AF356611

New Insights into the Diagnosis and Treatment of Temporomandibular Joint Diseases

New Insights into the Diagnosis and Treatment of Temporomandibular Joint Diseases

Editor

Luís Eduardo Almeida

Basel • Beijing • Wuhan • Barcelona • Belgrade • Novi Sad • Cluj • Manchester

Editor
Luís Eduardo Almeida
Surgical Sciences - Oral and
Maxillofacial Surgery
Marquette University School
of Dentistry
Milwuakee
United States

Editorial Office
MDPI
St. Alban-Anlage 66
4052 Basel, Switzerland

This is a reprint of articles from the Special Issue published online in the open access journal *Journal of Clinical Medicine* (ISSN 2077-0383) (available at: www.mdpi.com/journal/jcm/special_issues/50X815QLDX).

For citation purposes, cite each article independently as indicated on the article page online and as indicated below:

Lastname, A.A.; Lastname, B.B. Article Title. *Journal Name* **Year**, *Volume Number*, Page Range.

ISBN 978-3-7258-0404-7 (Hbk)
ISBN 978-3-7258-0403-0 (PDF)
doi.org/10.3390/books978-3-7258-0403-0

Contents

About the Editor . . . . **vii**

Preface . . . . **ix**

Louis Gerard Mercuri
Temporomandibular Joint Facts and Foibles
Reprinted from: *J. Clin. Med.* **2023**, *12*, 3246, doi:10.3390/jcm12093246 . . . . **1**

Luis Eduardo Almeida, Andrea Doetzer and Matthew L. Beck
Immunohistochemical Markers of Temporomandibular Disorders: A Review of the Literature
Reprinted from: *J. Clin. Med.* **2023**, *12*, 789, doi:10.3390/jcm12030789 . . . . **9**

**Karolina Lubecka, Kamila Checińska, Filip Bliźniak, Maciej Checiński, Natalia Turosz and
Adam Michcik et al.**
Intra-Articular Local Anesthetics in Temporomandibular Disorders: A Systematic Review and
Meta-Analysis
Reprinted from: *J. Clin. Med.* **2023**, *13*, 106, doi:10.3390/jcm13010106 . . . . **30**

**Maciej Checiński, Kamila Checińska, Natalia Turosz, Anita Brzozowska, Dariusz Chlubek
and Maciej Sikora**
Current Clinical Research Directions on Temporomandibular Joint Intra-Articular Injections: A
Mapping Review
Reprinted from: *J. Clin. Med.* **2023**, *12*, 4655, doi:10.3390/jcm12144655 . . . . **49**

Maciej Checiński, Kamila Checińska, Natalia Turosz, Maciej Sikora and Dariusz Chlubek
Intra-Articular Injections into the Inferior versus Superior Compartment of the
Temporomandibular Joint: A Systematic Review and Meta-Analysis
Reprinted from: *J. Clin. Med.* **2023**, *12*, 1664, doi:10.3390/jcm12041664 . . . . **68**

**Andrea Bargellini, Elena Mannari, Giovanni Cugliari, Andrea Deregibus, Tommaso
Castroflorio and Leila Es Sebar et al.**
Short-Term Effects of 3D-Printed Occlusal Splints and Conventional Splints on Sleep Bruxism
Activity: EMG–ECG Night Recordings of a Sample of Young Adults
Reprinted from: *J. Clin. Med.* **2024**, *13*, 776, doi:10.3390/jcm13030776 . . . . **87**

**Grzegorz Zieliński, Marcin Wójcicki, Michał Baszczowski, Agata Żyśko, Monika Litko-Rola
and Jacek Szkutnik et al.**
Influence of Soft Stabilization Splint on Electromyographic Patterns in Masticatory and Neck
Muscles in Healthy Women
Reprinted from: *J. Clin. Med.* **2023**, *12*, 2318, doi:10.3390/jcm12062318 . . . . **102**

**Francesco Inchingolo, Assunta Patano, Angelo Michele Inchingolo, Lilla Riccaldo, Roberta
Morolla and Anna Netti et al.**
Analysis of Mandibular Muscle Variations Following Condylar Fractures: A Systematic Review
Reprinted from: *J. Clin. Med.* **2023**, *12*, 5925, doi:10.3390/jcm12185925 . . . . **113**

Mathias Rentsch, Aleksandra Zumbrunn Wojczyńska, Luigi M. Gallo and Vera Colombo
Prevalence of Temporomandibular Disorders Based on a Shortened Symptom Questionnaire
of the Diagnostic Criteria for Temporomandibular Disorders and Its Screening Reliability for
Children and Adolescents Aged 7–14 Years
Reprinted from: *J. Clin. Med.* **2023**, *12*, 4109, doi:10.3390/jcm12124109 . . . . **129**

David Faustino Ângelo, Beatriz Mota, Ricardo São João, David Sanz and Henrique José Cardoso
Prevalence of Clinical Signs and Symptoms of Temporomandibular Joint Disorders Registered in the EUROTMJ Database: A Prospective Study in a Portuguese Center
Reprinted from: *J. Clin. Med.* **2023**, *12*, 3553, doi:10.3390/jcm12103553 . . . . **146**

Giuseppe Minervini, Rocco Franco, Maria Maddalena Marrapodi, Salvatore Crimi, Almir Badnjević and Gabriele Cervino et al.
Correlation between Temporomandibular Disorders (TMD) and Posture Evaluated trough the Diagnostic Criteria for Temporomandibular Disorders (DC/TMD): A Systematic Review with Meta-Analysis
Reprinted from: *J. Clin. Med.* **2023**, *12*, 2652, doi:10.3390/jcm12072652 . . . . **160**

About the Editor

Luís Eduardo Almeida

Dr. Luis Eduardo Almeida is a dynamic professional with an illustrious academic journey. He embarked on his pursuit of knowledge by earning a Doctorate in Dental Surgery (DDS) degree (1989–1993) from the prestigious Federal University of Parana State, UFPR, Brazil. His relentless dedication led him to achieve a Certificate in Oral and Maxillofacial Surgery (1995–1998) from the same institution.

Continuing his academic endeavors with fervor, he secured a Fellowship in Oral and Maxillofacial Surgery (1998–1999) at Northwestern University, Chicago, USA, followed by completing a Residency Program in the same field (2001–2003) at the same esteemed institution. He further honed his expertise by attaining a Master of Science in Health Sciences (2004–2006) and a Ph.D. in Health Sciences (2008–2013) from Pontifícia Universidade Católica do Paraná, PUC-PR, Brazil.

Presently, Dr. Luis Eduardo Almeida exudes enthusiasm as a guiding force at Marquette University – School of Dentistry. Serving as a Clinical Associate Professor in Surgical Sciences – Oral & Maxillofacial and Oral Surgery. His dynamic leadership as Director of the Predoctoral Program in Oral & Maxillofacial and Oral Surgery (since 2018) ignites a fervor for excellence in his students.

Certified by the Brazilian Dental Board and the Wisconsin Dental Board and Board eligible for the American Board of Oral and Maxillofacial Surgery, Dr. Luis Eduardo Almeida stands out as a pioneer in his field. His infectious enthusiasm extends beyond academia, as evidenced by his enriching experience as an Oral & Maxillofacial Surgeon in private practice (2003–2013).

In all his endeavors, Dr. Luis Eduardo Almeida embodies a spirit of enthusiasm, innovation, and excellence, leaving an indelible mark on the landscape of dentistry.

Preface

Welcome to the compilation of scholarly endeavors in "New Insights into the Diagnosis and Treatment of Temporomandibular Joint Diseases". This assemblage represents a focused exploration into the contemporary landscape of temporomandibular joint (TMJ) disorders, encapsulating a spectrum of research aimed at advancing our comprehension and therapeutic modalities.

Within the following pages, you will encounter a synthesis of empirical investigations delving into nuanced aspects of TMJ disorders. From innovative diagnostic methodologies to emerging therapeutic interventions, each contribution is underpinned by a commitment to expand the frontiers of knowledge in the field.

The crux of this compilation resides in its dedication to unraveling the intricacies of TMJ disorders. As an amalgamation of diverse studies, it strives to transcend conventional paradigms and redefine our understanding of these disorders. The scientific rigor and methodological precision embodied in each manuscript reflect an earnest pursuit of evidence-based insights.

TMJ research, by its very nature, demands an interdisciplinary approach. The convergence of methodologies from various scientific disciplines underscores the intricate interplay of factors contributing to TMJ disorders. From biomechanics to clinical outcomes, the breadth of topics covered in this compilation exemplifies the comprehensive perspective required for an exhaustive understanding.

Gratitude is extended to the contributing authors, whose scholarly contributions form the bedrock of this compilation. Their endeavors represent a collective stride toward unraveling the complex tapestry of TMJ disorders, propelling us into an era marked by refined diagnostic strategies and more efficacious treatment modalities.

Whether you are a seasoned researcher, clinician, or discerning reader with an interest in the scientific intricacies of TMJ disorders, we invite you to immerse yourself in the empirical richness that unfolds within these pages. May this compilation serve as a catalyst for further scientific inquiry, fostering a deeper understanding and engendering advancements in the diagnosis and treatment of TMJ disorders.

Luís Eduardo Almeida

Editor

Journal of
Clinical Medicine

Opinion

Temporomandibular Joint Facts and Foibles

Louis Gerard Mercuri 

Department of Orthopaedic Surgery, Rush University Medical Center, Chicago, IL 60612, USA;
louis_g_mercuri@rush.edu

Abstract: The purpose of this article is to dispel some of the major foibles associated with the etiology and management of TMJ disorders, while presenting some of the facts based on the scientific literature to date. To appreciate this kind of update, the reader must be an "out of the box thinker" which requires openness to new ways of seeing the world and a willingness to accept new concepts based on evolving evidence.

Keywords: temporomandibular joint; TMJ; temporomandibular joint disorders; TMD

check for
updates

Citation: Mercuri, L.G.
Temporomandibular Joint Facts and
Foibles. *J. Clin. Med.* **2023**, *12*, 3246.
https://doi.org/10.3390/jcm12093246

Academic Editors: Eiji Tanaka and
Luís Eduardo Almeida

Received: 19 April 2023
Revised: 25 April 2023
Accepted: 28 April 2023
Published: 1 May 2023

1. Introduction

In a 24 August 2015 New York Times op-ed entitled "The Case for Teaching Ignorance", Jamie Holmes wrote, "Presenting knowledge as more certain than it is discourages curiosity". There is no healthcare issue in dentistry that has proffered more quasi-scientific information than what is related to the etiology and management of temporomandibular joint (TMJ) disorders.

In a 1995 literature review commissioned by NIDR (now NIDCR) to determine the strength of the evidence regarding therapy for TMJ disorders, Antczak-Bouckoms found that the TMJ literature as of that date consisted primarily of uncontrolled clinical trials, case series, case reports, and simple descriptions of techniques; she concluded that such uncontrolled observations were subject to considerable bias and thus difficult to interpret. Further, she reckoned that "if treatment of TMD is going to follow the evidence-based trend in medicine, rather than opinion or pathophysiologic rationales, then more rigorously controlled clinical trials of most therapeutic options would be necessary" [1]. However, while much has changed, there are still a number of "uncontrolled clinical trials, case series, case reports, and simple descriptions of techniques" in the present day TMJ literature.

The purpose of this article is to dispel some of the major foibles associated with the etiology and management of TMJ disorders, while presenting some of the facts based on the scientific literature to date. To appreciate this kind of update, the reader must be an "out of the box thinker" which requires openness to new ways of seeing the world and a willingness to accept new concepts based on evolving evidence.

2. The Temporomandibular Joint

What is the TMJ besides being the articulation between the mandible and the base of the skull that allows the functions of mastication and speech as well as supporting deglutition and the upper airway? Do we want people to utilize cynical descriptions of the TMJ as being The Mouth Joint, The Money Joint, or The Mystery Joint? For surgeons, why is treating TMJ problems like looking for a good restaurant in an unfamiliar town where you do not know which joint to enter and which one to stay out of? (Seldin, E 1983).

As I see it, there are 10 foibles related to the TMJ that I feel must be addressed and investigated further, but even now there are many evidence-based facts to counteract them.

2.1. Problem #1

Patients, physicians, many dentists, and health care insurance providers do not consider the TMJ as being just another orthopedic joint. Orthopedics is the branch of medicine concerned with the diagnosis and treatment of acute, chronic, traumatic, and overuse injuries and other disorders of the musculoskeletal system. Is that not exactly what clinicians do who manage TMJ patients?

2.2. Problem #2

The etiology, diagnosis, and management focus has often been on only one joint component (e.g., the disc, masticatory muscles, teeth). The musculoskeletal system is a complex scheme made up not just of the bone and cartilage of the joints, articular discs, and menisci, but also the muscles that move the joints as well as the vascular, lymphatic, and neural components that nourish the joint components and prompt its movements [2].

2.3. Problem #3

Unfortunately, the TMJ has been compared anatomically to the hip, a ball-in-socket joint, as opposed to the knee, which has a similar range of motion with both sliding and lateral movements. Both joints have an interposed fibrocartilaginous structure, anatomically designated a "meniscus" in the knee, but described as a "disc" in the TMJ due to its anatomic shape and the nature of the supporting attachments. Intra-articular fibrocartilages are complete or incomplete plates of fibrocartilage that are attached to the joint capsule (the investing ligament) and that stretch across the joint cavity between a pair of conarticular surfaces. When complete, they are called "discs"; when incomplete, they are called "menisci" [3].

2.4. Problem #4

The etiology, diagnosis, and management focus for TMJ disorders has recently been fixated on the position of the disc. At first it was all about the occlusion of the teeth or the malposition of the mandibular condyle. However, when those mechanistic concepts were disproven, "internal derangement" of the articular disc became the target for both non-surgical and surgical treatments.

The term "internal derangement" was coined in 1814 by an English surgeon William Hey for meniscus problems in the knee [4]. The term was applied to the TMJ disc in 1826 by another English surgeon, Astley Cooper [5]. In 1887, Thomas Annandale, a Scottish surgeon, reported the repositioning of an "internally deranged" TMJ disc with horsehair sutures [6]. Soon thereafter, TMJ disc internal derangement became the prime suspect for joint noise, pain, and "locking", and this concept became the basis for performing TMJ discectomy [7].

Etiologic concepts of TMJ disorders shifted from these early versions of internal disc derangement to a strong emphasis on the teeth and muscles. This continued until the 1970s, when William Farrar and William McCarty developed the Normandy Study Group and resurrected internal derangement of the TMJ disc as being the main anatomical etiology of TMJ disorders [8].

Clyde Wilkes went a step further by stating that disc displacement led directly to condylar degeneration. Wilkes then proposed a classification and surgical management procedures for repositioning the disc [9]. This hypothesis led to the belief that internal derangements, or disc displacements, represented the basic pathological entity responsible for all the clinical signs and symptoms associated with the so-called TMJ pain dysfunction syndrome.

Clinical management of TMJ disorders took on new importance for surgeons because it was now assumed, based on Wilkes' assertions, that disc displacement was likely to progress to advanced degenerative states. This prompted the development of open surgical disc repositioning protocols by Dolwick [10], arthroscopic protocols by Sanders [11],

McCain [12], and Yang [13], as well as arthrocentesis protocols by Nitzan [14], all of which added to further understanding of the role of the TMJ disc in healthy and diseased states.

The author Dan Brown said, "Wide acceptance of an idea is not proof of its validity" [15]. This is true of many things, especially TMJ "internal derangement" and its progression to degeneration of the articulation.

De Bont stated that "disc displacement may represent extremes of normal variation" and "... internal derangement may be a sign of a range of conditions, rather than a single entity" [16].

Widmalm reported that "internal derangement does not always cause clinical symptoms; it may be contributory, but inflammatory changes in the joint capsule, synovial tissues and retro-discal tissues are likely more important for the development of symptoms than disc displacement or deformation per se [17]".

With the advent of MR imaging, investigators began to study the TMJs of random asymptomatic volunteers and found that 23–35% of these subjects had displaced discs [18–26]. This was analogous to studies of intervertebral disc displacement in asymptomatic low back pain subjects [27,28].

Since 23–35% of the population have abnormal disc–condyle relationships, it can be argued that MRI disc position does not have a direct correlation with clinical signs and symptoms. Thus, there is a need to understand how much of the biological variability of the disc–condyle relationship is really part of a natural course of joint wear and aging, and how much of it is the result of joint overload. A multivariate analysis did not show that TMJ pain correlates with a reduction in anterior disc displacement or mandibular condyle morphology [29].

Joint biomechanical instability is defined as the inability to maintain the functional relationship between the bones and associated soft tissues under normal physiological forces. When this occurs, it puts extra stress on the intra-articular and surrounding joint structures, leading to soft tissue damage or tears, joint degeneration, and pain [30]. Joints begin to degenerate or break down when the catabolic (destructive) processes exceed the anabolic (reparative) processes. The result is joint instability and disabling chronic musculoskeletal pain.

Histological examinations have confirmed that increased vascularity and inflammatory reactions are seen in the synovia and posterior disc attachment in patients who present with TMJ pathology. Synovial fluid is produced by fibroblast-like type B synovial cells. Physiologic changes in synovial fluid volume and content occur in response to trauma, inflammation, and bacterial, fungal, or viral penetrance. These changes in synovial fluid can increase friction, which can lead to joint degeneration [31,32].

In 2016, Israel concluded that "internal derangement" of the TMJ is not a disease, but a non-specific sign of tissue failure which, in some cases, may lead to biomechanical instability of the joint. This tissue failure is usually caused by TMJ mechanical overloading, leading to inflammatory/degenerative changes related to synovitis. The intra-articular changes associated with internal derangement can also be caused by a systemic arthropathy or a localized atypical arthropathy involving the TMJ [33].

Impairment of joint movement after repeated micro- or macro-trauma, including the TMJ, is the result of the development of synovitis within the joint capsule and Hilton's Orthopedic Law, i.e., "The nerves that innervate a joint, also innervate the muscles that move that joint, as well as the overlying skin" [34]. In the case of the TMJ, it is the trigeminal nerve (CN V). This reflex limits joint movement, thereby causing pain and possibly further damage.

Mobilization of the disc is of more importance for the reduction in signs and symptoms of internal derangement than anatomically repositioning it in relationship to the condylar head and the articular eminence. Therefore, disc function is more important than disc position. Scientific evidence regarding the effectiveness of TMJ disc repositioning remains scarce and needs further efforts to guide clinicians and patients when considering the clinical and surgical options to treat TMJ disc displacement [35].

Chantaracherd et al., in a cross-sectional study, found no association between TMJ disc status and TMD symptomatology as represented by pain, jaw function, and disability [36]. This suggests that TMJ disc displacements per se have minimal impact on patients' reported pain, function, and disability. This also suggests that treatments focusing on "correcting" those displacements (such as surgery) may have limited impact on patient-reported outcomes [37].

2.5. Problem #5

Dentists are taught to "cure" rather than "manage". If there is caries and it is removed and replaced with a restoration, the tooth is cured. If there is pulpal disease and endodontics is successfully performed, the tooth is cured. If there are non-restorable caries or advanced periodontal disease and the tooth is extracted, then those problems are cured.

However, there are few if any medical problems that are "cured"; instead, most are "managed". Hypertension, diabetes, musculoskeletal disorders, etc., are managed, not cured. TMJ disorders are musculoskeletal conditions, and therefore not "curable", but most of them are "manageable". This is an important concept for the clinician to accept and pass on to their TMJ patients so as not to give the patient the undeliverable and confounding expectations of a "cure" instead of the realistic and deliverable expectation of a "management" plan for their TMJ problem.

2.6. Problem #6

Patients, physicians, many dentists, and health care insurers are often using the wrong terminology. The commonly used term "Temporomandibular Disorders" (TMDs) is a collective label that is used to embrace several clinical problems that involve the masticatory musculature and the temporomandibular joint itself [38]. Therefore, like all musculoskeletal diseases, there are two major categories, namely, extra-articular muscle-based disorders and intra-articular joint-based disorders.

Most patients with signs and symptoms of an extra-articular muscle based TMJ disorder have masticatory and cervical myositis or myofascial pain that is amenable to non-invasive management. However, those with intra-articular TMJ disorders display demonstrable signs and symptoms of specific TMJ pathology, like other joints in the body, some of which can be managed conservatively while others may have indications for invasive management.

2.7. Problem #7

There are multiple confusing and often unrelated signs and symptoms that are reported by TMD patients, in addition to the frequent presence of other co-morbid painful conditions.

In 1985, Rugh and Solberg stated "...there is scant evidence to suggest that TMJ conditions are new diseases...rather, it appears that the recent interest is a result of redefining of old conditions, increasing education of professionals, increased public awareness, and a belief by clinicians that these signs and symptoms should be treated." [39]

As discussed in Problem #4, this kind of thinking led to the belief that disc displacement treatment is necessary to prevent progression of internal derangement disorders to degenerative joint disease. However, several studies have shown that TMJ disorders can often turn out to be self-limiting, or non-progressive conditions [40].

2.8. Problem #8

The dental profession has embraced the concept that the TMJ is a unique articulation because it has terminal ectodermal structures (teeth) and has fibrocartilage rather than hyaline cartilage covering the articular surfaces of the condyles and eminence. Therefore, dentistry focused diagnostic and therapeutic modalities on the occlusion and more recently focused on the intra-articular disc position.

The axial skeleton also has terminal ectodermal structures—finger and toenails. Is there any literature that directly correlates in the appendicular degenerative joint changes with abnormalities in finger or toenails?

Manfredini has written that multiple research findings support the absence of a disease-specific association between TMJ disorders and dental occlusion. There seems to be no solid ground to further hypothesize a role for dental occlusion in the pathophysiology of TMJ disorders. He encouraged clinicians to abandon the old gnathological paradigm in the management of TMJ disorders in practice [41].

The essential life functions of mastication, speech, airway support, and deglutition are supported by TMJ function and form. This places the TMJ complex under more cyclical loading and unloading than any other body joint over a lifetime. Therefore, the TMJ is considered to be load-bearing during masticatory function and even at rest or during full closure of the mouth. The fibrocartilaginous tissues, including the disc and articular cartilage, have important functions in stress distribution. Fibrocartilage is the strongest kind of cartilage, which is why the TMJ has fibro- rather than hyaline cartilage (the weakest of the three types of cartilage). Further, the bony surfaces of other highly stressed joints such as the sternoclavicular and symphysis pubis are covered with fibrocartilage [42–44].

2.9. Problem #9

Sophisms, bad science, opinions, and cults have developed around the diagnosis and management of TMJ disorders.

Why do some adhere to concepts that have proven to be scientifically invalid? When it comes to TMJ disorders, Mohl and Ohrbach believed the following were in play: early professional training and experience, reinforcement of familiar procedures, inertia, isolation in a private practice setting, insecurity, unfamiliarity with the literature, inability to assess scientific evidence, blind belief in "schools of thought" sponsored by charismatic gurus, and finally economics [45].

As an example, the hallmark of neuromuscular dentistry is the use of electronic diagnostic devices in the diagnosis and management of not only orofacial pain patients, but also for discovering problems in asymptomatic patients. Gonzales et al. stated that the results obtained from diagnostic testing should have a high probability of affecting either the correctness of the diagnosis, selection of the appropriate treatment, or both. They found that these electronic diagnostic devices did not provide any of this information [46].

In the American Association For Dental Research's 3 March 2010 Policy Statement on TMJ Disorders, they state that "...the consensus of recent scientific literature about currently available technological diagnostic devices for TMDs is that except for various imaging modalities, none of them shows the sensitivity and specificity required to separate normal subjects from TMD patients or to distinguish among TMD subgroups."

Further, this statement goes on to say that neuromuscular dentistry is not specifically recognized by the American Dental Association, although a variety of healthcare providers advertise themselves as TMJ specialists. As a result, many TMD treatments available today are based largely on beliefs, not on scientific evidence.

2.10. Problem #10

Misdiagnosis leading to improper and/or inappropriate management has led to iatrogenic disease.

The experience of the past 150 years in the diagnosis and management of chronic orofacial pain conditions has shown that a mechanistic, narrow approach is likely to produce iatrogenic harm, e.g., unnecessary root canal therapy, extractions, bite-changing restorations, TMJ surgery, etc. [47].

In an 18 November 2013 Wall Street Journal article entitled "The Biggest Mistakes Doctors Make", misdiagnosis was deemed the primary reason for many negative outcomes. As an example, if a clinician believes and espouses the idea that the underlying cause of most TMJ disorders is "internal derangement", every patient with signs and symptoms will be treated for this so-called "progressive disease". Since at least 23–35% who go on to an MRI will demonstrate a disc displacement that has nothing to do with their orofacial pain complaint, those findings may persuade the clinician to utilize unnecessary non-invasive, minimally invasive, or, worse yet, invasive treatment, all for the wrong reason. The literature is replete with such cases. Maslow's comment seems appropriate here: "When you only have a hammer, everything looks like a nail" [48].

3. Conclusions

Therefore, based on the above, the following recommendations are offered:

1. The TMJ is a complex orthopedic joint and therefore must be considered part of the body's musculoskeletal system.
2. The diagnosis and management of TMJ disorders must be based on the same medical orthopedic principles that are used for treating other joints in the body.
3. Clinicians and patients must understand that TMJ disorders are managed, not cured.
4. The etiologic role of synovitis in painful TMJ disorders must be acknowledged and elucidated.
5. Disc displacement is not a disease; instead, it is a non-specific sign of tissue failure leading to biomechanical instability of the joint, which may or may not be clinically significant.
6. TMJ disc function is more important than its position relative to the condyle.
7. Dental occlusion has no direct etiologic relevance to TMJ disorders.
8. Invasive management options must be reserved for patients with demonstrable intra-articular TMJ pathology.
9. Misdiagnosis and misinterpretation of clinical or imaging findings ultimately leads to iatrogenic TMJ problems.
10. The important thing is to never stop questioning. (Einstein, A.)

Funding: This research received no external funding.

Institutional Review Board Statement: Not applicable.

Informed Consent Statement: Not applicable.

Acknowledgments: Thanks to my mentor Charles S. Greene for reviewing this manuscript.

Conflicts of Interest: The author declares no conflict of interest.

References

1. Antczak-Bouckoms, A.A. Epidemiology of research for temporomandibular disorders. *J. Orofac. Pain* **1995**, *9*, 226–234.
2. Laskin, D.M. Temporomandibular Disorders: A Term Whose Time Has Passed! *J. Oral Maxillofac. Surg.* **2020**, *78*, 496–497. [CrossRef]
3. Britannica. Available online: https://www.britannica.com/science/joint-skeleton/Bicondylar-joint (accessed on 4 April 2023).
4. Hey, W. *Practical Observations in Surgery*; T Cadell & W Davis: London, UK, 1814.
5. Cooper, A.P. The knee. In *The Encyclopedia of Surgery*; Ashurst, J., Jr., Ed.; William Wood & Company: New York, NY, USA, 1889.
6. Annandale, T. On displacement of the inter-articular cartilage of the lower jaw and its treatment by operation. *Lancet* **1887**, *129*, 411. [CrossRef]
7. Lanz, A.B. Discus mandibularis. *Zenralbl. Chir.* **1909**, *36*, 289.
8. McCarty, W.L., Jr.; Farrar, W.B. Surgery for internal derangements of the temporomandibular joint. *J. Prosthet. Dent.* **1979**, *42*, 191–196. [CrossRef]
9. Wilkes, C.H. Internal derangement of the temporomandibular joint: Pathological variations. *Arch. Otolaryngol. Head Neck Surg.* **1989**, *115*, 469–477. [CrossRef]
10. Dolwick, M.F.; Katzberg, R.W.; Helms, C.A. Internal derangements of the temporomandibular joint: Fact or fiction? *J. Prosthet. Dent.* **1983**, *49*, 415–418. [CrossRef]
11. Sanders, B. Arthroscopic surgery of the temporomandibular joint: Treatment of internal derangement with persistent closed lock. *Oral Surg. Oral Med. Oral Pathol.* **1986**, *62*, 361–372. [CrossRef]
12. McCain, J.P.; de la Rua, H.; Le Blanc, W.G. Correlation of clinical, radiographic, and arthroscopic findings in internal derangements of the TMJ. *J. Oral Maxillofac. Surg.* **1989**, *47*, 913–921. [CrossRef]

13. Zhang, S.Y.; Liu, X.M.; Yang, C.; Cai, X.Y.; Chen, M.J.; Haddad, M.S.; Yun, B.; Chen, Z.Z. New arthroscopic disc repositioning and suturing technique for treating internal derangement of the temporomandibular joint: Part II—Magnetic resonance imaging evaluation. *J. Oral Maxillofac. Surg.* **2010**, *68*, 1813–1817. [CrossRef]

14. Nitzan, D.W. Arthrocentesis—Incentives for using this minimally invasive approach for temporomandibular disorders. *Oral Maxillofac. Surg. Clin. N. Am.* **2006**, *18*, 311–328. [CrossRef]

15. Brown, D. *The Lost Symbol*; Doubleday: New York, NY, USA, 2009; ISBN 978-0-385-50422-5.

16. de Bont, L.G.; Stegenga, B. Pathology of temporomandibular joint internal derangement and osteoarthrosis. *Int. J. Oral Maxillofac. Surg.* **1993**, *22*, 71–74. [CrossRef]

17. Widmalm, S.E.; Westesson, P.L.; Kim, I.K.; Pereira, F.J., Jr.; Lundh, H.; Tasaki, M.M.L. Temporomandibular Joint Pathosis Related to Age, Sex and Dentition in Autopsy Material. *Oral Surg. Oral Med. Oral Pathol.* **1994**, *78*, 416–425. [CrossRef]

18. Tasaki, M.M.; Westesson, P.L.; Isberg, A.M.; Ren, Y.F.; Tallents, R.H. Classification and prevalence of TMJ disc displacement in patients and asymptomatic volunteers. *Am. J. Orthod. Dentofac. Orthop.* **1996**, *109*, 249–262. [CrossRef]

19. Katzberg, R.W.; Westesson, P.L.; Tallents, R.H.; Drake, C.M. Anatomic disorders of the TMJ disc in asymptomatic subjects. *J. Oral Maxillofac. Surg.* **1996**, *54*, 147–153. [CrossRef]

20. Ribeiro, R.F.; Tallents, R.H.; Katzberg, R.W.; Murphy, W.C.; Moss, M.E.; Magalhaes, A.C.; Tavano, O. The prevalence of disc displacement in symptomatic and asymptomatic volunteers aged 6 to 25 years. *J. Orofac. Pain* **1997**, *11*, 37–47.

21. Larheim, T.A.; Westesson, P.; Sano, T. Temporomandibular joint disc displacement: Comparison in asymptomatic volunteers and patients. *Radiology* **2001**, *218*, 428–432. [CrossRef]

22. Oğütcen-Toller, M.; Taşkaya-Yilmaz, N.; Yilmaz, F. The evaluation of temporomandibular joint disc position in TMJ disorders using MRI. *Int. J. Oral Maxillofac. Surg.* **2002**, *31*, 603–607. [CrossRef]

23. Haiter-Neto, F.; Hollender, L.; Barclay, P.; Maravilla, K.R. Disk position and the bilaminar zone of the temporomandibular joint in asymptomatic young individuals by magnetic resonance imaging. *Oral Surg. Oral Med. Oral Pathol. Oral Radiol. Endod.* **2002**, *94*, 372–378. [CrossRef]

24. Schmitter, M.; Rammelsberg, P.; Hassel, A.; Schroeder, J.; Seneadza, V.; Balke, Z.; Essig, M. Evaluation of disk position and prevalence of internal derangement, in a sample of the elderly, by gadolinium-enhanced magnetic resonance imaging. *Oral Surg. Oral Med. Oral Pathol. Oral Radiol. Endod.* **2008**, *106*, 872–878. [CrossRef]

25. Peroz, I.; Seidel, A.; Griethe, M.; Lemke, A.J. MRI of the TMJ: Morphometric comparison of asymptomatic volunteers and symptomatic patients. *Quintessence Int.* **2011**, *42*, 659–667.

26. Poluha, R.L.; Canales, G.T.; Costa, Y.M.; Grossmann, E.; Bonjardim, L.R.; Conti, P.C.R. Temporomandibular joint disc displacement with reduction: A review of mechanisms and clinical presentation. *J. Appl. Oral Sci.* **2019**, *27*, e20180433. [CrossRef]

27. Jensen, M.C.; Brant-Zawadzki, M.N.; Obuchowski, N.; Modic, M.T.; Malkasian, D.; Ross, J.S. Magnetic resonance imaging of the lumbar spine in people without back pain. *N. Engl. J. Med.* **1994**, *331*, 69–73. [CrossRef]

28. Ketola, J.H.J.; Inkinen, S.I.; Karppinen, J.; Niinimäki, J.; Tervonen, O.; Nieminen, M.T. T2 -weighted magnetic resonance imaging texture as predictor of low back pain: A texture analysis-based classification pipeline to symptomatic and asymptomatic cases. *J. Orthop. Res.* **2021**, *39*, 2428–2438. [CrossRef]

29. Poluha, R.L.; Cunha, C.O.; Bonjardim, L.R.; Conti, P.C.R. Temporomandibular joint morphology does not influence the presence of arthralgia in patients with disk displacement with reduction: A magnetic resonance imaging-based study. *Oral Surg. Oral Med. Oral Pathol. Oral Radiol.* **2020**, *129*, 149–157. [CrossRef]

30. McGill, S.M.; Cholewicki, J. Biomechanical basis for stability: An explanation to enhance clinical utility. *J. Orthop. Sports Phys. Ther.* **2001**, *31*, 96–100. [CrossRef]

31. Kurita, K.; Westesson, P.L.; Sternby, N.H.; Eriksson, L.; Carlsson, L.E.; Lundh, H.; Toremalm, N.G. Histologic features of the temporomandibular joint disk and posterior disk attachment: Comparison of symptom-free persons with normally positioned disks and patients with internal derangement. *Oral Surg. Oral Med. Oral Pathol.* **1989**, *67*, 635–643. [CrossRef]

32. Holmlund, A.B.; Gynther, G.W.; Reinholt, F.P. Disk derangement and inflammatory changes in the posterior disk attachment of the temporomandibular joint. A histologic study. *Oral Surg. Oral Med. Oral Pathol.* **1992**, *73*, 9–12. [CrossRef]

33. Israel, H.A. Internal Derangement of the Temporomandibular Joint: New Perspectives on an Old Problem. *Oral Maxillofac. Surg. Clin. N. Am.* **2016**, *28*, 313–333. [CrossRef]

34. Hébert-Blouin, M.N.; Tubbs, R.S.; Carmichael, S.W.; Spinner, R.J. Hilton's law revisited. *Clin. Anat.* **2014**, *27*, 548–555. [CrossRef]

35. de Leeuw, R. Internal Derangements of the Temporomandibular Joint. *Oral Maxillofac. Surg. Clin. N. Am.* **2008**, *20*, 159–168. [CrossRef]

36. Chantaracherd, P.; John, M.T.; Hodges, J.S.; Schiffman, E.L. Temporomandibular joint disorders' impact on pain, function, and disability. *J. Dent. Res.* **2015**, *94* (Suppl. 3), 79S–86S. [CrossRef]

37. Schiffman, E.L.; Velly, A.M.; Look, J.O.; Hodges, J.S.; Swift, J.Q.; Decker, K.L.; Anderson, Q.N.; Templeton, R.B.; Lenton, P.A.; Kang, W.; et al. Effects of four treatment strategies for temporomandibular joint closed lock. *Int. J. Oral Maxillofac. Surg.* **2014**, *43*, 217–226. [CrossRef]

38. McNeill, C.; Mohl, N.D.; Rugh, J.D.; Tanaka, T.T. Temporomandibular disorders: Diagnosis, management, education, and research. *J. Am. Dent. Assoc.* **1990**, *120*, 253–263. [CrossRef]

39. Rugh, J.D.; Solberg, W.K. Oral health status in the United States: Temporomandibular disorders. *J. Dent. Educ.* **1985**, *49*, 398–406. [CrossRef]

40. Rugh, J.D.; Solberg, W.K. Psychological implications in temporomandibular joint dysfunction. *Oral Sci. Rev.* **1976**, *7*, 3–30.
41. Manfredini, D.; Lombardo, L.; Siciliani, G. Temporomandibular disorders and dental occlusion. A systematic review of association studies: End of an era? *J. Oral Rehabil.* **2017**, *44*, 908–923. [CrossRef]
42. Hylander, W.L.; Bays, R. An in vivo strain-gauge analysis of squamosal-dentary joint reaction force during mastication and incision in Macaca mulata and Macaca fascicularis. *Arch. Oral Biol.* **1979**, *24*, 689–697. [CrossRef]
43. Boyd, R.L.; Gibbs, C.H.; Mahan, P.E.; Richmond, A.F.; Laskin, J.L. Temporomandibular joint forces measured at the condyle of Macaca arctoides. *Am. J. Orthod. Dentofac. Orthop.* **1990**, *97*, 472–479. [CrossRef]
44. Beek, M.; Koolstra, J.H.; van Ruijven, L.J.; van Eijden, T.M.G.J. Three-dimensional finite element analysis of the human temporo-mandibular joint disc. *J. Biomech.* **2000**, *33*, 307–316. [CrossRef]
45. Mohl, N.D.; Ohrbach, R. The dilemma of scientific knowledge versus clinical management of TMD. *J. Prosth. Dent.* **1992**, *67*, 113–120. [CrossRef] [PubMed]
46. Gonzalez, Y.M.; Greene, C.S.; Mohl, N.D. Technological devices in the diagnosis of temporomandibular disorders. *Oral Maxillofac. Surg. Clin. N. Am.* **2008**, *20*, 211–220. [CrossRef] [PubMed]
47. Turp, J.C.; Hugger, A.; Sommer, C. Orofacial Pain—A Challenge and Chance Not Only for Dentistry. In *The Puzzle of Orofacial Pain: Integrated Research into Clinical Management*; Turp, J.C., Hugger, A., Sommer, C., Eds.; Karger: Basel, Switzerland, 2007.
48. Maslow, A.H. *The Psychology of Science: A Reconnaissance*; Harper & Row: New York, NY, USA, 1966.

Journal of
Clinical Medicine

Systematic Review

Immunohistochemical Markers of Temporomandibular Disorders: A Review of the Literature

Luis Eduardo Almeida [1,*], Andrea Doetzer [2] and Matthew L. Beck [1]

[1] Surgical Sciences Department, School of Dentistry, Marquette University, Milwaukee, WI 53233, USA
[2] Faculdade de Odontologia, Pontificia Universidade Catolica do Parana, Curitiba 80215-901, Brazil
* Correspondence: luiseduardo.almeida@marquette.edu

Abstract: Temporomandibular disorders (TMD) are a group of internal derangements encompassing dysfunction, displacement, degeneration of the temporomandibular joints and surroundings muscles of mastication, often accompanied by pain. Relationships between TMD and various chemical biomarkers have been examined throughout the years. This paper aims to gather evidence from the literature regarding other biomarkers and presenting them as one systematic review to investigate the potential links between TMD and different biochemical activity. To identify relevant papers, a comprehensive literature search was carried out in MEDLINE/PubMED, EMBASE, Web of Science and a manual search was performed in the International Journal of Oral and Maxillofacial Surgery, Journal of Oral and Maxillofacial surgery, and Journal of Cranio-Maxillo-Facial Surgery. The literature review produced extensive results relating to the biochemical and immunohistochemical markers of TMD. Many enzymes, inflammatory markers, proteoglycans, and hormones were identified and organized in tables, along with a brief description, study design, and conclusion of each study. Through this review, recurring evidence provides confidence in suggesting involvement of certain biomarkers that may be involved in this complex pathogenesis, in addition to pointing to differences in gender prevalence of TMD. However, more organized research on large human samples needs to be conducted to delve deeper into the understanding of how this disease develops and progresses.

Keywords: temporomandibular joint; temporomandibular disorders; biomarkers; immunohistochemistry; systematic review

Citation: Almeida, L.E.; Doetzer, A.; Beck, M.L. Immunohistochemical Markers of Temporomandibular Disorders: A Review of the Literature. *J. Clin. Med.* **2023**, *12*, 789. https://doi.org/10.3390/jcm12030789

Academic Editor: Mieszko Wieckiewicz

Received: 12 December 2022
Revised: 9 January 2023
Accepted: 13 January 2023
Published: 18 January 2023

1. Introduction

The temporomandibular joint (TMJ) is a ginglymoarthrodial joint acting as a union of the mandibular fossa of the temporal bone and the head of the mandibular condyle. It is used in the mastication processes via translational (gliding) and rotational (hinging) movements. Its primary components include the articular disc, the articular capsule, innervation and vascularization, ligaments, and the muscles of mastication [1].

The inner region of the TMJ contains an articular disc composed of fibrous connective tissue devoid of any blood vessels and nerves. It is attached to the capsular ligament anteriorly, posteriorly, medially, and laterally to separate the TMJ into two cavities. Posteriorly, loose connective tissue is attached to the connective tissue and is highly vascularized and innervated. Superiorly, the fibroelastic fascia attaches with elastic fibers to the tympanic plate. Inferiorly, at the posterior edge of the disc is the inferior retrodiscal lamina composed of fibrous, collagenous fibers. The disc is attached to the condylar formation at the medial and lateral aspects. The remaining body of the retrodiscal tissue is attached posteriorly to a large venous plexus that fills with blood as the condyle moves forward.

The TMJ is surrounded by the articular capsule, a capsular ligament with specialized endothelial cells that form a synovial lining which provides boundary and weeping lubrication. These cells provide synovial fluid, which act as a medium to prevent friction in the moving joint (boundary lubrication), as well as the ability of the articular surface to absorb a small amount of the synovial fluid in a compressed or nonmoving joint to eliminate friction (weeping lubrication) [1].

The supportive ligaments of the TMJ are collagenous and do not stretch, acting passively to restrain and limit border movements. These ligaments include the collateral, the capsular, and the temporomandibular ligaments. The collateral ligament serves to anchor the condyle. It is used to restrict movement of the disc away from the condyle and permits the hinge movement of the condyle. The capsular ligament surrounds the TMJ and attaches the temporal bone to the neck of condyle, resisting medial, lateral, and inferior forces that may separate or dislocate the articular surfaces. The temporomandibular ligament is composed of two portions: an oblique portion the connects the articular tubercle to the outer portion of the neck and resists excessive dropping of the condyle, and the horizontal portion that connects the articular tubercle to the tip of the condyle, limiting posterior movement of the condyle and disc. The accessory ligaments of the TMJ include the sphenomandibular and the stylomandibular, which prevent excess protrusive movements [1].

Temporomandibular disorders (TMD) are a group of internal derangements encompassing dysfunction, displacement, degeneration of the temporomandibular joints and surrounding muscles of mastication, often accompanied by pain. Currently, TMD is classified based on the articular disc and its mobility during mandibular movements. In early disease, the disc is displaced anteriorly, but returns to its normal position upon opening—termed anterior disc displacement with reduction (ADDwR). In later disease, the disc remains displaced even upon mandibular movements, creating a limitation in mobility and causing increased pain—termed anterior disc displacement without reduction (ADDwoR). Eventually, osteoarthritic changes may occur. The classic method of differentiating the stage of the disease was created by Wilkes. In his description, Wilkes uses clinical, radiographic and surgical findings to categorized cases into early, early/intermediate, intermediate, intermediate/late, and late stages [2].

It has been suggested that up to 25% of the population may experience symptoms of TMD, with a gender bias existing in which women are more affected by a ratio suggested to be greater than 2:1 [3]. Symptoms typically include clicking and popping of the joint, usually in early stages, limited range of motion and function, and orofacial pain. Current treatment modalities range from non-invasive options, such as physical therapy and occlusal splints, to minimally invasive options, such as trigger point injections and arthrocentesis, to invasive therapies, such as total joint replacement.

Despite our current understanding of TMD, the exact etiology of the disease remains unknown. It is crucial we develop our understanding of the progression and complications in order to diagnose and treat the disorder more effectively.

The relationships between TMD and various chemical biomarkers have been examined throughout the years. Matrix metalloproteinases (MMPs) are the major enzymes involved in extracellular matrix and basement membrane remodeling. Furthermore, they play a role in regulating chemokines, growth factors, proteases, among other molecules, in order to balance the inflammatory response [4]. Interleukins (IL) are a group of proteins that regulate cell growth, differentiation and motility. There are fifteen different interleukins known, and many of these are important factors in the inflammatory immune response [5]. These markers have the potential to be diagnostic and therapeutic targets for treating and managing TMD. MMPs and ILs are just a few of the immunohistochemical markers associated with TMD, and this paper aims to gather evidence from the literature regarding other common markers and presenting them as one systematic review to investigate the potential links between TMD and different biochemical activity.

2. Materials and Methods

2.1. Search Protocol

The MEDLINE/PubMed, EMBASE, and Web of Science databases were searched between January 2020 and March 2020. The manual search was performed in the three main journals of the field (International Journal of Oral and Maxillofacial Surgery, Journal of Oral and Maxillofacial Surgery, and Journal of Cranio-Maxillo-Facial Surgery), as well as from the reference list of studies included in this systematic review.

For the construction of the search strategy, MeSH (Medical Subject Heading) terms were used, which are considered descriptors of controlled subjects for searching MEDLINE, Web of Science, and Cochrane. To make the search more sensitive, vocabularies not controlled by the use of keywords were included. Entree terms were used for the search strategy in EMBASE.

The search was carried out by combining the terms ("MeSH" and keywords) for population and intervention, with the help of the Boolean operators "OR" and "AND". The strategy adopted for the MEDLINE/PubMed base was as follows: (TMJ OR temporomandibular joint OR TMD OR temporomandibular dysfunction OR temporomandibular joint dysfunction OR temporomandibular joint derangement) "AND" (immunohistochemistry OR immune antibody OR proteomic OR protein expression). In addition to these terms, the following keywords were used: disc displacement OR TMJ internal derangement OR TMD. The PRISMA 2020 statement was used as reference in reporting this systematic review. (Page, M.J.; McKenzie, J.: Bossuyt, P.: Boutron, I.; Hoffman, T.; Mulrow, C.; Shamseer, L.; Tetzlaff, J.; Akl, E.; Brennan, S.E.; et al. The Prisma 2020 Statement: An Updated Guideline for reporting Systematic Reviews. Syst. Rev. 2020, 10, 89.) The inclusion criteria was: human and animal research with a definitive clinical diagnosis or specific signs and symptoms for TMD, and protein expression results obtained through either immunohistochemistry OR immune antibody OR proteomic OR protein expression. The exclusion criteria were as follows: studies not in English, full text not available, studies not related to TMDs or where there was no mention of specific diagnoses or signs and symptoms, studies without the use of control to compare sample with and without protein expression.

A total of 107 studies were analyzed but only 86 were in accordance with PRISMA's criteria. The excluded studies either did not have a clear diagnosis of TMD or an adequate methodology. Three studies did not have the full text available, and one was not written in English. (Figure 1).

2.2. Data Analysis

The search protocol that was used was designed to be as inclusive as possible to encompass as many potential biomarkers that may be involved in the pathogenesis of TMD. As such, the methodology of the included studies varies widely, from utilization of different detection techniques in analyzing biomarker involvement, such as polymerase chain reaction (PCR), DNA extraction, immunohistochemical staining (IHC), among many others, to subject species—such as detection of biomarkers in human, rabbit or mouse tissues. With such variation in study design, a direct analysis of the data obtained through the included studies was unable to be conducted. Furthermore, the quality of each study, such as looking at statistical power, was not considered beyond the impact of the study based on factors such as sample size and the species from which the tissue was harvested in the study.

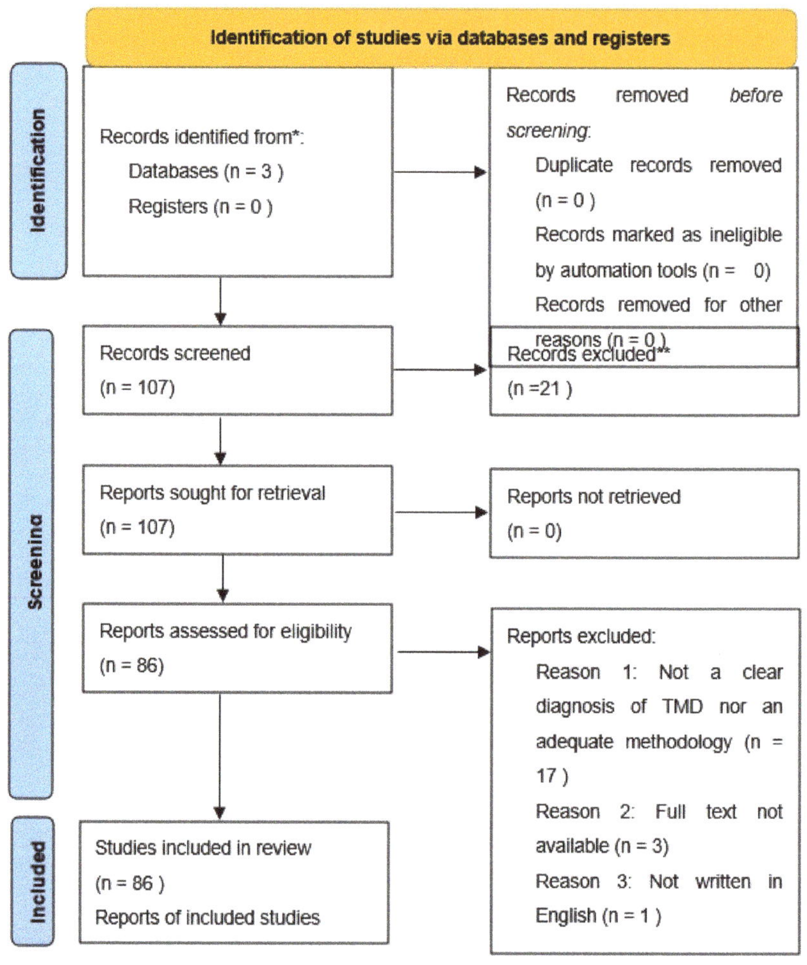

Figure 1. PRISMA 2020 flow diagram. * MEDLINE, Web of Science, and Cochrane, ** EMBASE.

3. Results

The literature review produced extensive results relating to the biochemical and immunohistochemical markers of TMD. The most studied are: enzymes, such as matrix metalloproteinases (MMP), aggrecanase, and cyclooxygenase (COX); cytokines and inflammatory markers, such as interleukins (IL), tumor necrosis factor- alpha (TNF-alpha), and vascular endothelial growth factor (VEGF); proteoglycans, such as tenascin, vimentin and fibronectin; hormones, such as estrogen, progesterone and relaxin; and miscellaneous markers that did not fit into the other categories, such as aquaporin, heat shock protein, and b-cell lymphoma-2 gene (BCL-2). Each study, along with a brief description, study design, and the conclusions, are organized in the tables below.

4. Discussion

4.1. Enzymes

Matrix metalloproteinases (MMPs) have been a large target of research attempts to understand the underlying pathogenesis of TMD. As previously mentioned, MMPs are a family of 26 proteins that are involved in extracellular matrix and basement membrane remodeling through the breakdown of collagen, gelatin, proteoglycans, and other components of the basement membrane. One of the first studies looking at MMPs in the discs of the TMJ was conducted by Kapila, using various methods, such as Western blotting and immunohistochemical staining, to identify the expression of MMPs (gelatinases, collagenases and stromelysin) in the discs of rabbit TMJ [4]. Many studies followed and showed the strongest evidence for the involvement of MMP-2 and MMP-9, with some evidence showing the involvement of MMP-7 (Kapila, [5], Tanaka, [6], Yoshida, [7], Loreto, [8,9], Almeida et al. [10]). Perroto [11], however, did not identify a significant difference in levels of MMP-13 between samples of human discs with TMD when compared to control samples. This points to the complex nature of basement-membrane-remodeling regulation involved in TMD. Evidence suggests that MMP-2 and MMP-9 are involved in the pathogenesis of TMD; however, more research needs to be conducted to support the involvement of the other suggested MMPs (MMP-1, MMP-7) and to definitively exclude those that have not been shown to be involved thus far (MMP-13). (Table 1).

Table 1. Seven Enzymes (MMPs, aggrecanase, cyclooxygenase).

Authors	IHC Marker	Study Design and Tissue Expression	Results	Conclusions
Kapila et al. (1995) [12]	MMP	Rabbit discs	MMPs, namely proMMP-9, proMMP-2, proMMP-1, and proMMP-3 were detected in TMJ articular discs.	Proteinases characterized in rabbit TMJ discs
Marchetti et al. (1999) [13]	MMP-2	Human discs	Positive immunoreaction pattern for MMP-2. Fibroblast-, chondroblast- and osteoblast-like cells displayed a positive cytoplasmic reaction. Structural modifications of the articular disc could be specific responses to changes in the function of the TMJ.	MMP-2 produced disc alteration
Quinn et al. (2000) [14]	Cyclooxygenase-2	Human synovial fluid	COX-2 is present in the TMJ synovial tissue and fluid from patients with internal derangement. Therefore, COX-2 antagonists may be indicated in the treatment of TMJ arthralgia.	COX-2 expression in internal derangements
Tanaka et al. (2001) [6]	MMPs	Human synovial fluid	Quantitative analysis showed that the degree of MMP-2 and MMP-9 expression was higher in human patients with ADDwoR than in human patients with ADDwR.	MMP-2 and MMP-9 associated with TMD
Yoshida et al. (2001) [7]	Cyclooxygenase-1,2	Human synovial membranes	COX-1 may be an important mechanism for maintaining normal homeostasis at the endothelial cells and fibroblasts with internal derangement of TMJ.	Cyclooxygenase-1 associated with TMJ homeostasis
Puzas et al. (2001) [15]	MMPs and Cytokines	Mice discs	The presence of a MMP-9 (92-kD gelatinase) in TMJ disc and articular cells likely function in the degradative process. Additionally, this enzyme is under the control of pro-inflammatory cytokines whereby TGFbeta and IL-1 stimulate and PGE(2) inhibits its activity.	MMP-9 associated with degenerative processes
Yoshida et al. (2002) [16]	Cyclooxygenase-2	Human disk and synovial membranes	There were obvious distinction of COX-2 immunoreactivity between the control specimens and internal derangement cases, in the region of posterior and/or anterior loose connective tissues. intensive COX-2 expression was detected in the synovial membrane of internal derangement cases.	COX-2 expression and TMD

Table 1. *Cont.*

Authors	IHC Marker	Study Design and Tissue Expression	Results	Conclusions
Yoshida et al. (2006) [17]	MMP and aggrecanase	Human synovial fluid	MMP-2, -9, and aggrecanase expression in the ID group were significantly higher than those in the normal group. Those with anterior disc displacement without reduction and severe OA showed significantly high expression of MMP-9 compared with other disease subgroups.	MMP-2, 9 and aggrecanase in TMD.
Yoshida et al. (2005) [18]	Aggrecanase	Human synovial fluid	Aggrecanase expression in TMJD group were significantly higher than that in the normal control group.	Aggrecanase associated with TMD
Hu et al. (2008) [8]	Urokinase-Type Plasminogen Activator and Urokinase-Type Plasminogen Activator Receptor	Human synovial fluid	uPA and uPAR in the synovial fluid may play a role in the pathogenesis of TMD, and the level of uPA and uPAR in synovial fluid of TMD could be used as a biochemical markers to reflect pathological degree of TMD.	uPA and uPAR associated with TMD
Matsumoto et al. (2008) [9]	Disintegrin and metalloproteinase with thrombospondin motifs 5 (ADAMTS-5)	Human discs	ADAMTS-5 is related to deformation and destruction of human TMJ discs affected by internal derangement.	ADAMTS-5 associated with TMD
Leonardi et al. (2010) [19]	Caspase 3	Human discs	Fatty degeneration is limited by apoptosis, with adipocytes being immunolabeled by caspase 3 antibody.	
Loreto et al. (2011) [20]	Caspase 3	Human discs	A greater proportion of caspase 3-positive cells were found in ADDwR and ADDwoR than in control discs.	Caspase 3 associated with TMD
Loreto et al. (2013) [11]	MMP-7 and MMP-9	Human discs	MMP upregulation in discs from patients contributes to disc damage.	MMP associated with TMD
Nascimento et al. (2013) [21]	MMP-2 and MMP-9	Rat trigeminal ganglion	MMP expression in the trigeminal ganglion was shown to vary during the phases of the inflammatory process. MMP-9 regulated the early phase and MMP-2 participated in the late phase of this process. Furthermore, increases in plasma extravasation in periarticular tissue and myeloperoxidase activity in the joint tissue, which occurred throughout the inflammation process, were diminished by treatment with DOX, a nonspecific MMP inhibitor.	MMP-2 and MMP-9 associated with process of TMJ inflammation
Almeida et al. (2015) [10]	MMP-2 and MMP-9	Human discs	MMP-2 expression was elevated in the disks of patients with displacement and without reduction. MMP-9 expression was not statistically elevated in the disks of patients with displacement and without reduction.	Elevation of MMP-2 in TMD.
Perroto et al. (2018) [22]	MMP-13	Human discs	MMP-13 is not significantly involved in collagen degradation.	Inflammatory cascade?

4.2. Cytokines and Inflammatory Markers

Our review of the literature found several different cytokines and inflammatory markers to be in association with TMD, including interleukins, CD antigens, TNF alpha,

TGF beta, VEGF, IGF, NFkB, IFN-y, capsaicin, and bradykinin. These markers are involved in many pathophysiological functions, including initiating and mediating inflammatory responses, activating immune cells, regulating apoptosis and angiogenesis, and regulating pain and fever.

As previously mentioned, the interleukins are a major family of cytokines that are extensively involved in the immune response. They are secreted by macrophages and T cells in order to modulate inflammation and fever, activate and maintain immune cells and bone marrow, produce immunoglobulins and acute phase proteins, and maintain granulomas. Various studies in our review found IL-1, IL-2, IL-6, IL-8, and IL-10 to be associated with TMD. Kim (2012) did not detect IL-4 and IL-5 in either the TMD group or the control group during a study of human synovial fluid [23]. Fu et al. (1995), Ogura et al. (2002), and Sato et al. (2003) found an association between IL-6 and TMD, while Caporal et al. (2017) found no significant difference when comparing TMD groups of ADDwR and ADDwoR, and with and without OA [24–27]. (Table 2).

Table 2. Cytokines and Inflammatory Markers (Interleukins, TNF alpha, FasL, VEGF, IGF, NFKB, IFN-y, capsaicin, BMP-2, bradykinin).

Authors	IHC Marker	Study Design and Tissue Expression	Results	Conclusions
Fu et al. (1995) [24]	IL-6	Human synovial fluid	Interleukin-6 level was greater than 100 U/mL in 13 of 18 patients with degenerative joint disease and in 5 of 12 patients with disc displacement. However, the interleukin-6 level was less than 100 U/mL (range, 20 to 75 U/mL) in all patients with masticatory muscle disorder.	IL-6 associated with acute TMD
Fujita et al. (1999) [28]	P substance	Human disc and synovial fluid	Expression of substance P seems to be closely related to histopathological changes of the human TMJ with internal derangement.	Substance P associated with internal derangement
Suzuki et al. (1999) [29]	BMP-2	Human discs	BMP-2 was predominantly localized in chondrocytes around the damaged areas of the articular disks. BMP-2 expression was also found in synovial cells and endothelial cells of blood vessels.	BMP-2 associated with TMD
Yoshida et al. (1999) [30]	CD34 antigen	Human disc and synovial membrane	CD34 is suggested to be correlated with the process of angiogenesis induced by internal derangement of the TMJ.	Vascular endothelium and angiogenesis
Yoshida et al. (1999) [30]	TGF-beta and Tenascin	Human synovial membranes	TGF-beta and tenascin were distributed in the affected synovial membrane of TMJ with internal derangement. These findings suggested that TGF-beta and tenascin might have a close relationship with synovitis, followed by tissue repair.	TGF-beta and tenascin in TMD
Leonardi et al. (2000) [31]	CD44 Standard Form	Human discs	The up-regulation of CD44H observed in some dysfunctional TMJ discs seems to indicate a prevention of apoptosis in fibroblast-like cells and an important role in phenotypical change of fibrochondrocytes into chondroblastlike cells, enabling the aggregation of chondroid tissue pericellular matrix components.	Prevention of apoptosis in TMJ discs

Table 2. *Cont.*

Authors	IHC Marker	Study Design and Tissue Expression	Results	Conclusions
Ogura et al. (2002) [25]	IL-1 Beta and IL-6	Human synovial cells	IL-1 beta increased IL-6 production in synovial cells. Enhanced production of IL-6, which is associated with bone resorption and inflammatory response, seems to be related to the progression of TMD.	IL-6 associated with TMD
Suzuki et al. (2002) [32]	IL-1 Beta and TNF-alpha	Human synovial fluids	IL-1beta and TNF-alpha may be involved with TMJ internal derangement and coordinately play a role in pathogenesis of TMJ internal derangement.	
Tobe et al. (2002) [33]	IL-1 Beta and IL-8	Human synovial cells	IL-1beta stimulated IL-8 production through an increase in IL-8 gene expression in HTS cells, which may be associated with the increase in infiltrating inflammatory cells seen in the synovial membrane of TMJ disorders.	IL-8 associated with TMD
Leonardi et al. (2003) [34]	VEGF	Human Disks	In disc specimens from internal derangement of the TMJ with significant tissue degeneration/regeneration, VEGF was consistently expressed.	VEGF expression in human disc with TMD
Sato et al. (2003) [26]	IL-6	Human synovial tissue	In synovial tissues from 21 of the 46 joints with internal derangement, interleukin 6 (IL-6) was expressed in the synovial lining cells and in the mononuclear cells infiltrating the periphery of the blood vessels.	IL-6 associated with TMD
Suzuki et al. (2003) [35]	Bradykinin	Human synovial tissues and fluids	Bradykinin was also detected in 19 patients' TMJ synovial fluids and the average of bradykinin concentration in the synovial fluids of patients was higher than that of the healthy controls. Although a statistically significant correlation was not observed, these findings support the hypothesis that bradykinin may also be involved in the pathogenesis of TMJ pain and synovitis.	Bradykinin associated with pathogenesis of TMD pain and synovitis.
Sato et al. (2003) [36]	VEGF	Human synovial tissues	VEGF may have an important role in the genesis of joint effusion.	VEGF and joint effusion
Tanimoto et al. (2004) [37]	TGF-beta	Rabbit synovial membranes	TGF-beta 1 enhances the expression of HAS2 mRNA in the TMJ synovial membrane fibroblasts and may contribute to the production of high-molecular-weight HA in the joint fluid.	TGF-beta associated with high molecular weight hyaluronan and more viscous fluid
Kaneyama et al. (2005) [38]	Tumor Necrosis Factor-Alpha, interleukin-6, interleukin-1beta, Soluble Tumor Necrosis Factor Receptors I and II, interleukin-6 Soluble Receptor, interleukin-1 Soluble Receptor Type II, interleukin-1 Receptor Antagonist	Human synovial fluid	The concentrations of TNF-alpha, IL-6, IL-1beta, sTNFR-I, and sTNFR-II were significantly higher in the synovial fluid of patients than in controls ($p < 0.05$). TNF-alpha level was positively correlated with those of IL-6, sTNFR-I, and sTNFR-II. In particular, there was a highly significant positive correlation between sTNFR-I and sTNFR-II.	All associated with TMD

Table 2. *Cont.*

Authors	IHC Marker	Study Design and Tissue Expression	Results	Conclusions
Yamaguchi et al. (2005) [39]	Hypoxia and interleukin-1beta and MMPs	Rabbit discs	The results showed that the combination of hypoxia and IL-1beta caused a significant increase in MMP-1, MMP-3, MMP-9 and MMP-13 mRNA.	Hypoxia and IL-1beta increases MMPs
Ogura et. Al. (2005) [40]	TNF-alpha	Human synovial membranes	Production of interleukin (IL)-8, growth-related oncogene (GRO)-alpha, monocyte chemoattractant protein (MCP)-1, and regulated upon activation normal T-cell expressed and secreted (RANTES) protein by synovial fibroblasts was increased by TNF-alpha.	Increased protein production of chemokines by synovial fibroblasts in response to TNF-alpha treatment.
Sato et. Al. (2007) [41]	IL-8	Human synovial tissues	IL-8 was up-regulated in inflamed synovial tissues in patients with internal derangement.	IL-8 associated with TMD with internal derangement
Matsumoto et al. (2006) [42]	Cytokines	Human synovial fluid	In synovial fluid samples, angiogenin (Ang), fibroblast growth factor (FGF)-9, insulin-like growth factor-binding protein (IGFBP)-3, interleukin (IL)-1alpha, IL-1beta, IL-8, inducible protein (IP)-10, macrophage inflammatory protein (MIP)-1beta, osteoprotegerin (OPG), transforming growth factor (TGF)-beta2, tissue inhibitor of metalloproteinase (TIMP)-1, TIMP-2, tumor necrosis factor (TNF)-beta and vascular endothelial growth factor (VEGF) were detectable. Furthermore, the expression levels of Ang, brain-derived neurotrophic factor (BDNF), FGF-4, FGF-9, IGFBP-2, IL-8, MIP-1beta, OPG, pulmonary and activation-regulated protein (PARC), TGF-beta2, TIMP-2 and VEGF were significantly associated with the presence of JE; among these, nine cytokines (Ang, BDNF, FGF-4, FGF-9, IGFBP-2, MIP-1beta, PARC, TGF-beta2 and TIMP-2) were hitherto not described in TMD.	Cytokine expression and TMD
Deschner et al. (2007) [43]	Insulin-like Growth Factor 1	Rat discs	Continuous biophysical strain seems to downregulate the expression of the IGF system and may reduce the potential of fibrocartilage for growth and repair.	IGF-1 associated with TMJ strain
Sato et al. (2007) [43]	IL-8	Human synovial tissue	IL-8 was up-regulated in inflamed synovial tissues in patients with internal derangement.	IL-8 associated with TMD
Ke et al. (2007) [44]	NF-kB and TNF-alpha	Human synovial fibroblasts	Activation of NF-kB is responsible for TNF-alpha-induced COX-2 expression in synovial fibroblasts from the TMJ.	NFkB, TNF-alpha, and COX-2 associated with TMD
Tojyo et al. (2008) [45]	Hypoxia and interleukin-1beta	Human synovial fluid and discs	The combination of hypoxia and interleukin-1beta caused a significant increase in tenascin-C protein and mRNA of synovial fibroblasts, but not in disc cells.	Hypoxia and interleukin-1beta increase tenascin-C protein

Table 2. *Cont.*

Authors	IHC Marker	Study Design and Tissue Expression	Results	Conclusions
Leonardi (2011) [46]	TRAIL and DR5- and CASP3-dependent apoptosis	Human discs	Apoptosis involvement in the angiogenesis as a self-limiting process in patients with temporomandibular joint.	Apoptosis activation
Kaneyama (2010) [23]	Cytokine Receptors	Human synovial fluid	Mean concentrations of cytokine receptors (tumor necrosis factor receptors I and II, interleukin (IL) 6 soluble receptor, IL-1 soluble receptor type II, and IL-1 receptor antagonist and protein)in the synovial fluid were significantly higher in the 30 joints with JE than in the 25 joints without JE.	Increase in cytokine receptor levels with TMD
Leonardi et al. (2011) [47]	TNF-Related Apoptosis-Inducing Ligand Expression	Human discs	Cell loss due to the involvement of TRAIL apoptotic pathway seems, in part, responsible for TMJ disk degeneration.	Apoptosis and TMD
Kim (2012) [48]	Granulocyte Macrophage Colony stimulating Factor (GM-CSF), interferon (INF), interleukin (IL)-1β, IL-2, IL-4, IL-5, IL-6, IL-8, IL-10 and tumor necrosis factor (TNF)-α	Human synovial fluid	Granulocyte Macrophage Colony stimulating Factor (GM-CSF), interferon (INF), interleukin (IL)-1β, IL-2, IL-6, IL-8, IL-10 and tumor necrosis factor (TNF)-α were detected in the TMD group, whereas no cytokines were detected in the control group. IL-4 and IL-5 were not detected in either the TMD group or in the control group.	Certain cytokines (noted left) associated with TMD
Camejo (2013) [49]	Fas Ligand (FasL)	Human discs	A higher area of in situ immunostaining of FasL was found in temporomandibular discs with ADDwR [1].	Apoptosis and TMD
Sicurezza (2013) [50]	β-Defensin	Human discs	The presence of β-defensin-4 in human TMJ discs affected by ADDwoR [2] was found, hypothesizing its possible role in articular bone disruption.	β-Defensin and disc degeneration
Caporal et al. (2017) [29]	IL-6	Human discs	No significant differences were observed between the groups ADDwR and ADDwoR, and with and without OA, in respect to the expression of IL-6.	No difference in IL-6 between two TMD groups
Nakagawa et al. (2017) [49]	IFN-γ and TNF-α	Human synovial fluid	TNF-α and IFN-γ function in a cooperative manner to regulate inflammatory chemokine expression in synovial fibroblasts.	TNF-α and IFN-γ associated with inflammatory regulation in TMJ.
Kaya et al. (2018) [50]	Chemerin	Human synovial fluid	Chemerin in synovial fluid may play a role as a predisposing factor and may represent a novel potential prognostic biochemical marker in the pathogenesis of TMJ disorders.	Chemerin associated with TMD
Sorenson et al. (2018) [51]	IL-1	Human synovial fluid	Articles that compared IL-1 concentrations in TMD vs. control groups found significant differences.	IL-1 associated with TMD
Luo et al. (2019) [52]	IL-37b	Human synovial fluid	IL-37b suppressed inflammation and inhibited osteoclast formation.	Anti-inflammatory effects of IL-37.

[1] Anterior Disc Displacement with Reduction; [2] Anterior Disc Displacement without Reduction.

4.3. Proteoglycans

The review of the literature found several different proteoglycans that are involved with TMD, such as tenascin, vimentin, fibronectin, decorin, biglycan, veriscan, and elastin. Proteoglycans are proteins located in the extracellular matrix, cell surface, or intracellular granules. They are involved in several processes, such as cell signaling and organization of the extracellular matrix [53]. Yoshida and Leonardi found that tenascin was associated with degenerated tissue, specifically in the portion of the TMJ synovial membrane affected by internal derangement [54]. Additionally, Toriva et al. (2006) [55] found that versican is associated with TMD and causes growth-related changes and regional differences in the TMJ discs of rats. The full table of proteoglycans and their respective results may be seen in the table above. (Table 3).

Table 3. Proteoglycans (Tenascin, vimentin, fibronectin, decorin, biglycan, versican, elastin).

Authors	IHC Marker	Study Design and Tissue Expression	Results	Conclusions
Yoshida et al. (1996) [56]	Tenascin	Human synovial membranes	Synovial cells in the synovial membrane produce tenascin in the diseased human temporomandibular joint	Tenascin associated with TMD
Yoshida et al. (1997) [28]	Tenascin	Human disc and synovial membranes	Tenascin is expressed specifically in the portion of the TMJ synovial membrane affected with internal derangement	Tenascin associated with degenerated tissue
Mizoguchi et al. (1998) [5]	Biglycan, Decorin and Large Chondroitin-Sulphate Proteoglycan	Rat discs	Staining for biglycan was intense in the posterior band. In contrast, staining for decorin was faint in the intermediate zone and the central part of the posterior band, moderate in the anterior and posterior attachments and most intense in the junction between the anterior band and attachment. Similarly, there was intense staining for large chondroitin-sulphate proteoglycan in the peripheral band.	Presence of biglycan, decorin and large chondroitin-sulphate proteoglycan in disc.
Kuwabara et al. (2002) [57]	Decorin and Biglycan	Rat discs	Regional differences in staining for decorin became prominent at 4, 8 and 16 weeks; decorin was more abundant in the peripheral area of the band than in the central area. In contrast, staining for biglycan was evenly distributed throughout the disc until 4 weeks, and after that became intense in the anterior and posterior bands.	Decorin and biglycan present in discs
Leonardi et al. (2002) [58]	Vimentin and Alpha-Smooth Muscle Actin	Human discs	Vimentin is expressed by all disc cell populations and it does not appear to be influenced by any disease condition of the disc; on the other hand the up-regulation of alpha-SM actin immunolabelling seems to be correlated with histopathological findings of tears and clefts.	Localization of Vimentin and Alpha-Smooth Muscle Actin in TMJ discs
Yoshida et al. (2002) [59]	Tenascin	Human synovial fluid and discs	Tenascin was produced specifically in synovial cells and vascular endothelial cells and. Fibroblasts were affected in the portion of TMJ with internal derangement.	Tenascin expression in TMD

Table 3. *Cont.*

Authors	IHC Marker	Study Design and Tissue Expression	Results	Conclusions
Kondoh et al. (2003) [60]	Type II Collagen	Human discs	The percentage of type II collagen in immunoreactive disc cells was significantly higher in the outer part (the articular surfaces) than in the inner part (the deep central areas) of the disc.	Increased collagen synthesis in TMJ discs
Yoshida et al. (2004) [54]	Vimentin	Human discs and synovial membrane	There was an obvious distinction of vimentin immunoreactivity between the control specimens and internal derangement cases, in the posterior and/or anterior loose connective tissues. In particular, intensive vimentin expression was detected in the hypertrophic synovial membrane of internal derangement cases.	Vimentin present in synovial membrane of TMD
Leonardi et al. (2004) [61]	Fibronectin	Human discs	The findings suggest that TMJ disc tissue can express fibronectin and that the expression is more pronounced in disc specimens of patients with internal derangements.	Fibronectin associated with internal derangements
Paegle et al. (2005) [62]	Proteoglycans aggrecan, versican, biglycan, decorin, fibromodulin and hyaluronan synthase 1	Human discs	Aggrecan expression was higher in patients with chronic closed lock. Within posterior disc attachment specimens, chronic closed lock showed a tendency for higher expression of biglycan and hyaluronan synthase 1.	Proteoglycans associated with chronic closed lock
Toriya et al. (2006) [55]	Versican-core protein of a large chondroitin sulphate proteoglycan	Rat discs	Growth-related changes and regional differences exist in the expression of versican in the TMJ discs of growing rats.	Versican associated with TMD
Moraes et al. (2008) [63]	Collagen Type IV	Human fetus discs	Marker of type IV collagen showed the presence of blood vessels in the central region of the temporomandibular disc.	Collagen present in disc formation
Li et al. (2008) [64]	Hyaluronan and hyaluronan synthase	Human synovial membrane	IL-1beta functions on regulating HAS expression and consequently promoting the secretion of HA in synovial lining cells from TMJ.	Hyaluronan present in synovial membranes
Natiella et al. (2009) [65]	Collagen Type I and Fibronectin	Human synovial fluid and discs	Disc specimens with advanced morphologic pathology showed significant labeling for fibronectin in 3 of 3 cases and for collagen I in 4 of 4 cases. There was no considerable difference in detection of either fibronectin or collagen I in TMJ synovial aspirates from patients with advanced disc pathology compared with controls.	Fibronectin found in pathologic TMJ discs
Matsumoto et al. (2010) [66]	Hyaluronan and hyaluronan synthase (HAS)	Human discs	Hyaluronan synthase-3 is related to the pathological changes of human TMJ discs affected by ID.	HAS associated with TMJ degradation
Kiga et al. (2010) [67]	Lumican, CD34 and vascular endothelial growth factor	Human discs	Assembly and regulation of collagen fibers. Lumican new collagen network by fibroblast-like cells. Presence of VEGF and CD34 inside the deformed disc.	Lumican expression associated with increased CD34 and VEGF in TMJ

Table 3. *Cont.*

Authors	IHC Marker	Study Design and Tissue Expression	Results	Conclusions
Fang et al. (2010) [68]	Chondromodulin-1 (ChM-1)	Rabbit discs	ChM-1 may play a role in the regulation of TMJ remodeling by preventing blood vessel invasion of the cartilage.	Anti-angiogenic factor in TMJ discs
Kiga (2011) [69]	Lumican and fibromodulin under interleukin-1 beta (IL-1 β)-stimulated conditions	Human discs	Lumican and fibromodulin display different behaviors and that lumican may promote regeneration of the TMJ after degeneration and deformation induced by IL-1 β.	IL-1 β induces a significant increase in lumican mRNA, but not in fibromodulin mRNA.
Leonardi et al. (2011) [70]	Lubricin	Human discs	A longstanding TMJ disc injury, affects lubricin expression in the TMJ disc tissue and not its surfaces; moreover, lubricin immunostaining is not correlated to TMJ disc histopathological changes.	No correlation between TMD and lubricin expression
Leonardi et al. (2012) [71]	Lubricin	Human discs	Lubricin may have a role in normal disc posterior attachment physiology through the prevention of cellular adhesion as well as providing lubrication during normal bilaminar zone function.	Lubricin associated with TMJ homeostasis
Hill et al. (2014) [72]	Lubricin	Rat synovial fluid	Lack of lubricin in the TMJ causes osteoarthritis-like degeneration that affects the articular cartilage as well as the integrity of multiple joint tissues.	Protective effects of lubricin on TMJ
Shinohara et al. (2014) [73]	Tenascin-C (TNC)	Mice synovial fluid and disc	TNC was expressed in the wounded TMJ disc and mandibular fossa, lack of TNC may reduce fibrous adhesion formation in the TMJ.	TNC associated with TMD and fibrous adhesion formation
Leonardi (2016) [74]	Lubricin	Human synovial fluid	Lubricin levels were inversely correlated with age and to Wilkes score. Lubricin decreases in synovial fluid with advanced disease.	Protective effects of lubricin on TMJ

4.4. Hormones

Our search of the literature resulted in seven studies looking at the involvement of three hormones in the pathogenesis of TMD. These studies examined relaxin, estrogen and progesterone in tissues from rabbit, mice and baboon discal tissues. Relaxin is a hormone that has been shown to be involved in matrix remodeling during pregnancy, and its effects are potentiated in the presence of estradiol [75]. A common finding among the studies conducted by Naqvi, Kapila, Hashem, and Park showed that an increased level of relaxin in disc samples was associated with increased levels of MMPs, such as collagenase and stromelysis, and a decrease in the levels of collagen and glycosaminoglycans (GAGs) [4,71,75]. Hashem, however, also found that the hormone progesterone prevented the effects of relaxin and estrogen, limiting the amount of matrix loss by preventing induction of MMPs [76].

An alternative methodology of looking at the hormonal involvement in the pathogenesis was utilized in some studies by examining the estrogen and/or relaxin receptor. Wang's study examined the level of estrogen receptors α and β, relaxin receptors LGR7 and LGR8, and progesterone receptor in various joints, and they found a higher number of hormone receptors in the TMJ of mice when compared to the knee joint [76]. In a study conducted by Puri, it was found that the level of estrogen-α receptors in the TMJ, however, could be influenced by inflammation—an increase in inflammation was correlated with a decrease in estrogen-α receptors [77]. Together, these help illustrate the multifactorial etiology that makes understanding this disorder so difficult.

These findings support the idea that hormonal influences are a key factor in the gender disparity associated with TMD. With studies showing a large number of hormonal receptors located in the joint, and with studies showing an increase in relaxin and estrogen associated with increased MMP activity and decreased collagen and GAGs, evidence points toward hormones playing a key role in this disparity. Although these studies suggest this, however, it is important to note that all studies were carried out on different species, and more research should be conducted with human samples for stronger support. Similarly, future studies should focus on specific receptors and hormones to identify which hormones have the largest impact on the pathogenesis, as well as look at the role of inflammation. (Table 4).

Table 4. Hormones (estrogen, progesterone, relaxin).

Authors	IHC Marker	Study Design and Tissue Expression	Results	Conclusions
Naqvi et al. (2004) [76]	Relaxin and β-estradiol	Rabbit discs	Relaxin and β-estradiol plus relaxin induced the MMPs collagenase-1 and stromelysin-1 in fibrocartilaginous explants, accompanied by a loss of GAGs and collagen but not altering the synthesis of GAGs.	Relaxin associated with degenerative effects on TMJ
Hashem et al. (2006) [77]	Relaxin, β-estradiol, and progesterone alone or in various combinations.	Rabbit discs	Collagen caused by β-estradiol, relaxin, or β-estradiol + relaxin causes loss of disc glycosaminoglycans, Progesterone prevented relaxin- or β-estradiol-mediated loss of these molecules.	Estradiol and relaxin causes loss of glycosaminoglycans, and progesterone prevents it.
Wang et al. (2008) [78]	Estrogen receptors α, β, relaxin receptors LGR7 and LGR8, and progesterone receptor	Mice discs	TMJ cells had higher ER-α (>2.8-fold), ER-β (>2.2-fold), LGR7 (>3-fold) and PR (>1.8-fold), and lower LGR8 (0.5-fold).	Estrogen, relaxin, and progesterone within TMJ discs
Kapila et al. (2009) [11]	β-estradiol or relaxin and progesterone/ estrogen receptors (ER)-α and −β, relaxin-1 receptor (RXFP1, LGR7), and INSL3 receptor (RXFP2, LGR8)	Rabbit discs	Relaxin produces a dose-dependent induction of tissue-degrading enzymes of the matrix metalloproteinase family, specifically MMP-1, MMP-3, MMP-9, and MMP-13 in cell isolates and tissue explants from TMJ fibrocartilage. The induction of these MMPs is accompanied by loss of collagen and glycosaminoglycans, which was blocked by a pan-MMP inhibitor. Progesterone attenuated the induction of MMPs.	Estrogen, relaxin, and progesterone within TMJ discs
Puri et al. (2009) [75]	Estrogen receptor α	Rat disc and synovial fluid	The number of ER α -positive cells in the TMJ was not affected by inflammation or 17 beta-estradiol with exception of the retrodiscal tissue	Low estrogen receptor in normal TMJ
McDaniel et al. (2014) [79]	Estrogen and progesterone	Baboon discs	Treatment of baboon TMJ disc cells with estrogen led to reduced PRG4 promoter activity and mRNA expression in vitro.	Negative regulation of PRG4 by estrogen.
Park et al. (2019) [80]	Estrogen and progesterone	Mouse discs	Administration of Estrogen but not Progesterone caused a significant loss of TMJ collagen and glycosaminoglycans, accompanied by amplification of ERα and specific increases in MMP9 and MMP13 expression.	E2-mediated upregulation of MMP9 and MMP13.

4.5. Miscellaneous

Many of the studies that resulted from our search of the current literature included biomarkers that do not fit the preceding categories. Leonardi et. al. looked at the levels of heat shock protein 27 (HSP-27) in human disc tissues and found that an increased level of the protein was associated with TMD, while the level of HSP-27 was virtually undetectable in samples from fetal tissue and normal joint tissue [79]. Heat shock proteins act within cells to protect against protein denaturation; however, they have been suggested to be involved in the pathogenesis of many diseases with elevated levels for prolonged periods of time.

Huang et al. looked at the presence of BCL-2 and BAX in disc tissue from rabbit samples [80]. BCL-2 is a protooncogene that is anti-apoptotic in nature, while BAX is a pro-apoptotic gene. Immunohistochemical staining of the disc samples revealed the presence of receptors for both BCL-2 and BAX, which suggests the possibility of their involvement in the regulation of the apoptotic process of TMD [80]. With respect to tumor suppressors, Castorina et. al. looked at the presence of P53 expression (along with VEGF) in human disc samples from subjects with TMD and found a positive correlation between pathological changes of the tissue and expression of the tumor suppressor protein P53, as well as levels of VEGF [81]. These studies are added evidence of the complex nature of the regulation of the apoptotic pathway in TMD.

Aquaporins are membrane channel proteins that allow water to pass through cell membranes to help regulate osmotic hemostasis [82]. They are found as part of normal physiology, with their presence identified in processes in the kidney, gastrointestinal tract, among other tissues, but have also been identified as part of various pathologic processes. In two studies carried out by Loreto, human disc samples were utilized to examine the presence of aquaporin-1 channels [53,83]. In the first study, it was found that these channels are part of normal homeostatic regulation of the TMJ; however, in the second study, it was found that the channel protein was upregulated in disc samples from subjects with TMD when compared with samples without TMD [82,84].

It has been previously shown that angiogenesis plays an important role in the progression of TMD, and, with that, the involvement of VEGF. Dickkopf-related protein (DKK-1) is a protein that inhibits the Wnt/beta-cantenin pathway and has been previously shown to play a role in physiologic and pathologic processes thorugh its role in promoting angiogenesis and recruitment of endothelial cells [85]. In a study conducted by Jiang, human synovial fibroblasts were examined for the presence of DKK-1 and found a high level of expression in the synovial fluid in samples obtained from subjects with TMD [84].

Elastin-derived proteins (EDPs) are bioactive peptides that arise with the degradation of elastin, a major component of the extracellular matrix [86]. An increase in elastin degradation and turnover has been shown to be a part of many pathological processes and through induction of inflammatory markers. Kobayachi used immunohistochemical staining to examine levels of EDPs in human synovial fluid and found an association between EDP, IL-6 and MMP-12, suggesting a role in EDPs in regulation of the inflammatory cascade in the TMJ [86].

Finally, in a study conducted by Fujimura, human synovial fluid was analyzed using electrophoresis in an attempt to look at synovial protein patterns [87]. The group found a significantly higher concentration of total protein in the synovial fluid of samples of patients with various stages of TMD when compared to healthy samples, especially in the amount of proteins with a higher molecular weight. Even when comparing the differences of protein concentration within the stages, a more advanced (osteoarthritis) stages were associated with a higher concentration of protein than earlier stages. These results support the idea that there is an increase in the level of inflammation as TMD progresses to higher stages. (Table 5).

Table 5. Miscellaneous (aquaporin, DKK1, etc).

Authors	IHC Marker	Study Design and Tissue Expression	Results	Conclusions
Leonardi et al. (2002) [87]	Heat shock protein 27	Human discs	HSP-27 upregulates in internal derangement specimens with major histopathological changes; it is not expressed or only weakly expressed in TMJ discs of fetuses and normal TMJ discs.	HSP-27 upregulation in TMD
Huang et al. (2004) [82]	Bcl-2 and Bax	Rabbit discs	Chondrocyte cytoplasm in the disc exhibited a high intensity for Bcl-2, while Bax activity was only sporadically observed. Bcl-2 and Bax proteins are present in TMJ cartilage and their expression patterns suggest that these oncoproteins are involved in chondrocyte survival or death via apoptotic pathways.	Apoptotic pathways within TMJ discs
Fujimura et al. (2006) [84]	Synovial fluid proteins	Human synovial fluid	Approximately 22 different protein bands with molecular weights ranging from 14 to 700 kd were clearly discernible on electrophoresis. The relative amounts of specific proteins in the SF of the TMD group were also different from those in the AS group ($p < 0.05$). The major difference in total protein concentration appeared to be due to the increased abundance of relatively high molecular weight proteins (>140 kd) in the TMD patients as compared to the AS group.	Increased concentration of synovial proteins associated with increased stage of TMD.
Loreto (2012) [85]	Aquaporin	Human discs	AQP1 is normally expressed in the TMJ disc and confirm a role for it in the maintenance of TMJ homeostasis.	Aquaporin acts on TMJ homeostasis
Loreto (2012) [85]	Aquaporin	Human discs	Aquaporin-1 is expressed and upregulated in temporomandibular joint with anterior disc displacement (both with and without reduction).	Channel protein involved in plasma membrane water permeability
Jiang (2015) [81]	Dickkopf-related Protein 1 (DKK-1)	Human synovial fibroblasts	DKK-1 is associated with angiogenesis in the synovial fluid of patients with TMD.	DKK-1 associated with TMD
Kobayashi (2017) [53]	Elastin-derived peptides	Human synovial fluid	Upregulation of IL-6 and MMP-12 expression by EDPs may be mediated through elastin-binding proteins (EBP) and a protein kinase A signaling cascade.	Elastin-derived peptides modulate the inflammatory cascade
Castorina et al. (2019) [83]	P53 and VEGF	Human discs	P53 and VEGF expression in TMJ discs with internal derangement correlate with degeneration.	P53 and VEGF associated with TMD

5. Conclusions

The pathogenesis underlying the development and progression temporomandibular joint disorder is a very complex biochemical process. As evidenced by this review, much research has been conducted to identify the different underlying involvement of biological markers. Through this review, recurring evidence provides confidence in suggesting the involvement of matrix metalloproteinases that may be involved in basement membrane and disc pathogenesis, inflammatory markers that may be modulating pain and TMJ tissue breakdown, as well as point to differences in gender prevalence of TMD. Moreover, proteoglycans and several proteins were found to be involved in the inflammatory and apoptotic cascade contributing to TMD progression and severity. Evidence in the current literature has allowed us to increase the comprehension of the underlying pathophysiology of TMD; however, as mentioned throughout this paper, more organized research on larger human samples needs to be conducted in order to delve deeper into the understanding of how this disease develops and progresses.

Author Contributions: Conceptualization, L.E.A. and A.D.; methodology, L.E.A.; software, L.E.A.; validation, L.E.A., A.D. and M.L.B.; formal analysis, L.E.A.; investigation, L.E.A.; resources, L.E.A.; data curation, L.E.A.; writing—original draft preparation, L.E.A.; writing—review and editing, M.L.B.; visualization, M.L.B.; supervision, L.E.A.; project administration, L.E.A. All authors have read and agreed to the published version of the manuscript.

Funding: This research received no external funding.

Institutional Review Board Statement: Not applicable.

Informed Consent Statement: Not applicable.

Data Availability Statement: Data can be found on PubMed.

Acknowledgments: We would like to acknowledge the Marquette University School of Dentistry for their support with this research.

Conflicts of Interest: The authors declare no conflict of interest.

References

1. Okeson, J.S. General Considerations in the Treatment of Temporomandibular Disorders. In *Management of Temporomandibular Disorders and Occlusion*, 6th ed.; Elsevier: Amsterdam, The Netherlands, 2008.
2. Murphy, M.K.; MacBarb, R.F.; Wong, M.E.; Athanasiou, K.A. Temporomandibular Disorders: A Review of Etiology, Clinical Management, and Tissue Engineering Strategies. *Int. J. Oral Maxillofac. Implant.* **2013**, *28*, e393–e414. [CrossRef]
3. Yadav, S.; Yang, Y.; Dutra, E.H.; Robinson, J.L.; Wadhwa, S. Temporomandibular Joint Disorders in Older Adults. *J. Am. Geriatr. Soc.* **2018**, *66*, 1213–1217. [CrossRef] [PubMed]
4. Yoshida, H.; Fujita, S.; Iizuka, T.; Yoshida, T.; Sakakura, T. The specific expression of tenascin in the synovial membrane of the temporomandibular joint with internal derangement: An immunohistochemical study. *Histochem. Cell Biol.* **1997**, *107*, 479–484. [CrossRef] [PubMed]
5. Mizoguchi, I.; Scott, P.G.; Dodd, C.M.; Rahemtulla, F.; Sasano, Y.; Kuwabara, M.; Satoh, S.; Saitoh, S.; Hatakeyama, Y.; Kagayama, M.; et al. An immunohistochemical study of the localization of biglycan, decorin and large chondroitin-sulphate proteoglycan in adult rat temporomandibular joint disc. *Arch. Oral Biol.* **1998**, *43*, 889–898. [CrossRef] [PubMed]
6. Tanaka, A.; Kumagai, S.; Kawashiri, S.; Takatsuka, S.; Nakagawa, K.; Yamamoto, E.; Matsumoto, N. Expression of matrix metalloproteinase-2 and -9 in synovial fluid of the temporomandibular joint accompanied by anterior disc displacement. *J. Oral Pathol. Med.* **2001**, *30*, 59–64. [CrossRef]
7. Yoshida, H.; Fukumura, Y.; Fujita, S.; Nishida, M.; Iizuka, T. The distribution of cyclooxygenase-1 in human temporomandibular joint samples: An immunohistochemical study. *J. Oral Rehabil.* **2001**, *28*, 511–516. [CrossRef]
8. Hu, L.; Liang, X.-H.; Zhu, G.-Q.; Hu, J.; Shi, Z.-D. Expression of urokinase-type plasminogen activator and urokinase-type plasminogen activator receptor in synovial fluid of patients with temporomandibular disorders. *Zhonghua Kou Qiang Yi Xue Za Zhi = Zhonghua Kouqiang Yixue Zazhi = Chin. J. Stomatol.* **2008**, *43*, 160–163. Available online: https://www.ncbi.nlm.nih.gov/pubmed/18788551 (accessed on 25 October 2022).
9. Matsumoto, T.; Tojyo, I.; Kiga, N.; Hiraishi, Y.; Fujita, S. Expression of adamts-5 in deformed human temporomandibular joint discs. *Histol. Histopathol.* **2008**, *23*, 1485–1493. [CrossRef]
10. Almeida, L.E.; Caporal, K.; Ambros, V.; Azevedo, M.; Noronha, L.; Leonardi, R.; Trevilatto, P.C. Immunohistochemical expression of matrix metalloprotease-2 and matrix metalloprotease-9 in the disks of patients with temporomandibular joint dysfunction. *J. Oral Pathol. Med.* **2014**, *44*, 75–79. [CrossRef]
11. Loreto, C.; Leonardi, R.; Musumeci, G.; Pannone, G.; Castorina, S. An ex vivo study on immunohistochemical localization of MMP-7 and MMP-9 in temporomandibular joint discs with internal derangement. *Eur. J. Histochem.* **2013**, *57*, e12. [CrossRef]
12. Kapila, S.; Wang, W.; Uston, K. Matrix Metalloproteinase Induction by Relaxin Causes Cartilage Matrix Degradation in Target Synovial Joints. *Ann. N. Y. Acad. Sci.* **2009**, *1160*, 322–328. [CrossRef]
13. Marchetti, C.; Cornaglia, I.; Casasco, A.; Bernasconi, G.; Baciliero, U.; Stetler-Stevenson, W. Immunolocalization of gelatinase-A (matrix metalloproteinase-2) in damaged human temporomandibular joint discs. *Arch. Oral Biol.* **1999**, *44*, 297–304. [CrossRef] [PubMed]
14. Quinn, J.H.; Kent, J.N.; Moise, A.; Lukiw, W.J. Cyclooxygenase-2 in synovial tissue and fluid of dysfunctional temporomandibular joints with internal derangement. *J. Oral Maxillofac. Surg.* **2000**, *58*, 1229–1232. [CrossRef] [PubMed]
15. Puzas, J.E.; Landeau, J.M.; Tallents, R.; Albright, J.; Schwarz, E.M.; Landesberg, R. Degradative pathways in tissues of the temporomandibular joint. Use of in vitro and in vivo models to characterize matrix metalloproteinase and cytokine activity. *Cells Tissues Organs* **2001**, *169*, 248–256. [CrossRef] [PubMed]

16. Yoshida, H.; Fukumura, Y.; Fujita, S.; Nishida, M.; Iizuka, T. The expression of cyclooxygenase-2 in human temporomandibular joint samples: An immunohistochemical study. *J. Oral Rehabil.* **2002**, *29*, 1146–1152. [CrossRef] [PubMed]
17. Yoshida, K.; Takatsuka, S.; Hatada, E.; Nakamura, H.; Tanaka, A.; Ueki, K.; Nakagawa, K.; Okada, Y.; Yamamoto, E.; Fukuda, R. Expression of matrix metalloproteinases and aggrecanase in the synovial fluids of patients with symptomatic temporomandibular disorders. *Oral Surg. Oral Med. Oral Pathol. Oral Radiol. Endodontol.* **2006**, *102*, 22–27. [CrossRef]
18. Yoshida, K.; Takatsuka, S.; Tanaka, A.; Hatada, E.; Nakamura, H.; Nakagawa, K.; Okada, Y. Aggrecanase analysis of synovial fluid of temporomandibular joint disorders. *Oral Dis.* **2005**, *11*, 299–302. [CrossRef]
19. Leonardi, R.; Migliore, M.R.; Almeida, L.E.; Trevilatto, P.C.; Loreto, C. Limited fatty infiltration due to apoptosis in human degenerated temporomandibular joint disks: An immunohistochemical study. *J. Craniofac. Surg.* **2010**, *21*, 1508–1511. [CrossRef]
20. Loreto, C.; Almeida, L.E.; Trevilatto, P.; Leonardi, R. Apoptosis in displaced temporomandibular joint disc with and without reduction: An immunohistochemical study. *J. Oral Pathol. Med.* **2011**, *40*, 103–110. [CrossRef]
21. Nascimento, G.C.; Rizzi, E.; Gerlach, R.F.; Leite-Panissi, C.R.A. Expression of mmp-2 and mmp-9 in the rat trigeminal gangli-on during the development of temporomandibular joint inflammation. *Braz. J. Med. Biol. Res.* **2013**, *46*, 956–967. [CrossRef]
22. Perotto, J.H.; Camejo, F.D.A.; Doetzer, A.D.; Almeida, L.E.; Azevedo, M.; Olandoski, M.; Noronha, L.; Trevilatto, P.C. Expression of MMP-13 in human temporomandibular joint disc derangement and osteoarthritis. *Cranio®* **2017**, *36*, 161–166. [CrossRef] [PubMed]
23. Kim, Y.-K.; Kim, S.-G.; Kim, B.-S.; Lee, J.-Y.; Yun, P.-Y.; Bae, J.-H.; Oh, J.-S.; Ahn, J.-M.; Kim, J.-S.; Lee, S.-Y. Analysis of the cytokine profiles of the synovial fluid in a normal temporomandibular joint: Preliminary study. *J. Cranio-Maxillofac. Surg.* **2012**, *40*, e337–e341. [CrossRef] [PubMed]
24. Fu, K.-Y.; Ma, X.; Zhang, Z.; Pang, X.; Chen, W. Interleukin-6 in synovial fluid and HLA-DR expression in synovium from patients with temporomandibular disorders. *J. Orofac. Pain* **1995**, *9*, 131–137. Available online: https://www.ncbi.nlm.nih.gov/pubmed/7488982 (accessed on 25 October 2022). [PubMed]
25. Ogura, N.; Tobe, M.; Sakamaki, H.; Kujiraoka, H.; Akiba, M.; Abiko, Y.; Nagura, H. Interleukin-1β induces interleukin-6 mRNA expression and protein production in synovial cells from human temporomandibular joint. *J. Oral Pathol. Med.* **2002**, *31*, 353–360. [CrossRef]
26. Sato, J.; Segami, N.; Nishimura, M.; Demura, N.; Yoshimura, H.; Yoshitake, Y.; Nishikawa, K. Expression of interleukin 6 in synovial tissues in patients with internal derangement of the temporomandibular joint. *Br. J. Oral Maxillofac. Surg.* **2003**, *41*, 95–101. [CrossRef]
27. Camejo, F.D.A.; Azevedo, M.; Ambros, V.; Caporal, K.S.T.; Doetzer, A.D.; Almeida, L.E.; Olandoski, M.; Noronha, L.; Trevilatto, P.C. Interleukin-6 expression in disc derangement of human temporomandibular joint and association with osteoarthrosis. *J. Cranio-Maxillofac. Surg.* **2017**, *45*, 768–774. [CrossRef]
28. Yoshida, H.; Fujita, S.; Nishida, M.; Iizuka, T. The expression of substance P in human temporomandibular joint samples: An immunohistochemical study. *J. Oral Rehabil.* **1999**, *26*, 338–344. [CrossRef]
29. Suzuki, T.; Bessho, K.; Segami, N.; Nojima, T.; Iizuka, T. Bone morphogenetic protein-2 in temporomandibular joints with internal derangement. *Oral Surg. Oral Med. Oral Pathol. Oral Radiol. Endodontol.* **1999**, *88*, 670–673. [CrossRef]
30. Yoshida, H.; Yoshida, T.; Iizuka, T.; Sakakura, T.; Fujita, S. The expression of transforming growth factor beta (TGF-beta) in the synovial membrane of human temporomandibular joint with internal derangement: A comparison with tenascin expression. *J. Oral Rehabil.* **1999**, *26*, 814–820. [CrossRef]
31. Leonardi, R.; Villari, L.; Piacentini, C.; Bernasconi, G.; Baciliero, U.; Travali, S. Cd44 standard form (cd44h) expression and distribution in dysfunctional human temporomandibular joint discs. *Int. J. Oral Maxillofac. Surg.* **2000**, *29*, 296–300. Available online: https://www.ncbi.nlm.nih.gov/pubmed/11030403 (accessed on 25 October 2022). [CrossRef]
32. Suzuki, T.; Segami, N.; Nishimura, M.; Nojima, T. Co-expression of interleukin-1beta and tumor necrosis factor alpha in synovial tissues and synovial fluids of temporomandibular joint with internal derangement: Comparision with histological grading of synovial inflammation. *J. Oral Pathol. Med.* **2002**, *31*, 549–557. [CrossRef]
33. Tobe, M.; Ogura, N.; Abiko, Y.; Nagura, H. Interleukin-1β stimulates interleukin-8 production and gene expression in synovial cells from human temporomandibular joint. *J. Oral Maxillofac. Surg.* **2002**, *60*, 741–747. [CrossRef]
34. Leonardi, R.; Muzio, L.L.; Bernasconi, G.; Caltabiano, C.; Piacentini, C. Expression of vascular endothelial growth factor in human dysfunctional temporomandibular joint discs. *Arch. Oral Biol.* **2003**, *48*, 185–192. [CrossRef] [PubMed]
35. Suzuki, T.; Segami, N.; Nishimura, M.; Sato, J.; Nojima, T. Bradykinin expression in synovial tissues and synovial fluids obtained from patients with internal derangement of the temporomandibular joint. *Cranio®* **2003**, *21*, 265–270. [CrossRef] [PubMed]
36. Sato, J.; Segami, N.; Nishimura, M.; Kaneyama, K.; Demura, N.; Yoshimura, H. Relation between the expression of vascular endothelial growth factor in synovial tissues and the extent of joint effusion seen on magnetic resonance imaging in patients with internal derangement of the temporomandibular joint. *Br. J. Oral Maxillofac. Surg.* **2003**, *41*, 88–94. [CrossRef] [PubMed]
37. Tanimoto, K.; Suzuki, A.; Ohno, S.; Honda, K.; Tanaka, N.; Doi, T.; Yoneno, K.; Ohno-Nakahara, M.; Nakatani, Y.; Ueki, M.; et al. Effects of TGF-β on Hyaluronan Anabolism in Fibroblasts Derived from the Synovial Membrane of the Rabbit Temporomandibular Joint. *J. Dent. Res.* **2004**, *83*, 40–44. [CrossRef]

38. Kaneyama, K.; Segami, N.; Sun, W.; Sato, J.; Fujimura, K. Analysis of tumor necrosis factor-α, interleukin-6, interleukin-1β, soluble tumor necrosis factor receptors I and II, interleukin-6 soluble receptor, interleukin-1 soluble receptor type II, interleukin-1 receptor antagonist, and protein in the synovial fluid of patients with temporomandibular joint disorders. *Oral Surg. Oral Med. Oral Pathol. Oral Radiol. Endodontol.* **2004**, *99*, 276–284. [CrossRef]
39. Yamaguchi, A.; Tojyo, I.; Yoshida, H.; Fujita, S. Role of hypoxia and interleukin-1β in gene expressions of matrix metalloproteinases in temporomandibular joint disc cells. *Arch. Oral Biol.* **2005**, *50*, 81–87. [CrossRef]
40. Ogura, N.; Tobe, M.; Sakamaki, H.; Nagura, H.; Abiko, Y.; Kondoh, T. Tumor necrosis factor-alpha increases chemokine gene expression and production in synovial fibroblasts from human temporomandibular joint. *J. Oral Pathol. Med.* **2005**, *34*, 357–363. [CrossRef]
41. Sato, J.; Segami, N.; Nishimura, M.; Yoshitake, Y.; Kaneyama, K.; Kitagawa, Y. Expression of interleukin 8 in synovial tissues in patients with internal derangement of the temporomandibular joint and its relationship with clinical variables. *Oral Surg. Oral Med. Oral Pathol. Oral Radiol. Endodontol.* **2007**, *103*, 467–474. [CrossRef]
42. Matsumoto, K.; Honda, K.; Ohshima, M.; Yamaguchi, Y.; Nakajima, I.; Micke, P.; Otsuka, K. Cytokine profile in synovial fluid from patients with internal derangement of the temporomandibular joint: A preliminary study. *Dentomaxillofacial Radiol.* **2006**, *35*, 432–441. [CrossRef]
43. Deschner, J.; Rath-Deschner, B.; Reimann, S.; Bourauel, C.; Götz, W.; Jepsen, S.; Jäger, A. Regulatory effects of biophysical strain on rat TMJ discs. *Ann. Anat.-Anat. Anz.* **2007**, *189*, 326–328. [CrossRef] [PubMed]
44. Ke, J.; Long, X.; Liu, Y.; Zhang, Y.; Li, J.; Fang, W.; Meng, Q. Role of NF-κB in TNF-α-induced COX-2 Expression in Synovial Fibroblasts from Human TMJ. *J. Dent. Res.* **2007**, *86*, 363–367. [CrossRef] [PubMed]
45. Tojyo, I.; Yamaguchi, A.; Nitta, T.; Yoshida, H.; Fujita, S.; Yoshida, T. Effect of hypoxia and interleukin-1? on expression of tenascin-C in temporomandibular joint. *Oral Dis.* **2007**, *14*, 45–50. [CrossRef] [PubMed]
46. Leonardi, R.; Almeida, L.E.; Rusu, M.; Sicurezza, E.; Palazzo, G.; Loreto, C. Tumor Necrosis Factor-Related Apoptosis-Inducing Ligand Expression Correlates to Temporomandibular Joint Disk Degeneration. *J. Craniofac. Surg.* **2011**, *22*, 504–508. [CrossRef]
47. Camejo, F.D.A.; Almeida, L.E.; Doetzer, A.D.; Caporal, K.S.T.; Ambros, V.; Azevedo, M.; Alanis, L.R.A.; Olandoski, M.; Noronha, L.; Trevilatto, P.C. FasL expression in articular discs of human temporomandibular joint and association with osteoarthrosis. *J. Oral Pathol. Med.* **2013**, *43*, 69–75. [CrossRef]
48. Sicurezza, E.; Loreto, C.; Musumeci, G.; Almeida, L.E.; Rusu, M.; Grasso, C.; Leonardi, R. Expression of β-defensin 4 on temporomandibular joint discs with anterior displacement without reduction. *J. Cranio-Maxillofac. Surg.* **2013**, *41*, 821–825. [CrossRef]
49. Ohta, K.; Naruse, T.; Kato, H.; Ishida, Y.; Nakagawa, T.; Ono, S.; Shigeishi, H.; Takechi, M. Differential regulation by IFN-γ on TNF-α-induced chemokine expression in synovial fibroblasts from temporomandibular joint. *Mol. Med. Rep.* **2017**, *16*, 6850–6857. [CrossRef]
50. Kaya, G.; Yavuz, G.Y.; Kızıltunç, A. Expression of chemerin in the synovial fluid of patients with temporomandibular joint disorders. *J. Oral Rehabil.* **2018**, *45*, 289–294. [CrossRef]
51. Sorenson, A.; Hresko, K.; Butcher, S.; Pierce, S.; Tramontina, V.; Leonardi, R.; Loreto, C.; Bosio, J.; Almeida, L.E. Expression of Interleukin-1 and temporomandibular disorder: Contemporary review of the literature. *Cranio®* **2017**, *36*, 268–272. [CrossRef]
52. Luo, P.; Feng, C.; Jiang, C.; Ren, X.; Gou, L.; Ji, P.; Xu, J. IL-37b alleviates inflammation in the temporomandibular joint cartilage via IL-1R8 pathway. *Cell Prolif.* **2019**, *52*, e12692. [CrossRef] [PubMed]
53. Kobayashi, K.; Jokaji, R.; Miyazawa-Hira, M.; Takatsuka, S.; Tanaka, A.; Ooi, K.; Nakamura, H.; Kawashiri, S. Elastin-derived peptides are involved in the processes of human temporomandibular disorder by inducing inflammatory responses in synovial cells. *Mol. Med. Rep.* **2017**, *16*, 3147–3154. [CrossRef]
54. Yoshida, H.; Fukumura, Y.; Nishida, M.; Fujita, S.; Iizuka, T. The immunohistochemical distribution of vimentin in human temporomandibular joint samples. *J. Oral Rehabil.* **2004**, *31*, 47–51. [CrossRef] [PubMed]
55. Toriya, N.; Takuma, T.; Arakawa, T.; Abiko, Y.; Sasano, Y.; Takahashi, I.; Sakakura, Y.; Rahemtulla, F.; Mizoguchi, I. Expression and localization of versican during postnatal development of rat temporomandibular joint disc. *Histochem. Cell Biol.* **2005**, *125*, 205–214. [CrossRef]
56. Yoshida, H.; Yoshida, T.; Iizuka, T.; Sakakura, T.; Fujita, S. An immunohistochemical and in situ hybridization study of the expression of tenascin in synovial membranes from human temporomandibular joints with internal derangement. *Arch. Oral Biol.* **1996**, *41*, 1081–1085. [CrossRef] [PubMed]
57. Kuwabara, M.; Takuma, T.; Scott, P.G.; Dodd, C.M.; Mizoguchi, I. Biochemical and immunohistochemical studies of the protein expression and localization of decorin and biglycan in the temporomandibular joint disc of growing rats. *Arch. Oral Biol.* **2002**, *47*, 473–480. [CrossRef]
58. Leonardi, R.; Villari, L.; Piacentini, C.; Bernasconi, G.; Travali, S.; Caltabiano, C. Immunolocalization of vimentin and alpha-smooth muscle actin in dysfunctional human temporomandibular joint disc samples. *J. Oral Rehabil.* **2002**, *29*, 282–286. [CrossRef]
59. Yoshida, H.; Fujita, S.; Nishida, M.; Iizuka, T.; Yoshida, T.; Sakakura, T. The expression of tenascin mRNA in human temporomandibular joint specimens. *J. Oral Rehabil.* **2002**, *29*, 765–769. [CrossRef]

60. Kondoh, T.; Hamada, Y.; Iino, M.; Takahashi, T.; Kikuchi, T.; Fujikawa, K.; Seto, K. Regional differences of type II collagen synthesis in the human temporomandibular joint disc: Immunolocalization study of carboxy-terminal type II procollagen peptide (chondrocalcin). *Arch. Oral Biol.* **2003**, *48*, 621–625. [CrossRef]
61. Leonardi, R.; Michelotti, A.; Farella, M.; Caltabiano, R.; Lanzafame, S. Fibronectin Upregulation in Human Temporomandibular Joint Disks with Internal Derangement. *J. Craniofac. Surg.* **2004**, *15*, 678–683. [CrossRef]
62. Paegle, D.; Holmlund, A.; Hjerpe, A. Expression of proteoglycan mRNA in patients with painful clicking and chronic closed lock of the temporomandibular joint. *Int. J. Oral Maxillofac. Surg.* **2005**, *34*, 656–658. [CrossRef] [PubMed]
63. De Moraes, L.O.; Lodi, F.R.; Gomes, T.S.; Marques, S.R.; Fernandes, J.A., Jr.; Oshima, C.T.; Alonso, L.G. Immunohisto-chemical expression of collagen type iv antibody in the articular disc of the temporomandibular joint of human fetuses. *Ital. J. Anat. Embryol.* **2008**, *113*, 91–95. Available online: https://www.ncbi.nlm.nih.gov/pubmed/18702236 (accessed on 25 October 2022).
64. Li, J.; Long, X.; Ke, J.; Meng, Q.-G.; Lee, C.C.W.; Doocey, J.M.; Zhu, F. Regulation of HAS expression in human synovial lining cells of TMJ by IL-1β. *Arch. Oral Biol.* **2008**, *53*, 60–65. [CrossRef] [PubMed]
65. Natiella, J.R.; Burch, L.; Fries, K.M.; Upton, L.G.; Edsberg, L.E. Analysis of the Collagen I and Fibronectin of Temporomandibular Joint Synovial Fluid and Discs. *J. Oral Maxillofac. Surg.* **2009**, *67*, 105–113. [CrossRef]
66. Matsumoto, T.; Inayama, M.; Tojyo, I.; Kiga, N.; Fujita, S. Expression of hyaluronan synthase 3 in deformed human temporo-mandibular joint discs: In vivo and in vitro studies. *Eur. J. Histochem.* **2010**, *54*, e50. [CrossRef] [PubMed]
67. Kiga, N.; Tojyo, I.; Matsumoto, T.; Hiraishi, Y.; Shinohara, Y.; Fujita, S. Expression of lumican related to CD34 and VEGF in the articular disc of the human temporomandibular joint. *Eur. J. Histochem.* **2010**, *54*, e34. [CrossRef] [PubMed]
68. Fang, W.; Friis, T.E.; Long, X.; Xiao, Y. Expression of chondromodulin-1 in the temporomandibular joint condylar cartilage and disc. *J. Oral Pathol. Med.* **2009**, *39*, 356–360. [CrossRef] [PubMed]
69. Kiga, N.; Tojyo, I.; Matsumoto, T.; Hiraishi, Y.; Shinohara, Y.; Makino, S.; Fujita, S. Expression of lumican and fibromodulin following interleukin-1 beta stimulation of disc cells of the human temporomandibular joint. *Eur. J. Histochem.* **2011**, *55*, e11. [CrossRef]
70. Leonardi, R.; Almeida, L.E.; Loreto, C. Lubricin immunohistochemical expression in human temporomandibular joint disc with internal derangement. *J. Oral Pathol. Med.* **2011**, *40*, 587–592. [CrossRef]
71. Leonardi, R.; Rusu, M.; Loreto, F.; Loreto, C.; Musumeci, G. Immunolocalization and expression of lubricin in the bilaminar zone of the human temporomandibular joint disc. *Acta Histochem.* **2012**, *114*, 1–5. [CrossRef]
72. Hill, A.; Duran, J.; Purcell, P. Lubricin Protects the Temporomandibular Joint Surfaces from Degeneration. *PLoS ONE* **2014**, *9*, e106497. [CrossRef] [PubMed]
73. Shinohara, Y.; Okamoto, K.; Goh, Y.; Kiga, N.; Tojyo, I.; Fujita, S. Inhibition of fibrous adhesion formation in the temporomandibu-lar joint of tenascin-C knockout mice. *Eur. J. Histochem.* **2014**, *58*, 2337. [CrossRef] [PubMed]
74. Leonardi, R.; Perrotta, R.; Almeida, L.; Loreto, C.; Musumeci, G. Lubricin in synovial fluid of mild and severe temporomandibular joint internal derangements. *Med. Oral Patol. Oral Y Cirugía Bucal.* **2016**, *21*, e793–e799. [CrossRef] [PubMed]
75. Puri, J.; Hutchins, B.; Bellinger, L.L.; Kramer, P.R. Estrogen and inflammation modulate estrogen receptor alpha expression in specific tissues of the temporomandibular joint. *Reprod. Biol. Endocrinol.* **2009**, *7*, 155. Available online: https://www.ncbi.nlm.nih.gov/pubmed/20043825 (accessed on 25 October 2022). [CrossRef]
76. Naqvi, T.; Duong, T.T.; Hashem, G.; Shiga, M.; Zhang, Q.; Kapila, S. Relaxin's induction of metalloproteinases is associated with the loss of collagen and glycosaminoglycans in synovial joint fibrocartilaginous explants. *Thromb. Haemost.* **2005**, *7*, 1–11. [CrossRef]
77. Hashem, G.; Zhang, Q.; Hayami, T.; Chen, J.; Wang, W.; Kapila, S. Relaxin and β-estradiol modulate targeted matrix degradation in specific synovial joint fibrocartilages: Progesterone prevents matrix loss. *Thromb. Haemost.* **2006**, *8*, R98. [CrossRef]
78. Wang, W.; Hayami, T.; Kapila, S. Female hormone receptors are differentially expressed in mouse fibrocartilages. *Osteoarthr. Cartil.* **2009**, *17*, 646–654. [CrossRef]
79. McDaniel, J.S.; Babu, R.A.S.; Navarro, M.M.; LeBaron, R.G. Transcriptional regulation of proteoglycan 4 by 17β-estradiol in immortalized baboon temporomandibular joint disc cells. *Eur. J. Oral Sci.* **2014**, *122*, 100–108. [CrossRef]
80. Park, Y.; Chen, S.; Ahmad, N.; Hayami, T.; Kapila, S. Estrogen Selectively Enhances TMJ Disc but Not Knee Meniscus Matrix Loss. *J. Dent. Res.* **2019**, *98*, 1532–1538. [CrossRef]
81. Jiang, S.-J.; Li, W.; Li, Y.-J.; Fang, W.; Long, X. Dickkopf-related protein 1 induces angiogenesis by upregulating vascular endothelial growth factor in the synovial fibroblasts of patients with temporomandibular joint disorders. *Mol. Med. Rep.* **2015**, *12*, 4959–4966. [CrossRef]
82. Huang, Q.; Singh, B.; Sharawy, M. Immunohistochemical analysis of Bcl-2 and Bax oncoproteins in rabbit craniomandibular joint. *Arch. Oral Biol.* **2003**, *49*, 143–148. [CrossRef] [PubMed]
83. Fujimura, K.; Segami, N.; Yoshitake, Y.; Tsuruoka, N.; Kaneyama, K.; Sato, J.; Kobayashi, S. Electrophoretic separation of the synovial fluid proteins in patients with temporomandibular joint disorders. *Oral Surg. Oral Med. Oral Pathol. Oral Radiol. Endodontol.* **2005**, *101*, 463–468. [CrossRef] [PubMed]
84. Loreto, C.; Castro, E.L.; Musumeci, G.; Loreto, F.; Rapisarda, G.; Rezzani, R.; Castorina, S.; Leonardi, R.; Rusu, M.C. Aquaporin 1 expression in human temporomandibular disc. *Acta Histochem.* **2012**, *114*, 744–748. [CrossRef]

85. Loreto, C.; Galanti, C.; Almeida, L.E.; Leonardi, R.; Pannone, G.; Musumeci, G.; Carnazza, M.L.; Caltabiano, R. Expression and localization of aquaporin-1 in temporomandibular joint disc with internal derangement. *J. Oral Pathol. Med.* **2012**, *41*, 642–647. [CrossRef] [PubMed]
86. Leonardi, R.; Caltabiano, M.; Cascone, P.; Loreto, C. Expression of Heat Shock Protein 27 (HSP27) in Human Temporomandibular Joint Discs of Patients with Internal Derangement. *J. Craniofac. Surg.* **2002**, *13*, 713–717. [CrossRef] [PubMed]
87. Castorina, S.; Lombardo, C.; Castrogiovanni, P.; Musumeci, G.; Barbato, E.; Almeida, L.E.; Leonardi, R. P53 and vegf expres-sion in human temporomandibular joint discs with internal derangement correlate with degeneration. *J. Biol. Regul. Homeost. Agents* **2019**, *33*, 1657–1662. [CrossRef] [PubMed]

Journal of
Clinical Medicine

Systematic Review

Intra-Articular Local Anesthetics in Temporomandibular Disorders: A Systematic Review and Meta-Analysis

Karolina Lubecka [1], Kamila Chęcińska [2], Filip Bliźniak [1], Maciej Chęciński [1], Natalia Turosz [3], Adam Michcik [4], Dariusz Chlubek [5,*] and Maciej Sikora [5,6]

[1] Department of Oral Surgery, Preventive Medicine Center, Komorowskiego 12, 30-106 Cracow, Poland; lubeckarolina@gmail.com (K.L.); fblizniak@gmail.com (F.B.); maciej@checinscy.pl (M.C.)

[2] Department of Glass Technology and Amorphous Coatings, Faculty of Materials Science and Ceramics, AGH University of Science and Technology, Mickiewicza 30, 30-059 Cracow, Poland; checinska@agh.edu.pl

[3] Institute of Public Health, Jagiellonian University Medical College, Skawińska 8, 31-066 Cracow, Poland; natalia.turosz@gmail.com

[4] Department of Maxillofacial Surgery, Medical University of Gdansk, Mariana Smoluchowskiego 17, 80-214 Gdańsk, Poland; adammichcik@gumed.edu.pl

[5] Department of Biochemistry and Medical Chemistry, Pomeranian Medical University, Powstańców Wielkopolskich 72, 70-111 Szczecin, Poland; sikora-maciej@wp.pl

[6] Department of Maxillofacial Surgery, Hospital of the Ministry of Interior, Wojska Polskiego 51, 25-375 Kielce, Poland

* Correspondence: dchlubek@pum.edu.pl

Abstract: This systematic review with meta-analysis was conducted to evaluate the effectiveness of local anesthetic administration into temporomandibular joint cavities in relieving pain and increasing mandibular mobility. Randomized controlled trials were included with no limitation on report publication dates. Final searches were performed on 15 October 2023, using engines provided by the US National Library, Bielefeld University, and Elsevier Publishing House. The risk of bias was assessed using the Cochrane Risk of Bias 2 tool. Articular pain and mandible abduction values and their mean differences were summarized in tables and graphs. Eight studies on a total of 252 patients evaluating intra-articular administration of articaine, bupivacaine, lidocaine, and mepivacaine were included in the systematic review. None of the eligible studies presented a high risk of bias in any of the assessed domains. An analgesic effect of intra-articular bupivacaine was observed for up to 24 h. In the long-term follow-up, there were no statistically significant changes in quantified pain compared to both the baseline value and the placebo group, regardless of the anesthetic used (articaine, bupivacaine, and lidocaine). There is no scientific evidence on the effect of intra-articular administration of local anesthesia on the range of motion of the mandible. Therefore, in the current state of knowledge, the administration of local anesthetics into the temporomandibular joint cavities can only be considered as a short-term pain relief measure.

Keywords: temporomandibular joint disorders; intra-articular injections; bupivacaine; lidocaine; articaine; mepivacaine

Citation: Lubecka, K.; Chęcińska, K.; Bliźniak, F.; Chęciński, M.; Turosz, N.; Michcik, A.; Chlubek, D.; Sikora, M. Intra-Articular Local Anesthetics in Temporomandibular Disorders: A Systematic Review and Meta-Analysis. *J. Clin. Med.* **2024**, *13*, 106. https://doi.org/10.3390/jcm13010106

Academic Editor: Luís Eduardo Almeida

Received: 24 November 2023
Revised: 17 December 2023
Accepted: 21 December 2023
Published: 24 December 2023

1. Introduction

1.1. Background

The pain associated with mastication results mainly from abnormal function of the temporomandibular joints or masticatory muscles [1–4]. In the course of physical examination, it is possible to distinguish muscle pain from joint pain, which guides further diagnosis and treatment [5–7]. The severity of articular pain is measured primarily on a visual analog scale [8,9]. Painful reduction of jaw mobility, and thus difficulty with food intake, is a significant factor in deteriorating the patient's quality of life [10–14]. In cases of severe articular pain with limited mouth opening, the main cause may be difficult to determine, and the therapy undertaken is sometimes empirical [1,15]. The range of methods for treating articular pain and limited mandibular mobility is very wide and combination therapies are often used [3–7,10]. Depending on the specific diagnosis and the severity of the symptoms, psychotherapy, physiotherapy, systemic pharmacotherapy, splint therapy, dry needling and intramuscular injections, intra-articular injections, arthrocentesis, arthroscopy, open joint surgery, and joint replacement are used [3–7,10]. Due to their minimal invasiveness, intra-articular administration of drugs, hyaluronic acid, and blood products are the subject of current scientific research [5,10]. Too slow or ineffective treatment of pain induces the search for ad hoc relief. One of the obvious solutions to relieving persistent pain is to start drug therapy [16]. An alternative to systemic analgesic pharmacotherapy is the local administration of drugs. The professional literature indicates the possibility of performing nearby nerve blocks, intra-articular administration of anti-inflammatory drugs (corticosteroids and nonsteroidal anti-inflammatory drugs), analgesics (opioids), or local anesthetics [17–22].

1.2. Rationale

Local anesthetics are well-researched, inexpensive, widely used, and easily available [23,24]. Intra-articular administration of local anesthetics is one of the recognized, albeit controversial, orthopedic procedures [25,26]. According to the latest reports, cytotoxicity of bupivacaine in intra-articular injections is suspected, and there have been many reports of chondrolysis after shoulder arthroscopy in which intra-articular injections of anesthetics were used [25,27]. Injection of local anesthetics into the temporomandibular joint cavities is not commonly performed [21,28]. The growing number of scientific publications on this topic allows for a first cross-sectional evaluation and encourages a critical assessment of the effectiveness of the discussed therapy [21,28].

1.3. Objectives

The primary objective of this systematic review with meta-analysis is to compare the effectiveness of local anesthetic administration in temporomandibular joint cavities in relieving articular pain compared to placebo or other substances. An analogous comparison with regard to the change in the range of mandibular mobility was adopted as a secondary objective.

2. Materials and Methods

This study was conducted in accordance with the Preferred Reporting Items for Systematic Reviews and Meta-Analyses guidelines and reported in the International Prospective Register of Systematic Reviews database under the number: CRD42023484735 [29,30].

2.1. Eligibility Criteria

The review included randomized controlled trials of the injection of local anesthetics into the temporomandibular joints. The inclusion of studies on healthy volunteers was intended to ensure the comprehensiveness of the systematic review. The outcome criterion was taken into account when qualifying for the meta-analysis, but failure to meet it did not exclude the report from inclusion in the systematic review. Data on changes in articular pain intensity or mandible mobility were required. No time frame limits were applied. Details are presented in Table 1.

Table 1. Eligibility criteria.

	Criteria for Inclusion	Criteria for Exclusion
Problem	Patients diagnosed with temporomandibular disorders or healthy volunteers	Cadaver studies
Intervention	Local anesthetic intra-articular injection	None
Comparison	Injection with the omission or replacement of the local anesthetic	None
Outcomes	Articular pain severity or mandibular mobility range	Unquantifiable results
Settings	Randomized trials	Less than 5 patients per group

2.2. Information Sources

This systematic review was conducted using three of the leading medical database search engines: (1) the US National Library of Medicine PubMed, (2) the German Bielefeld Academic Search Engine, and (3) the Dutch Elsevier Scopus [31–33]. All final searches were conducted on the same day, 15 October 2023.

2.3. Search Strategy

The search strategy was based on the Problem and Intervention eligibility criteria. It was implemented in the form of a single query, common to all search engines: "temporomandibular joint" AND (injection OR injections) AND ("local anesthetic" OR "local anaesthetic" OR benzocaine OR procaine OR chloroprocaine OR lidocaine OR prilocaine OR tetracaine OR bupivacaine OR cinchocaine OR ropivacaine).

2.4. Selection Process

Records identified during the medical database search were transferred to the Rayyan automation tool (Qatar Computing Research Institute, Doha, Qatar and Rayyan Systems, Cambridge, MA, USA) [34]. This tool identified potential duplicates, which were manually verified and, if confirmed, removed (M.C. and K.C.). Then, continuing the use of Rayyan, the same researchers performed a screening based on titles and abstracts. Records identified unanimously as not meeting the Problem or Intervention criteria were discarded. In cases of discrepancies in the assessment, a given record was left for full-text verification of the report. Eligibility determined on the basis of the full content of the reports was initially assessed by two researchers (M.C. and K.C.) and, in case of doubt, discussed among the entire team until consensus was reached.

2.5. Data Collection Process

The data needed for synthesis were extracted (M.C. and K.C.) without the use of automation tools, based only on the content of the reports. The data collected from the content of the reports were initially entered into a summary table, and its refined version was placed in the Results section of this paper.

2.6. Data Items

The researchers collected the following data items from the content of primary study reports: (1) initial severity of articular pain; (2) final severity of articular pain; (3) initial range of mandible abduction; and (4) final range of mandible abduction. Joint pain expressed on a visual analog scale (VAS) was preferred, and in the absence of this variable, a numerical rating scale (NRS) was accepted and converted proportionally to 0–10 if necessary. The quantified pain values were unified on a scale of 0–10. For several different measurements of mandible abduction range, preference was given in the following order: (1) maximum unassisted mouth opening; (2) maximum mouth opening without pain; and (3) maximum manually assisted mouth opening. These data were collected for both the study and control groups.

Additionally, the following data were extracted: (1) first author and year of publication of the report; (2) the number of patients in the study and control groups; (3) type of local anesthetic used in the study group; (4) dose of the single-administered preparation; (5) number of injections and interval between injections; (6) interventions in control groups; and (7) description of co-interventions in study and control groups.

2.7. Study Risk of Bias Assessment

A bias risk assessment was performed (K.L. and F.B.) using the revised Cochrane risk-of-bias tool for randomized trials (RoB 2) and visualized using the Robvis tool (c7c1bdd) [35,36].

2.8. Effect Measures

For the purposes of synthesizing and presenting the results, mean differences were calculated for (1) articular pain and (2) the range of mandible abduction. A MedCalc tool was used (MedCalc Software (22.016), Ostend, Belgium) [37].

2.9. Synthesis Methods

All studies with a risk of bias lower than high were qualified for synthesis. Synthesis was performed by combining data extracted from reports and mean difference results in a summary table. The synthesis results were presented in charts using Google Workspace tools (Google LLC, Mountain View, CA, USA, (Version: 16 October 2023 Scheduled Release)).

2.10. Reporting Bias Assessment

In the case of missing data, this fact was noted, but the series was not discarded. No further reporting bias assessments were undertaken.

2.11. Certainty Assessment

The summary of findings was tabulated with the risk of bias in the source reports provided.

3. Results

3.1. Study Selection

The selection process identified 23 reports describing the administration of local anesthetics into the temporomandibular joint cavities. Of these, 15 were excluded from the full-text review due to the lack of a control group or the inability to assess the effect of local anesthetics despite control groups present (Table 2) [38–52]. Non-qualified reports are addressed in the Discussion section. Eight reports presented randomized trials comparing interventions differing only in the intra-articular administration of local anesthetic (Table 3) [53–60]. The data from their content were extracted, synthesized, and analyzed in the Results section of this review. The detailed selection process is illustrated in a flow diagram (Figure 1).

Table 2. Reports excluded at the eligibility stage.

First Author, Publication Year	Title	DOI or PMID Number (If the Former Was Not Assigned)	Reason for Exclusion
Bhargava, 2023 [38]	A Comparative Preliminary Randomized Clinical Study to Evaluate Heavy Bupivacaine Dextrose Prolotherapy (HDP) and Autologous Blood Injection (ABI) for Symptomatic Temporomandibular Joint Hypermobility Disorder.	10.1007/s12663-022-01738-x	Ineligible comparison
Shan, 2023 [39]	Platelet-rich plasma and hyaluronic acid for injection treatment of temporomandibular joint degeneration in Affiliated Stomatological Hospital of Guangzhou Medical University	10.57760/sciencedb.o00013.00022	Ineligible comparison
Prakash, 2022 [40]	Intra-articular platelet-rich plasma injection versus hydrocortisone with local anesthetic injections for temporomandibular disorders.	10.6026/97320630018991	Ineligible comparison
Dasukil, 2021 [41]	Efficacy of Prolotherapy in Temporomandibular Joint Disorders: An Exploratory Study.	10.1007/s12663-020-01328-9	No comparison
Louw, 2019 [42]	Treatment of Temporomandibular Dysfunction With Hypertonic Dextrose Injection (Prolotherapy): A Randomized Controlled Trial With Long-term Partial Crossover.	10.1016/j.mayocp.2018.07.023	Ineligible comparison

Table 2. *Cont.*

First Author, Publication Year	Title	DOI or PMID Number (If the Former Was Not Assigned)	Reason for Exclusion
Gupta, 2018 [43]	Comparison between intra-articular platelet-rich plasma injection versus hydrocortisone with local anesthetic injections in temporomandibular disorders: A double-blind study.	10.4103/njms.njms_69_16	Ineligible comparison
Refai, 2017 [44]	Long-term therapeutic effects of dextrose prolotherapy in patients with hypermobility of the temporomandibular joint: a single-arm study with 1-4 year follow-up	10.1016/j.bjoms.2016.12.002	No comparison
Chakraborty, 2016 [45]	Ultrasound-Guided Temporomandibular Joint Injection for Chronic Posthemimandibulectomy Jaw Pain	10.1213/xaa.0000000000000384	No comparison (case report)
Zhou, 2014 [46]	Modified dextrose prolotherapy for recurrent temporomandibular joint dislocation.	10.1016/j.bjoms.2013.08.018	No comparison
Samiee, 2011 [47]	Temporomandibular joint injection with corticosteroid and local anesthetic for limited mouth opening.	10.2334/josnusd.53.321	No comparison
Refai, 2011 [48]	The efficacy of dextrose prolotherapy for temporomandibular joint hypermobility: A preliminary prospective, randomized, double-blind, placebo-controlled clinical trial	10.1016/j.joms.2011.02.128	Ineligible comparison
Guarda Nardini, 2002 [49]	Treatment of temporomandibular joint closed-lock using intra-articular injection of mepivacaine with immediate resolution durable in time (six months follow-up)	PMID: 11845117	No comparison
Sato, 1997 [50]	Effect of lavage with injection of sodium hyaluronate for patients with nonreducing disk displacement of the temporomandibular joint.	10.1016/s1079-2104(97)90337-1	Ineligible comparison
Kamada, 1993 [51]	Changes in synovial fluid N-acetyl-beta-glucosaminidase activity in the human temporomandibular joint with dysfunction.	PMID: 8182502	Ineligible comparison
Danzig, 1992 [52]	Effect of an anesthetic injected into the temporomandibular joint space in patients with TMD.	PMID: 1298765	No comparison

Table 3. Reports included in the systematic review.

First Author, Publication Year	Title	DOI or PMID Number (If the Former Was Not Assigned)
Zarate, 2020 [53]	Dextrose Prolotherapy Versus Lidocaine Injection for Temporomandibular Dysfunction: A Pragmatic Randomized Controlled Trial.	10.1089/acm.2020.0207
Ziegler, 2010 [54]	Analgesic effects of intra-articular morphine in patients with temporomandibular joint disorders: a prospective, double-blind, placebo-controlled clinical trial.	10.1016/j.joms.2009.04.049
Ayesh, 2007 [55]	Effects of local anesthetics on somatosensory function in the temporomandibular joint area.	10.1007/s00221-007-0893-4
Zuniga, 2007 [56]	The Analgesic Efficacy and Safety of Intra-Articular Morphine and Mepivacaine Following Temporomandibular Joint Arthroplasty	10.1016/j.joms.2007.04.001
Tjakkes, 2007 [57]	The effect of intra-articular injection of ultracain in the temporomandibular joint in patients with preauricular pain—A randomized prospective double-blind placebo-controlled crossover study	10.1097/ajp.0b013e31802f0950
Lobbezoo, 2003 [58]	Effects of TMJ anesthesia and jaw gape on jaw-stretch reflexes in humans	10.1016/s1388-2457(03)00155-x

Table 3. *Cont.*

First Author, Publication Year	Title	DOI or PMID Number (If the Former Was Not Assigned)
Furst, 2001 [59]	The use of intra-articular opioids and bupivacaine for analgesia following temporomandibular joint arthroscopy: a prospective, randomized trial.	10.1053/joms.2001.25820
Gu, 1998 [60]	Visco-supplementation therapy in internal derangement of temporomandibular joint.	PMID: 11245058

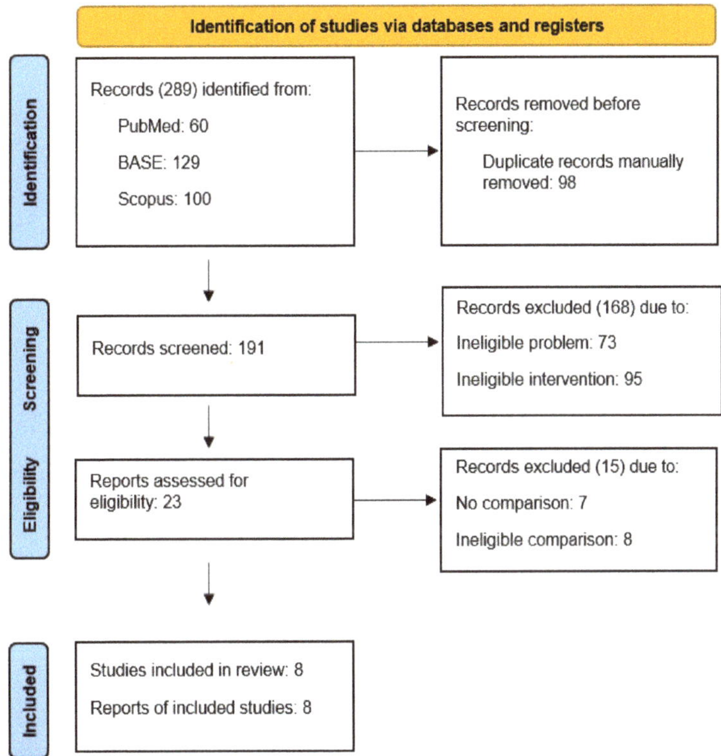

Figure 1. PRISMA flow diagram.

3.2. Study Characteristics

Eight eligible randomized clinical trials were conducted on a total of 252 patients [53–60]. Articaine (4%, one study), bupivacaine (0.25–0.5%, three studies), lidocaine (0.2–2%, two studies), and mepivacaine (1–3%, two studies) were used. In the study groups, 1 to 3 or as-needed interventions were performed at intervals of 2 to 42 days, which gave a total dose of anesthetic from 2.5 to 40 mg. Six of the studies reported results for placebo groups (Table 4) [54–59].

3.3. Risk of Bias in Studies

The risk of bias results obtained using the Cochrane risk-of-bias tool for randomized trials (RoB 2) are illustrated in Figures 2 and 3. None of the studies identified any risks related to the randomization process or deviations from the intended interventions. There were some concerns in the remaining domains but no high risk of bias was noted. Therefore, none of the studies were rejected at this stage.

Table 4. Study characteristics.

First Author	Diagnosis	Number of Patients (Study/Controls)	Local Anesthetic Type and Dose	Number of Injections/Interval (Days)	Total Anesthetic Dose	Interventions in Control Groups	Co-Interventions Common to the Study Group and Controls
Zarate	TMDs	15/14	0.2% lidocaine, 1 mL	3/28	6 mg	0.2% lidocaine + 20% dextrose, 1 mL injection	N/A
Ziegler	TMDs	12/36	0.5% bupivacaine, 2 mL	3/2	30 mg	(1) 2 mL of 0.9% saline, (2) 5 mg of morphine in X mL of 0.9% saline, or (3) 10 mg of morphine in X mL of 0.9% saline injection	N/A
Ayesh	Healthy volunteers	14/14 (contralateral control)	0.25% bupivacaine, 1 mL	1/N/A	2.5 mg	0.9% saline injection	N/A
Zuniga	Post-arthroplasty status	10/25	3% mepivacaine, 1 mL	1/N/A	30 mg	(1) 0.9% saline injection (placebo), (2) 0.1% morphine, 1 mL or (3) 0.1% morphine + 3% mepivacaine, 1 mL	N/A
Tjakkes	TMDs	20/20 (crossover control)	4% articaine + 1:200,000 pinephrine, 0.5 mL	2/14	40 mg	0.9% saline, 0.5 mL	Application of EMLA topical anesthesia 45 min before injection
Lobbezoo	Healthy volunteers	6/5	1% mepivacaine, 1 mL	2/14–42	20 mg	0.9% saline, 1 mL	N/A
Furst	Post-arthroscopy state	24/8	0.5% bupivacaine, 2 mL	1/N/A	10 mg	(1) 0.9% saline, 3 mL (2) 0.2% morphine, 1 mL or (3) 0.2% morphine, 1 mL + 0.5% bupivacaine, 2 mL	Postoperative application of morphine (4 mg, i.v.) and acetaminophen 325 mg + codeine 15 mg
Gu	TMDs	43/20	2% lidocaine, 1 mL	1 (more if needed)/N/A	20 mg	1% hyaluronic acid, 0.3–1 mL	Infiltration anesthesia of the preauricular area (2% lidocaine), articular cavity irrigation (0.9% saline, 5 mL)

TMDs—temporomandibular disorders; N/A—not applicable.

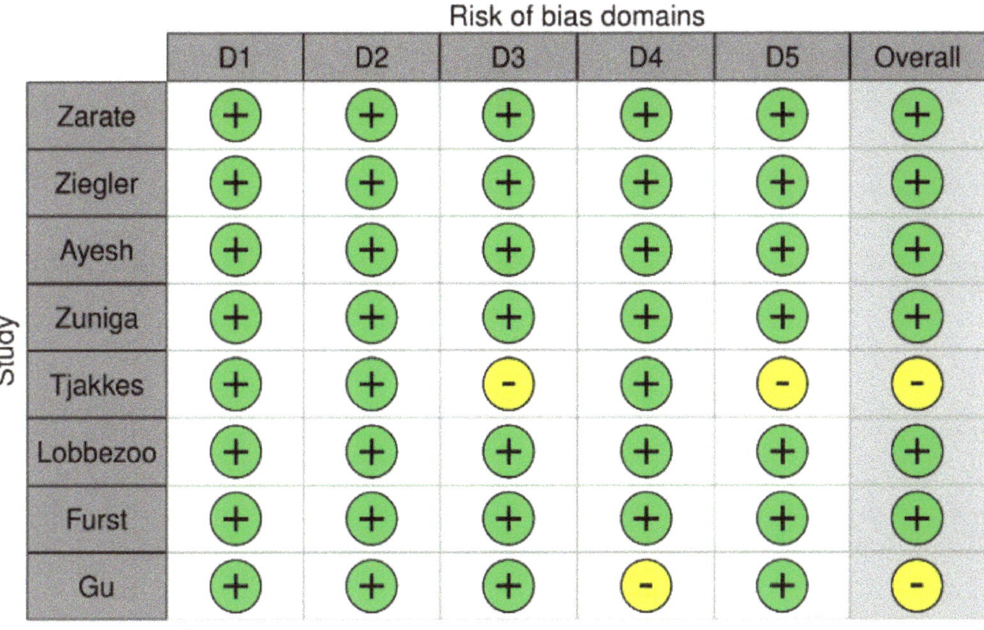

Figure 2. Risk of bias in studies: traffic light plot.

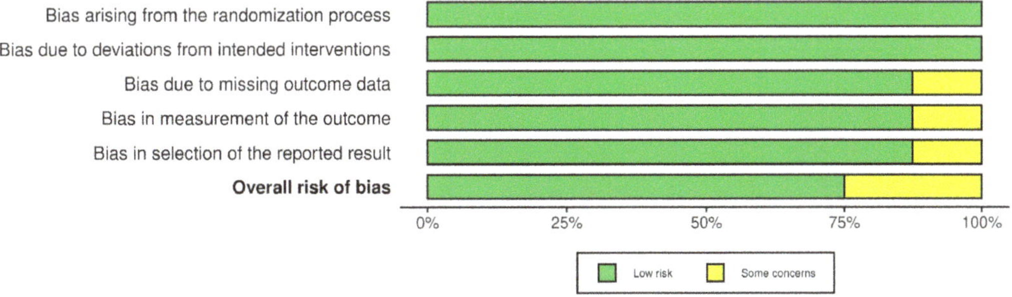

Figure 3. Risk of bias in studies: summary plot.

3.4. Results of Individual Studies

3.4.1. Pain Intensity

The initial values of articular pain in the study and placebo groups and their change over time are presented in Table 5. The qualified reports by Zuniga et al. and Gu et al. described the change in the severity of articular pain, but the results could not be quantified [56,60]. In the study by Ziegler et al., the intervention was performed three times; only the initial pain values and those during the observation period after the first and before the second intervention were entered into the table. Due to the lack of standard deviations provided, the standard error of the calculated mean differences could not be determined for the study by Ziegler et al. [54]. In the study by Ayesh et al., partial pain values were not recorded, which made further processing of the results impossible [55]. In the studies by Zarate et al. and Tjakkes et al., statistically significant differences in articular pain values compared to the initial values were observed over a period of 2 weeks to a year [53,57]. Tjakkes et al. reported numerical results of changes in the intensity of joint pain for the group receiving a local anesthetic versus the placebo group. Fourteen days after the intervention, articular pain decreased in the treated group and increased in the control group. The difference in mean was -1.0 ± 0.9. However, this difference was not statistically significant ($p = 0.28$). Pain severity during this period was lower at every measurement for the local anesthetic groups [57]. Furst et al. reported only post-intervention pain values. In the study by these authors, no statistically significant differences in pain intensity were observed in any of the patient groups in the period from 4 to 24 h after the intervention [59].

3.4.2. Mandibular Abduction

The range of mandibular abduction was measured only in some of the studies included in this review. Zarate et al. examined the maximum pain-free mandibular abduction, which they defined as the interincisal distance at the opening to the point of discomfort [53]. Tjakkes et al. examined maximum unassisted mandibular abduction, which they defined as the interincisal distance with maximum mouth opening on request [57]. Lobbezzo et al. took this parameter into account but as constant values for which they determined the tension of the masticatory muscles [58]. The study by Gu et al. presented only unquantified results, which made further processing impossible [60]. Therefore, Table 6 presents only the results from the studies by Zarate et al. and Tjakkes et al., which, due to their paucity, cannot be further processed [53,57]. The extent of mandibular abduction after 3 months of follow-up in the study by Zarate et al. did not differ significantly ($p = 0.59$) between the lidocaine and lidocaine plus dextrose groups. In both groups, there was an increase in these values compared to the initial ones, but it was not statistically significant [53]. During a two-week follow-up period, Tjakkes et al. did not observe a statistically significant difference between the abduction gain among the patient groups ($p = 0.10$) [57].

Table 5. Results of individual studies in the VAS/NRS articular pain domain over time. Values (with standard deviations where known) and mean differences (with standard errors where calculable) are provided.

First Author	Intervention Group	Sample Size	Initial	2 h	4 h	6 h	8 h	12 h	16 h	20 h	1 Day	2 Days	2 Weeks	1 Month	2 Months	3 Months	1 Year
Zarate	Lidocaine	21 joints	7.2 ± 0.8	N/S	N/S	N/S	N/S	N/S	N/S	N/S	N/S	N/S	N/S	5.4 ± 2.1 -1.8 ± 0.5*	4.6 ± 2.2 -2.6 ± 0.5*	4.3 ± 2.6 -2.9 ± 0.6*	4.6 ± 2.5 -2.6 ± 0.6*
	Lidocaine + dextrose	22 joints	7.2 ± 1.1	N/S	N/S	N/S	N/S	N/S	N/S	N/S	N/S	N/S	N/S	4.4 ± 2.4 -2.8 ± 0.6*	4.4 ± 2.4 -2.8 ± 0.6*	2.9 ± 2.6 -4.3 ± 0.6*	2.4 ± 2.6 -4.8 ± 0.6*
Ziegler	Bupivacaine 0.5%	12 joints	10	1.0 9.0	N/S	6.4 3.6	N/S	7.9 2.1	N/S	N/S	8.0 2.0	8.0 2.0	N/S	N/S	N/S	N/S	N/S
	Morphine 10 mg	12 joints	10	1.8 8.2	N/S	3.6 6.4	N/S	3.86.2	N/S	N/S	3.8 6.2	3.8 6.2	N/S	N/S	N/S	N/S	N/S
	Morphine 5 mg	12 joints	10	3.2 6.8	N/S	4.0 6.0	N/S	4.0 6.0	N/S	N/S	3.9 6.1	3.9 6.1	N/S	N/S	N/S	N/S	N/S
	Placebo (0.9% saline)	12 joints	10	6.2 3.8	N/S	7.1 2.9	N/S	8.1 1.9	N/S	N/S	8.5 1.5	8.51.5	N/S	N/S	N/S	N/S	N/S
Tjakkes	Articaine	20 patients	6.6 ± 2.3	N/S	N/S	N/S	N/S	N/S	N/S	N/S	N/S	N/S	5.3 ± 3.1 -1.3 ± 2.4*	N/S	N/S	N/S	N/S
	Placebo (0.9% saline)	20 patients	5.8 ± 2.2	N/S	N/S	N/S	N/S	N/S	N/S	N/S	N/S	N/S	6.3 ± 2.7 0.5 ± 1.5*	N/S	N/S	N/S	N/S
Furst	Bupivacaine	8 joints	N/S	2.2 ± 2.6 (baseline)	2.4 ± 1.9 0.2 ± 1.1†	2.1 ± 2.2 -0.1 ± 1.2†	1.3 ± 1.0 -0.9 ± 1.0†	1.9 ± 1.4 -0.3 ± 1.0†	2.8 ± 2.0 0.6 ± 1.2†	2.1 ± 1.4 -0.1 ± 1.0†	3.2 ± 3.2 1.0 ± 1.5†	N/S	N/S	N/S	N/S	N/S	N/S
	Bupivacaine + morphine	8 joints	N/S	5.5 ± 2.9 (baseline)	6.8 ± 2.0 1.3 ± 1.2†	7.2 ± 1.5 1.7 ± 1.2†	6.2 ± 1.8 0.7 ± 1.2†	4.8 ± 2.5 -0.7 ± 1.3†	4.2 ± 2.4 -1.3 ± 1.3†	4.1 ± 1.6 -1.4 ± 1.2†	3.9 ± 1.4 -1.6 ± 1.1†	N/S	N/S	N/S	N/S	N/S	N/S
	Morphine	8 joints	N/S	3.7 ± 2.4 (baseline)	3.4 ± 3.5 -0.3 ± 1.5†	4.2 ± 3.6 0.5 ± 1.5†	3.6 ± 2.0 -0.1 ± 1.1†	3.8 ± 2.7 0.1 ± 1.3†	4.8 ± 1.9 1.1 ± 1.1†	2.7 ± 2.2 -1.0 ± 1.2†	2.4 ± 2.1 -1.3 ± 1.1†	N/S	N/S	N/S	N/S	N/S	N/S
	Placebo (0.9% saline)	8 joints	N/S	5.6 ± 1.9 (baseline)	7.2 ± 2.0 1.6 ± 1.0†	6.8 ± 2.4 1.2 ± 1.1†	7.6 ± 2.0 2.0 ± 1.0†	6.7 ± 2.4 1.1 ± 1.1†	4.7 ± 1.4 -0.9 ± 0.8†	4.0 ± 2.6 -1.6 ± 1.1†	4.4 ± 2.6 -1.2 ± 1.1†	N/S	N/S	N/S	N/S	N/S	N/S

N/S—not specified; *—statistically significant ($p < 0.05$); †—no statistical significance ($p \geq 0.05$).

3.5. Results of Syntheses

Below, graphical summaries of the results of the mean values of articular pain intensity for the groups receiving anesthetics (black lines) and the placebo groups (gray lines) are presented. Due to the large number of time points for the initial observation period, the same data are illustrated multiple times but at different scales in Figures 4–6. The following figures illustrate the differences in pain intensity between the study groups and placebo on two observation time scales (Figures 7 and 8). Attempts to fit linear regression models were unsuccessful; therefore, mean values are illustrated (dashed lines).

3.6. Certainty of Evidence

The key results of this systematic review are summarized in Table 7. The articular pain results are supported by four and the mouth opening range results by two randomized clinical trials, all free of high risk of bias in any of the domains assessed.

Table 6. Results of individual studies in the mandibular abduction domain (in millimeters). Values (with standard deviations where known) and mean differences (with standard errors where calculable) are provided.

First Author	Intervention Group	Sample Size	Initial	2 Weeks	3 Months
Zarate	Lidocaine	14 patients	42.4 ± 9.3	N/S	47.8 ± 7.8 5.4 ± 3.2 †
	Lidocaine + dextrose	15 patients	38.7 ± 10.6	N/S	43.4 ± 9.8 4.7 ± 3.7 †
Tjakkes	Articaine	20 patients	N/S	N/S 2.0 ± 2.9	N/S
	Placebo (0.9% saline)	20 patients	N/S	N/S 0.4 ± 2.9	N/S

N/S—not specified; †—no statistical significance ($p \geq 0.05$).

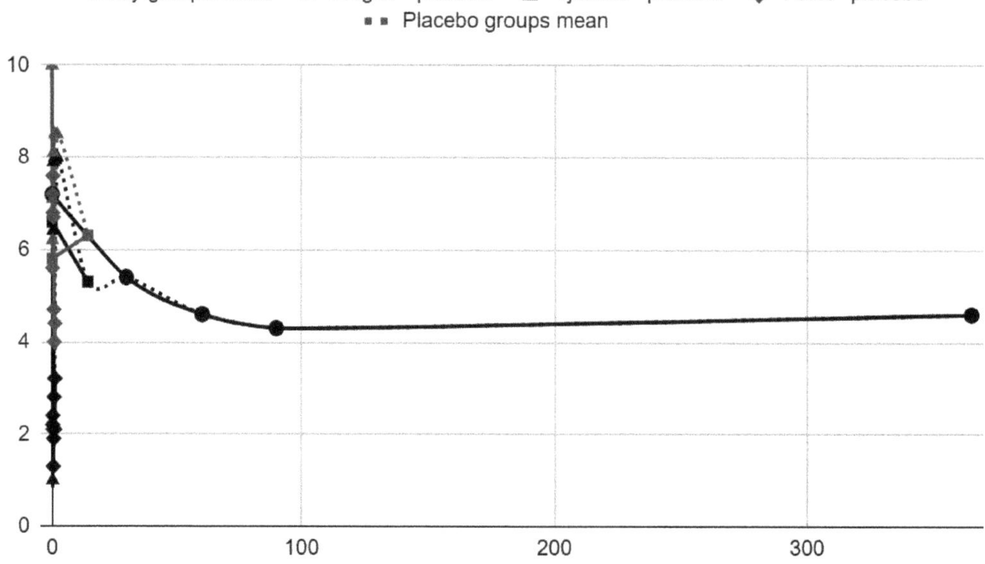

Figure 4. VAS/NRS pain intensity over time in days.

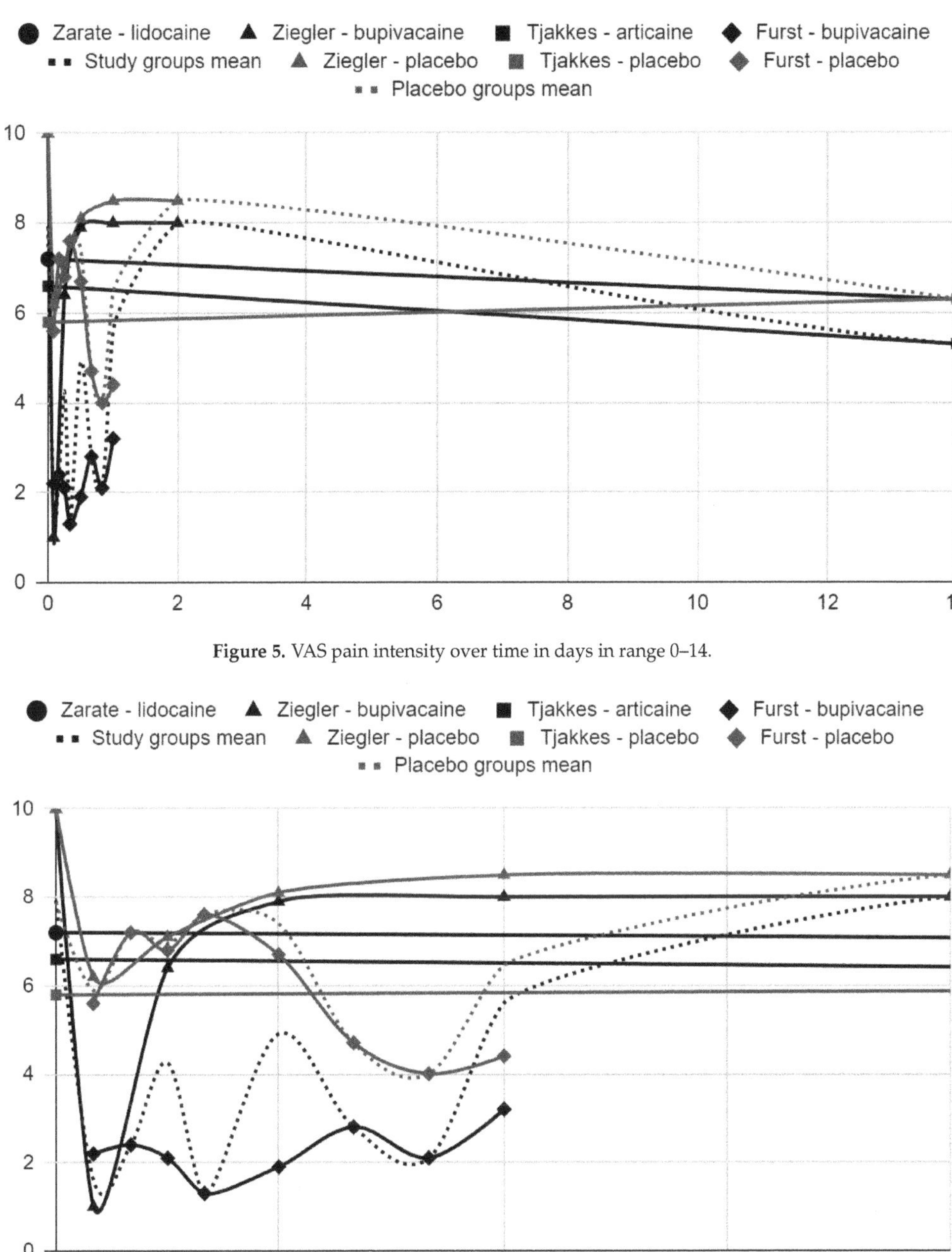

Figure 5. VAS pain intensity over time in days in range 0–14.

Figure 6. VAS pain intensity over time in days in range 0–2.

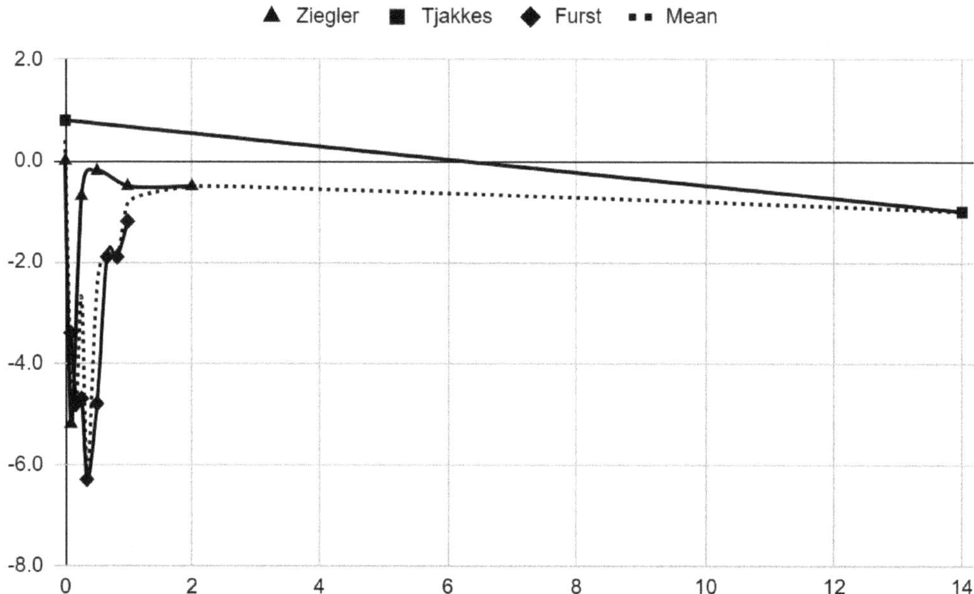

Figure 7. Differences in VAS pain between treatment and placebo groups over time in days.

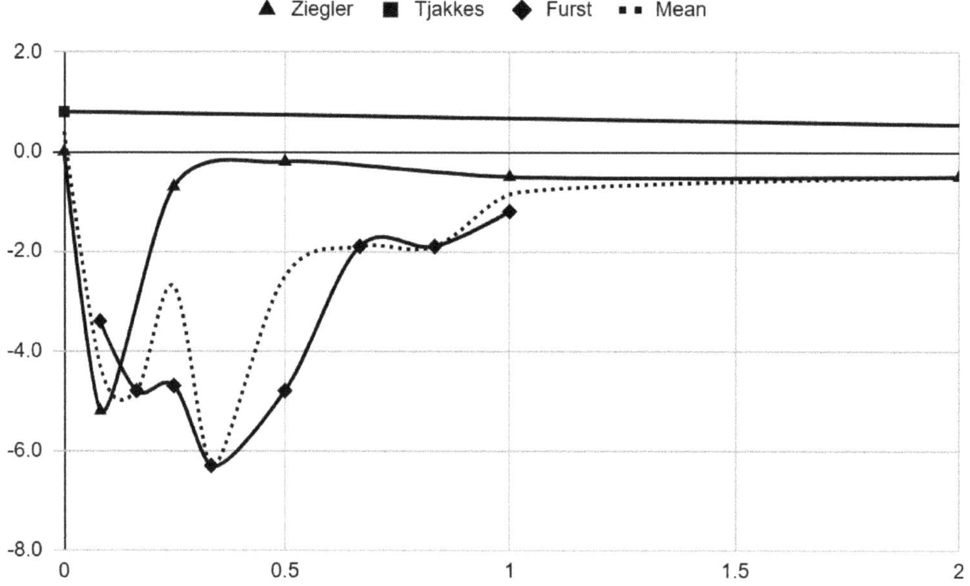

Figure 8. Differences in VAS pain between treatment and placebo groups over time in days in range 0–2.

Table 7. Summary of findings.

Domain	Level of Evidence	Number of Studies	Total Sample Sizes in Local Anesthetic and Control Groups	Risk of Bias in Studies	Findings
Articular pain	Randomized controlled trials	4	61/40	From low to some concerns	- On the first day, bupivacaine provides a noticeably better analgesic effect than placebo. However, there is no data to prove the statistical significance of these differences. - In the period from 24 h to 2 weeks, there are no statistically significant differences between local anesthetics and the placebo. - No further observations were made with placebo control groups.
Mandibular mobility	Randomized controlled trials	2	34/20	From low to some concerns	- No statistically significant difference in jaw abduction was observed between the articaine and placebo groups 2 weeks after the intervention. - No statistically significant change in the range of mouth opening was observed after 3 months of observation of the group that received intra-articular lidocaine.

4. Discussion

4.1. General Interpretation of the Results in the Context of Other Evidence

4.1.1. Randomized Controlled Trials

The administration of local anesthetics into the cavities of the temporomandibular joints seems to be justified only in the context of temporary relief of articular pain. The use of bupivacaine appears to provide an immediate analgesic effect that lasts for up to 24 h [54,59]. This may be sufficient to transfer the patient to a higher-reference center and undergo other types of treatment. In the long-term follow-up, none of the local anesthetics resulted in statistically significant improvement in pain either from baseline or compared to the placebo groups [53,57].

Based on the collected material, the intra-articular administration of local anesthetics does not seem to have any effect on the range of motion of the mandible. However, these conclusions are supported by only two clinical studies, which do not cover the first day [53,57].

4.1.2. Ineligible Control Group Studies

In studies in which (a) the control differed not only in the absence of a local anesthetic or (b) in which all groups received an anesthetic, it was impossible to assess the impact of the substances in question on the treatment outcome. Local anesthetics were administered in combination with (1) hypertonic dextrose as a standard prolotherapy protocol (3 studies), (2) corticosteroid giving a worse effect than platelet-rich plasma without anesthetic (2 studies), or (3) sodium hyaluronate as a joint rinsing agent before viscosupplementation (2 studies). Lidocaine was also used for joint cavity rinsing before collecting synovial fluid for laboratory tests (1 study) [38–40,42,43,48,50,51].

4.1.3. No Control Group Studies

Uncontrolled studies make it possible to determine changes in disease severity indicators during treatment, in relation to their initial values. However, they present difficulty in indicating which of the components of the therapy accounted for success. The following studies included combinations of local anesthetics with hypertonic dextrose and corticosteroids. In the hypertonic dextrose reports of Dasukil et al., Refai, and Zhou et al. it was unanimously assessed that prolotherapy brought the desired effect in the form of resolution of dislocations. The first two studies also presented a decrease in the intensity of pain during treatment [41,44,46].

The corticosteroid reports of Chakraborty et al. and Samiee et al. described intra-articular injections for treating post-hemimandibulectomy pain (single case) and mandibular immobility, respectively. In both situations, therapeutic success was achieved, expressed by pain relief and an increase in the range of mandible abduction, respectively [45,47].

The only two identified studies that describe the intra-articular administration of local anesthetics without therapeutically active additives are reports by Guarda Nardini et al. and Danzig et al. In the first of them, the physical administration of fluid under pressure into the joint cavity may have been of considerable importance, which in combination resulted in a significant increase in the range of jaw mobility and almost complete disappearance of articular pain in the context of other studies. The study of Danzig et al. showed immediate relief of pain after intra-articular administration of lidocaine in a group of 23 patients [49,52].

4.2. Potential Chondrotoxicity

The effect of intra-articular injections on the cartilage of the temporomandibular joints is currently an actively discussed topic [61]. Attempts are made to remove inflammatory mediators by performing arthrocentesis, introducing anti-inflammatory mediators in autologous blood concentrates, and autografting stem cells [13,62–66]. These interventions are chondroprotective and even regenerative in nature [67,68]. In the context of this state of advancement of injection techniques, the administration of substances with chondrotoxic potential seems to be unjustified.

The selection process inadvertently identified one experimental study on the administration of local anesthetics to the temporomandibular joint cavities. In a 2022 report by Asan et al., cytotoxicity of administration of 1 mL of lidocaine, bupivacaine, or articaine into the cavities of rabbit temporomandibular joints was indicated [69]. In the post-mortem examination, thinning and unevenness of articular cartilage and a reduced amount of collagen were observed compared to the placebo group receiving physiological saline solution. It was shown that the weakest adverse effects among the study groups were observed after the administration of articaine. The described results cannot be directly interpolated to humans due to the different body weights and volumes of joint cavities for the same volume (1 mL) of the agent administered in most of the identified clinical studies.

A 2023 report by Zhang et al. compiles data from various articles assessing the chondrotoxicity of local anesthetics (bupivacaine, ropivacaine, lidocaine, and mepivacaine) administered intra-articularly into shoulder and knee joints during arthroscopic surgery [70]. Joint chondrolysis was observed postoperatively due to significant disruption of chondrocyte cultures by local anesthetics. Bupivacaine toxicity was the highest due to its longest half-life among the tested local anesthetics; consequently, it is not recommended by Zhang et al. for intra-articular administration. Therefore, we call into question the safety of intra-articular administration of local anesthetics in temporomandibular disorders and encourage a review of experimental studies focused on this problem.

4.3. Limitations of the Evidence

The randomized controlled trials included in this review were characterized by high heterogeneity in terms of: (1) the type of local anesthetic used (bupivacaine, lidocaine, mepivacaine, and articaine); (2) number of intra-articular administrations (1 to 3 or as needed); (3) intervals between interventions (from 2 to 42 days); and (4) the total dose of the drug (from 2.5 to 40 mg).

4.4. Limitations of the Review Processes

The search query was based on English keywords, which made it impossible to identify and include fully foreign-language reports.

5. Conclusions

(1)   Bupivacaine administered into the temporomandibular joint provided temporary pain relief, which lasted no longer than 24 h. In longer follow-ups, no statistically significant analgesic effectiveness was noted.

(2)   There is no evidence of a statistically significant improvement in the range of jaw mobility after the intra-articular administration of local anesthetics.

(3)   Local anesthetics administered intra-articularly have chondrotoxic potential, which requires verification in a separate systematic review.

Author Contributions: Conceptualization, M.C. and M.S.; Methodology, K.C. and M.C.; Software, K.L. and F.B.; Validation, N.T., A.M. and D.C.; Formal Analysis, K.L. and M.C.; Investigation, K.L., K.C., F.B., M.C. and A.M.; Resources, K.L., K.C., F.B. and M.C.; Data Curation, K.L., K.C., F.B. and M.C.; Writing—Original Draft Preparation, K.L., K.C., F.B., M.C., N.T. and M.S.; Writing—Review and Editing, M.C., N.T., A.M., D.C. and M.S.; Visualization, F.B., N.T. and M.C.; Supervision, D.C. and M.S.; Project Administration, D.C and M.S. All authors have read and agreed to the published version of the manuscript.

Funding: This research received no external funding.

Institutional Review Board Statement: Not applicable.

Informed Consent Statement: Not applicable.

Data Availability Statement: All data are contained within the article.

Conflicts of Interest: The authors declare no conflict of interest.

References

1.   Schiffman, E.; Ohrbach, R.; Truelove, E.; Look, J.; Anderson, G.; Goulet, J.-P.; List, T.; Svensson, P.; Gonzalez, Y.; Lobbezoo, F.; et al. Diagnostic Criteria for Temporomandibular Disorders (DC/TMD) for Clinical and Research Applications: Recommendations of the International RDC/TMD Consortium Network* and Orofacial Pain Special Interest Group†. *J. Oral Facial Pain Headache* **2014**, *28*, 6–27. [CrossRef] [PubMed]

2.   International Classification of Orofacial Pain, 1st Edition (ICOP). *Cephalalgia* **2020**, *40*, 129–221. [CrossRef] [PubMed]

3.   Nitecka-Buchta, A.; Walczynska-Dragon, K.; Batko-Kapustecka, J.; Wieckiewicz, M. Comparison between Collagen and Lidocaine Intramuscular Injections in Terms of Their Efficiency in Decreasing Myofascial Pain within Masseter Muscles: A Randomized, Single-Blind Controlled Trial. *Pain Res. Manag.* **2018**, *2018*, e8261090. [CrossRef] [PubMed]

4.   Nowak, Z.; Chęciński, M.; Nitecka-Buchta, A.; Bulanda, S.; Ilczuk-Rypuła, D.; Postek-Stefańska, L.; Baron, S. Intramuscular Injections and Dry Needling within Masticatory Muscles in Management of Myofascial Pain. Systematic Review of Clinical Trials. *Int. J. Environ. Res. Public Health* **2021**, *18*, 9552. [CrossRef] [PubMed]

5.   Sikora, M.; Czerwińska-Niezabitowska, B.; Chęciński, M.A.; Sielski, M.; Chlubek, D. Short-Term Effects of Intra-Articular Hyaluronic Acid Administration in Patients with Temporomandibular Joint Disorders. *J. Clin. Med.* **2020**, *9*, 1749. [CrossRef] [PubMed]

6.   Pihut, M.; Ferendiuk, E.; Szewczyk, M.; Kasprzyk, K.; Wieckiewicz, M. The Efficiency of Botulinum Toxin Type A for the Treatment of Masseter Muscle Pain in Patients with Temporomandibular Joint Dysfunction and Tension-Type Headache. *J. Headache Pain* **2016**, *17*, 29. [CrossRef]

7.   Urbański, P.; Trybulec, B.; Pihut, M. The Application of Manual Techniques in Masticatory Muscles Relaxation as Adjunctive Therapy in the Treatment of Temporomandibular Joint Disorders. *Int. J. Environ. Res. Public Health* **2021**, *18*, 12970. [CrossRef]

8.   Stefanovski, V.; Daskalova, B.; Mladenovski, M.; Lazarevski, G.; Panchevska, S. Visual Analog Scale for Pain Analysis in Patients with Temporomandibular Dysfunction. *Acad. Med. J.* **2022**, *2*, 128–134. [CrossRef]

9.   Emshoff, R.; Bertram, S.; Emshoff, I. Clinically Important Difference Thresholds of the Visual Analog Scale: A Conceptual Model for Identifying Meaningful Intraindividual Changes for Pain Intensity. *Pain* **2011**, *152*, 2277–2282. [CrossRef]

10.   Sikora, M.; Sielski, M.; Chęciński, M.; Chęcińska, K.; Czerwińska-Niezabitowska, B.; Chlubek, D. Patient-Reported Quality of Life versus Physical Examination in Treating Temporomandibular Disorders with Intra-Articular Platelet-Rich Plasma Injections: An Open-Label Clinical Trial. *Int. J. Environ. Res. Public Health* **2022**, *19*, 13299. [CrossRef]

11.   Aktaş, A.; Ilgaz, F.; Serel Arslan, S. Dietary Intakes of Individuals with Temporomandibular Disorders: A Comparative Study. *J. Oral Rehabil.* **2023**, *50*, 655–663. [CrossRef] [PubMed]

12.   Nasri-Heir, C.; Touger-Decker, R. Temporomandibular Joint Disorders and the Eating Experience. *Dent. Clin. N. Am.* **2023**, *67*, 367–377. [CrossRef] [PubMed]

13. Sikora, M.; Sielski, M.; Chęciński, M.; Nowak, Z.; Czerwińska-Niezabitowska, B.; Chlubek, D. Repeated Intra-Articular Administration of Platelet-Rich Plasma (PRP) in Temporomandibular Disorders: A Clinical Case Series. *J. Clin. Med.* **2022**, *11*, 4281. [CrossRef] [PubMed]

14. Bitiniene, D.; Zamaliauskiene, R.; Kubilius, R.; Leketas, M.; Gailius, T.; Smirnovaite, K. Quality of Life in Patients with Temporomandibular Disorders. A Systematic Review. *Stomatologija* **2018**, *20*, 3–9. [PubMed]

15. Kopacz, Ł.; Ciosek, Ż.; Gronwald, H.; Skomro, P.; Ardan, R.; Lietz-Kijak, D. Comparative Analysis of the Influence of Selected Physical Factors on the Level of Pain in the Course of Temporomandibular Joint Disorders. *Pain Res. Manag.* **2020**, *2020*, 1036306. [CrossRef] [PubMed]

16. Andre, A.; Kang, J.; Dym, H. Pharmacologic Treatment for Temporomandibular and Temporomandibular Joint Disorders. *Oral Maxillofac. Surg. Clin. N. Am.* **2022**, *34*, 49–59. [CrossRef] [PubMed]

17. Xie, Y.; Zhao, K.; Ye, G.; Yao, X.; Yu, M.; Ouyang, H. Effectiveness of Intra-Articular Injections of Sodium Hyaluronate, Corticosteroids, Platelet-Rich Plasma on Temporomandibular Joint Osteoarthritis: A Systematic Review and Network Meta-Analysis of Randomized Controlled Trials. *J. Evid. -Based Dent. Pract.* **2022**, *22*, 101720. [CrossRef]

18. Turosz, N.; Chęcińska, K.; Chęciński, M.; Michcik, A.; Chlubek, D.; Sikora, M. Adverse Events of Intra-Articular Temporomandibular Joint Injections: A Systematic Search and Review. *Pomeranian J. Life Sci.* **2023**, *69*, 48–54. [CrossRef]

19. Montinaro, F.; Nucci, L.; d'Apuzzo, F.; Perillo, L.; Chiarenza, M.C.; Grassia, V. Oral Nonsteroidal Anti-Inflammatory Drugs as Treatment of Joint and Muscle Pain in Temporomandibular Disorders: A Systematic Review. *CRANIO®* **2022**, 1–10. [CrossRef]

20. Goiato, M.C.; da Silva, E.V.F.; de Medeiros, R.A.; Túrcio, K.H.L.; Dos Santos, D.M. Are Intra-Articular Injections of Hyaluronic Acid Effective for the Treatment of Temporomandibular Disorders? A Systematic Review. *Int. J. Oral Maxillofac. Surg.* **2016**, *45*, 1531–1537. [CrossRef]

21. Chęciński, M.; Chęcińska, K.; Turosz, N.; Brzozowska, A.; Chlubek, D.; Sikora, M. Current Clinical Research Directions on Temporomandibular Joint Intra-Articular Injections: A Mapping Review. *J. Clin. Med.* **2023**, *12*, 4655. [CrossRef] [PubMed]

22. Gopalakrishnan, V.; Nagori, S.A.; Roy Chowdhury, S.K.; Saxena, V. The Use of Intra-Articular Analgesics to Improve Outcomes after Temporomandibular Joint Arthrocentesis: A Review. *Oral Maxillofac. Surg.* **2018**, *22*, 357–364. [CrossRef] [PubMed]

23. Ogle, O.E.; Mahjoubi, G. Local Anesthesia: Agents, Techniques, and Complications. *Dent. Clin. N. Am.* **2012**, *56*, 133–148. [CrossRef] [PubMed]

24. Becker, D.E.; Reed, K.L. Local Anesthetics: Review of Pharmacological Considerations. *Anesth. Prog.* **2012**, *59*, 90–102. [CrossRef] [PubMed]

25. Erkocak, O.F.; Aydın, B.K.; Celik, J.B. Intraarticular Local Anesthetics: Primum Non Nocere. *Knee Surg. Sports Traumatol. Arthrosc.* **2012**, *20*, 2124. [CrossRef] [PubMed]

26. Moiniche, S.; Mikkelsen, S.; Wetterslev, J.; Dahl, J. A Systematic Review of Intra-Articular Local Anesthesia for Postoperative Pain Relief after Arthroscopic Knee Surgery. *Reg. Anesth. Pain Med.* **1999**, *24*, 430–437. [CrossRef] [PubMed]

27. Campo, M.M.; Kerkhoffs, G.M.M.J.; Sierevelt, I.N.; Weeseman, R.R.; Van Der Vis, H.M.; Albers, G.H.R. A Randomised Controlled Trial for the Effectiveness of Intra-Articular Ropivacaine and Bupivacaine on Pain after Knee Arthroscopy: The DUPRA (DUtch Pain Relief after Arthroscopy)-Trial. *Knee Surg. Sports Traumatol. Arthrosc.* **2012**, *20*, 239–244. [CrossRef]

28. Chęciński, M.; Chęcińska, K.; Nowak, Z.; Sikora, M.; Chlubek, D. Treatment of Mandibular Hypomobility by Injections into the Temporomandibular Joints: A Systematic Review of the Substances Used. *J. Clin. Med.* **2022**, *11*, 2305. [CrossRef]

29. Page, M.J.; McKenzie, J.E.; Bossuyt, P.M.; Boutron, I.; Hoffmann, T.C.; Mulrow, C.D.; Shamseer, L.; Tetzlaff, J.M.; Akl, E.A.; Brennan, S.E.; et al. The PRISMA 2020 Statement: An Updated Guideline for Reporting Systematic Reviews. *BMJ* **2021**, *88*, n71. [CrossRef]

30. PROSPERO—International Prospective Register of Systematic Reviews. Available online: https://www.crd.york.ac.uk/prospero/ (accessed on 14 November 2023).

31. PubMed Overview. Available online: https://pubmed.ncbi.nlm.nih.gov/ (accessed on 14 November 2023).

32. BASE—Bielefeld Academic Search Engine | What Is BASE? Available online: https://www.base-search.net/about/en/index.php (accessed on 4 May 2023).

33. Scopus | Abstract and Citation Database | Elsevier. Available online: https://www.elsevier.com/products/scopus (accessed on 14 November 2023).

34. Ouzzani, M.; Hammady, H.; Fedorowicz, Z.; Elmagarmid, A. Rayyan—A Web and Mobile App for Systematic Reviews. *Syst. Rev.* **2016**, *5*, 210. [CrossRef]

35. Sterne, J.A.C.; Savović, J.; Page, M.J.; Elbers, R.G.; Blencowe, N.S.; Boutron, I.; Cates, C.J.; Cheng, H.-Y.; Corbett, M.S.; Eldridge, S.M.; et al. RoB 2: A Revised Tool for Assessing Risk of Bias in Randomised Trials. *BMJ* **2019**, *366*, l4898. [CrossRef] [PubMed]

36. McGuinness, L.A.; Higgins, J.P.T. Risk-of-bias VISualization (Robvis): An R Package and Shiny Web App for Visualizing Risk-of-bias Assessments. *Res. Synth. Methods* **2021**, *12*, 55–61. [CrossRef] [PubMed]

37. Schoonjans, F. MedCalc Statistical Software. Available online: https://www.medcalc.org/ (accessed on 14 November 2023).

38. Bhargava, D.; Sivakumar, B.; Bhargava, P.G. A Comparative Preliminary Randomized Clinical Study to Evaluate Heavy Bupivacaine Dextrose Prolotherapy (HDP) and Autologous Blood Injection (ABI) for Symptomatic Temporomandibular Joint Hypermobility Disorder. *J. Maxillofac. Oral Surg.* **2023**, *22*, 110–118. [CrossRef] [PubMed]

39. Shan, Y.S. Platelet-Rich Plasma and Hyaluronic Acid for Injection Treatment of Temporomandibular Joint Degeneration in Affiliated Stomatological Hospital of Guangzhou Medical University 2023, 113664 bytes, 1 Files. Available online: https://www.scidb.cn/en/detail?dataSetId=a55c9b031d4a4da6ba28cf4d2526b9fe (accessed on 15 October 2023).

40. Prakash, J. Intra-Articular Platelet-Rich Plasma Injection versus Hydrocortisone with Local Anesthetic Injections for Temporo Mandibular Disorders. *Bioinformation* **2022**, *18*, 991–997. [CrossRef] [PubMed]

41. Dasukil, S.; Shetty, S.K.; Arora, G.; Degala, S. Efficacy of Prolotherapy in Temporomandibular Joint Disorders: An Exploratory Study. *J. Maxillofac. Oral Surg.* **2021**, *20*, 115–120. [CrossRef]

42. Louw, W.F.; Reeves, K.D.; Lam, S.K.H.; Cheng, A.-L.; Rabago, D. Treatment of Temporomandibular Dysfunction With Hypertonic Dextrose Injection (Prolotherapy): A Randomized Controlled Trial With Long-Term Partial Crossover. *Mayo Clin. Proc.* **2019**, *94*, 820–832. [CrossRef]

43. Gupta, S.; Sharma, A.; Purohit, J.; Goyal, R.; Malviya, Y.; Jain, S. Comparison between Intra-Articular Platelet-Rich Plasma Injection versus Hydrocortisone with Local Anesthetic Injections in Temporomandibular Disorders: A Double-Blind Study. *Natl J. Maxillofac. Surg.* **2018**, *9*, 205. [CrossRef]

44. Refai, H. Long-Term Therapeutic Effects of Dextrose Prolotherapy in Patients with Hypermobility of the Temporomandibular Joint: A Single-Arm Study with 1–4 Years' Follow up. *Br. J. Oral Maxillofac. Surg.* **2017**, *55*, 465–470. [CrossRef]

45. Chakraborty, A.; Datta, T.; Lingegowda, D.; Khemka, R. Ultrasound-Guided Temporomandibular Joint Injection for Chronic Posthemimandibulectomy Jaw Pain. *A&A Case Rep.* **2016**, *7*, 203–206. [CrossRef]

46. Zhou, H.; Hu, K.; Ding, Y. Modified Dextrose Prolotherapy for Recurrent Temporomandibular Joint Dislocation. *Br. J. Oral Maxillofac. Surg.* **2014**, *52*, 63–66. [CrossRef]

47. Samiee, A.; Sabzerou, D.; Edalatpajouh, F.; Clark, G.T.; Ram, S. Temporomandibular Joint Injection with Corticosteroid and Local Anesthetic for Limited Mouth Opening. *J. Oral Sci.* **2011**, *53*, 321–325. [CrossRef] [PubMed]

48. Refai, H.; Altahhan, O.; Elsharkawy, R. The Efficacy of Dextrose Prolotherapy for Temporomandibular Joint Hypermobility: A Preliminary Prospective, Randomized, Double-Blind, Placebo-Controlled Clinical Trial. *J. Oral Maxillofac. Surg.* **2011**, *69*, 2962–2970. [CrossRef] [PubMed]

49. Guarda Nardini, L.; Tito, R.; Beltrame, A. Treatment of temporo-mandibular joint closed-lock using intra-articular injection of mepivacaine with immediate resolution durable in time (six months follow-up). *Minerva Stomatol.* **2002**, *51*, 21–28.

50. Sato, S.; Ohta, M.; Ohki, H.; Kawamura, H.; Motegi, K. Effect of Lavage with Injection of Sodium Hyaluronate for Patients with Nonreducing Disk Displacement of the Temporomandibular Joint. *Oral Surg. Oral Med. Oral Pathol. Oral Radiol. Endodontol.* **1997**, *84*, 241–244. [CrossRef] [PubMed]

51. Kamada, A.; Fujita, A.; Kakudo, K.; Okazaki, J.; Ida, M.; Sakaki, T. Changes in Synovial Fluid N-Acetyl-Beta-Glucosaminidase Activity in the Human Temporomandibular Joint with Dysfunction. *J. Osaka Dent. Univ.* **1993**, *27*, 107–111. [PubMed]

52. Danzig, W.; May, S.; McNeill, C.; Miller, A. Effect of an Anesthetic Injected into the Temporomandibular Joint Space in Patients with TMD. *J. Craniomandib. Disord.* **1992**, *6*, 288–295. [PubMed]

53. Zarate, M.A.; Frusso, R.D.; Reeves, K.D.; Cheng, A.-L.; Rabago, D. Dextrose Prolotherapy Versus Lidocaine Injection for Temporomandibular Dysfunction: A Pragmatic Randomized Controlled Trial. *J. Altern. Complement. Med.* **2020**, *26*, 1064–1073. [CrossRef]

54. Ziegler, C.M.; Wiechnik, J.; Mühling, J. Analgesic Effects of Intra-Articular Morphine in Patients with Temporomandibular Joint Disorders: A Prospective, Double-Blind, Placebo-Controlled Clinical Trial. *J. Oral Maxillofac. Surg.* **2010**, *68*, 622–627. [CrossRef]

55. Ayesh, E.E.; Ernberg, M.; Svensson, P. Effects of Local Anesthetics on Somatosensory Function in the Temporomandibular Joint Area. *Exp. Brain Res.* **2007**, *180*, 715–725. [CrossRef]

56. Zuniga, J.R.; Ibanez, C.; Kozacko, M. The Analgesic Efficacy and Safety of Intra-Articular Morphine and Mepivacaine Following Temporomandibular Joint Arthroplasty. *J. Oral Maxillofac. Surg.* **2007**, *65*, 1477–1485. [CrossRef]

57. Tjakkes, G.-H.E.; Tenvergert, E.M.; de Bont, L.G.M.; Stegenga, B. The Effect of Intra-Articular Injection of Ultracain in the Temporomandibular Joint in Patients with Preauricular Pain: A Randomized Prospective Double-Blind Placebo-Controlled Crossover Study. *Clin. J. Pain* **2007**, *23*, 233–236. [CrossRef] [PubMed]

58. Lobbezoo, F.; Wang, K.; Aartman, I.H.A.; Svensson, P. Effects of TMJ Anesthesia and Jaw Gape on Jaw-Stretch Reflexes in Humans. *Clin. Neurophysiol.* **2003**, *114*, 1656–1661. [CrossRef] [PubMed]

59. Furst, I.M.; Kryshtalskyj, B.; Weinberg, S. The Use of Intra-Articular Opioids and Bupivacaine for Analgesia Following Temporomandibular Joint Arthroscopy: A Prospective, Randomized Trial. *J. Oral Maxillofac. Surg.* **2001**, *59*, 979–983; discussion 983–984. [CrossRef] [PubMed]

60. Gu, Z.; Wu, Q.; Zhang, Y.; Zhang, Z.; Sun, K. Visco-Supplementation Therapy in Internal Derangement of Temporomandibular Joint. *Chin. Med. J. Engl.* **1998**, *111*, 656–659. [PubMed]

61. Derwich, M.; Mitus-Kenig, M.; Pawlowska, E. Mechanisms of Action and Efficacy of Hyaluronic Acid, Corticosteroids and Platelet-Rich Plasma in the Treatment of Temporomandibular Joint Osteoarthritis—A Systematic Review. *Int. J. Mol. Sci.* **2021**, *22*, 7405. [CrossRef] [PubMed]

62. Guarda-Nardini, L.; De Almeida, A.M.; Manfredini, D. Arthrocentesis of the Temporomandibular Joint: Systematic Review and Clinical Implications of Research Findings. *J. Oral Facial Pain Headache* **2021**, *35*, 17–29. [CrossRef] [PubMed]

63. Siewert-Gutowska, M.; Pokrowiecki, R.; Kamiński, A.; Zawadzki, P.; Stopa, Z. State of the Art in Temporomandibular Joint Arthrocentesis-A Systematic Review. *J. Clin. Med.* **2023**, *12*, 4439. [CrossRef]

64. Chęciński, M.; Chęcińska, K.; Turosz, N.; Kamińska, M.; Nowak, Z.; Sikora, M.; Chlubek, D. Autologous Stem Cells Transplants in the Treatment of Temporomandibular Joints Disorders: A Systematic Review and Meta-Analysis of Clinical Trials. *Cells* **2022**, *11*, 2709. [CrossRef]

65. Liapaki, A.; Thamm, J.R.; Ha, S.; Monteiro, J.L.G.C.; McCain, J.P.; Troulis, M.J.; Guastaldi, F.P.S. Is There a Difference in Treatment Effect of Different Intra-Articular Drugs for Temporomandibular Joint Osteoarthritis? A Systematic Review of Randomized Controlled Trials. *International J. Oral Maxillofac. Surg.* **2021**, *50*, 1233–1243. [CrossRef]

66. Haddad, C.; Zoghbi, A.; El Skaff, E.; Touma, J. Platelet-rich Plasma Injections for the Treatment of Temporomandibular **Joint** Disorders: A Systematic Review. *J. Oral Rehabil.* **2023**, *50*, joor.13545. [CrossRef]

67. Köhnke, R.; Ahlers, M.O.; Birkelbach, M.A.; Ewald, F.; Krueger, M.; Fiedler, I.; Busse, B.; Heiland, M.; Vollkommer, T.; Gosau, M.; et al. Temporomandibular Joint Osteoarthritis: Regenerative Treatment by a Stem Cell Containing Advanced Therapy Medicinal Product (ATMP)—An In Vivo Animal Trial. *Int. J. Mol. Sci.* **2021**, *22*, 443. [CrossRef] [PubMed]

68. Coskun, U.; Candirli, C.; Kerimoglu, G.; Taskesen, F. Effect of Platelet-Rich Plasma on Temporomandibular Joint Cartilage Wound Healing: Experimental Study in Rabbits. *J. Cranio-Maxillofac. Surg.* **2019**, *47*, 357–364. [CrossRef] [PubMed]

69. Asan, C.Y.; Ağyüz, G.; Canpolat, D.G.; Demirbas, A.E.; Asan, M.; Yay, A.; Ülger, M.; Karakükcü, Ç. Chondrotoxic Effects of Intra-Articular Injection of Local Anaesthetics in the Rabbit Temporomandibular Joint. *Int. J. Oral Maxillofac. Surg.* **2022**, *51*, 1337–1344. [CrossRef] [PubMed]

70. Zhang, K.; Li, M.; Yao, W.; Wan, L. Cytotoxicity of Local Anesthetics on Bone, Joint, and Muscle Tissues: A Narrative Review of the Current Literature. *J. Pain Res.* **2023**, *16*, 611–621. [CrossRef]

Journal of
Clinical Medicine

Review

Current Clinical Research Directions on Temporomandibular Joint Intra-Articular Injections: A Mapping Review

Maciej Chęciński [1], Kamila Chęcińska [2], Natalia Turosz [3], Anita Brzozowska [4], Dariusz Chlubek [5,*] and Maciej Sikora [5,6]

1   Department of Oral Surgery, Preventive Medicine Center, Komorowskiego 12, 30-106 Cracow, Poland; maciej@checinscy.pl
2   Department of Glass Technology and Amorphous Coatings, Faculty of Materials Science and Ceramics, AGH University of Science and Technology, Mickiewicza 30, 30-059 Cracow, Poland; checinska@agh.edu.pl
3   Institute of Public Health, Jagiellonian University Medical College, Skawińska 8, 31-066 Cracow, Poland; natalia.turosz@gmail.com
4   Preventive Medicine Center, Komorowskiego 12, 30-106 Kraków, Poland; brzanita@gmail.com
5   Department of Biochemistry and Medical Chemistry, Pomeranian Medical University, Powstańców Wielkopolskich 72, 70-111 Szczecin, Poland; sikora-maciej@wp.pl
6   Department of Maxillofacial Surgery, Hospital of the Ministry of Interior, Wojska Polskiego 51, 25-375 Kielce, Poland
*   Correspondence: dchlubek@pum.edu.pl

Abstract: This mapping review aims to identify and discuss current research directions on intracavitary temporomandibular joints (TMJs) injections. The inclusion criteria allowed studies published in the last full six years, based on patients diagnosed with temporomandibular joint disorders (TMDs), treated by TMJ intra-articular injections. Medical databases covered by the Association for Computing Machinery, Bielefeld Academic Search Engine, PubMed, and Elsevier Scopus engines were searched. The results were visualized with tables, charts, and diagrams. Of the 2712 records identified following the selection process, 152 reports were qualified for review. From January 2017, viscosupplementation with hyaluronic acid (HA) was the best-documented injectable administered into TMJ cavities. However, a significant growing trend was observed in the number of primary studies on centrifuged blood preparations administrations that surpassed the previously leading HA from 2021.

Keywords: temporomandibular joint; temporomandibular disorders; intra-articular injections; viscosupplementation; blood preparations; mesenchymal stem cells

Citation: Chęciński, M.; Chęcińska, K.; Turosz, N.; Brzozowska, A.; Chlubek, D.; Sikora, M. Current Clinical Research Directions on Temporomandibular Joint Intra-Articular Injections: A Mapping Review. *J. Clin. Med.* **2023**, *12*, 4655. https://doi.org/10.3390/jcm12144655

Academic Editor: Eiji Tanaka

Received: 23 May 2023
Revised: 27 June 2023
Accepted: 11 July 2023
Published: 13 July 2023

1. Introduction

1.1. Background

The temporomandibular joints (TMJs) connect the mandible to the temporal bones. These joints are essential to the proper functioning of the stomatognathic system, including opening and closing the mouth, chewing, and speaking [1]. Rotation and slide in TMJs are palpable on both sides in the preauricular area during abduction and adduction of the mandible [2,3]. Each TMJ consists of the mandibular condyle, the articular fossa of the temporal bone, and the cartilage disc that separates the two bones and cushions them during movement [1,4]. The joint is surrounded by a network of muscles, ligaments, and nerves that help stabilize and control its function [1,5].

Temporomandibular disorders (TMDs) are a collective term for a group of conditions manifested by abnormal function of the temporomandibular joints (TMJs) [6,7]. According to the meta-analysis by Valesan et al., the overall prevalence of TMDs in the adult population is approximately 31% [6] The causes of TMDs are seen primarily in malocclusions, morphological abnormalities, and post-traumatic changes within TMJs, and masticatory muscle dysfunction [5,8–10]. The causes of TMDs should also be sought in general deterioration of health (including psychological burden) and limited access to medical care, which could be observed with an increase in the frequency of TMDs diagnoses during the COVID-19 pandemic, according to the study by Haddad et al., to about 42% [8,9]. However, Ginszt et al. showed that there is a certain mechanical effect of wearing medical masks on muscle activity, in particular the anterior part of the temporalis muscle, which may also be important for the increase in the incidence of TMDs [10]. TMDs can manifest as articular and/or muscular pain, acoustic symptoms from TMJs, and reduced chewing quality [5,11,12]. Amongst TMDs treatment methods are biofeedback, cognitive–behavioral therapy, physiotherapy, oral drug therapy, splint therapy, changing the occlusive conditions, and minimally invasive, arthroscopic, or open surgery [13–17].

Minimally invasive intra-articular manipulations are currently considered a viable alternative in the management of TMDs, especially when more conservative treatments fail to provide relief from TMDs symptoms [18–20]. These techniques include arthrocentesis and intra-articular injections [20,21]. Arthrocentesis consists in rinsing the joint cavity with infusion fluids using two- and one-needle methods [20,21]. Intra-articular injections involve injecting the drug directly into the TMJ cavity [22,23]. Intra-articular injections are indicated to relieve joint pain, suppress inflammation and improve joint function [22,23].

Various substances are administered intra-articularly, including corticosteroids (CSs), hyaluronic acid (HA), and blood products such as platelet-rich plasma (PRP) or injectable platelet-rich fibrin (I-PRF) [17,18,24]. CSs are known for their strong anti-inflammatory effect. Supplementation of the main component of synovial fluid, HA, improves the mobility of joint surfaces relative to each other [24,25]. PRP and I-PRF, differing in composition resulting from the preparation, have the added advantage of being highly safe due to their autogenous nature [17,24,25].

1.2. Rationale

Scientific articles published in recent years indicate a sudden increase in the number of substances administered intra-articularly, and surgical technique is constantly improving. The growing number of primary research papers demonstrates the increasing popularity of intra-articular injections. Therefore, it seems reasonable to frequently update the state of knowledge about injection techniques in the treatment of TMDs. To the knowledge of the authors of this paper, no systematic map on this subject has been published to date.

1.3. Objectives

This mapping review aims to identify and discuss current research directions on intracavitary TMJs injections.

2. Materials and Methods

The systematic map was prepared by: (1) defining eligibility criteria; (2) developing a search strategy; (3) searching medical databases using leading engines; (4) selecting reports according to predetermined criteria; (5) assessing the research level of evidence; (6) synthesizing the results; (7) presenting the main research directions.

2.1. Eligibility Criteria

The eligibility criteria were established in accordance with the PICOS methodology (Table 1) [26–28]. Studies based on patients diagnosed with TMDs were included. Due to the different etiology and treatment, patients with TMDs as a manifestation of a general disease, e.g., rheumatoid arthritis or juvenile idiopathic arthritis, were excluded. Cadaver, animal, or in vitro studies were omitted as not including patients. Systematic reviews and meta-analyses based on eligible studies were included. Interventions containing the administration or administrations into the temporomandibular joint cavity were included. Additional interventions of a different kind were allowed, such as physiotherapy, pharmacotherapy, splint therapy, etc. Arthrocentesis alone, without intra-articular administration of any substance, was excluded. More invasive intra-articular manipulations, i.e., arthroscopy or open surgery, were disqualified. Due to the inclusion of studies with varying levels of evidence, the criterion of comparison was not applicable. Changes in any TMDs severity index were allowed as an outcome. Single case reports and any series less than 4 cases were rejected. In order to demonstrate the current directions of research, reports published in the last full 6 years, i.e., from 1 January 2017, to final searches conducted on 13 March 2023, were included.

Table 1. Review eligibility criteria.

Domain	Inclusion Criteria	Exclusion Criteria
Population	Patients diagnosed with TMDs	TMDs as a systemic disease component
Intervention	TMJ intra-articular injection	Arthrocentesis alone or more invasive interventions, e.g., arthroscopy
Comparison	Any or none	Not applicable
Outcomes	TMDs severity assessment	Not applicable
Settings	Reports based on 4 or more cases	Reports published before 2017

2.2. Search Strategy

The search strategy was based on terms identifying TMJ and injections. In its basic form, the query was:

"(temporomandibular OR TMJ OR TMJs) AND (injection OR injections OR puncture OR punctures OR administration OR administrations)".

The following search engines were used: (1) Association for Computing Machinery: Guide to Computing Literature (ACM; 3,470,491 records) [29]; (2) Bielefeld Academic Search Engine (BASE; 320,685,924 records) [30]; (3) National Library of Medicine: PubMed (NLM; over 35,000,000 records) [31]; (4) Elsevier Scopus (ES; over 87,000,000 records) [32,33]. For each search engine, the necessary query modifications were made to ensure the validity of the search (Table A1). Filters were used to exclude studies published before 2017, where possible.

2.3. Selection Process

Reports were selected for the systematic map in two stages by two authors (M.C. and A.B.) using Rayyan tool [34]. Screening consisted of including abstracts according to PICOS criteria. Acceptance by any of the judges resulted in the promotion of the record to the eligibility stage. In case of discrepancies regarding inclusion, decisions were made by consensus, with the casting vote of the third investigator (K.C.).

2.4. Qualification of Reports Due to the Study Design

The information on the design of the studies included in the review was extracted from the source reports by two authors (M.C. and K.C.) and unified using the Oxford Centre for Evidence-Based Medicine 2011 Levels of Evidence scale [35]. Systematic reviews involving randomized controlled trials were qualified as Level 1. Levels 2–4 were assigned to randomized controlled, non-randomized controlled, and uncontrolled trials, respectively.

2.5. Syntheses

The results of this mapping review were tabularized and illustrated by an organizational chart representing the research directions forks falling within the eligibility criteria described above. The numbers of individual articles in particular forks were presented with a bar, bubble, and column charts, with trend lines indicating the dominant directions of primary research on the last one.

3. Results

Of the 2712 records identified, 152 reports were ultimately qualified for the mapping review, with 32, 53, 28, and 39 reports in levels of evidence from 1 to 4, respectively

(Figures 1–5, Table A2) [18,19,22–25,36–171]. In the selection process, a total of 1407 duplicates were rejected, mainly due to overlapping search engines. At the screening stage, 1119 entries not related to TMJs injection treatment were excluded; these were present due to the intentionally unrestrictive choice of keywords in the queries. At the very end of the selection, 34 articles (mainly case reports) were rejected in the course of the full-text evaluation in accordance with the adopted inclusion and exclusion criteria.

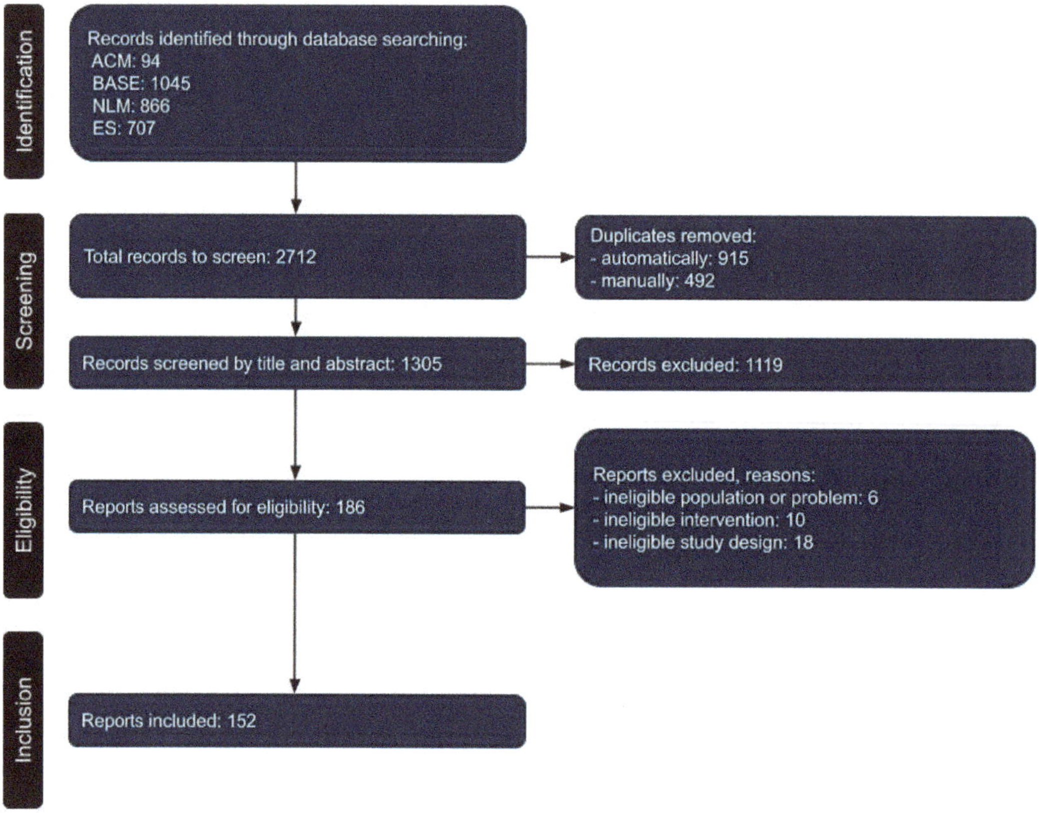

Figure 1. Flow diagram of the selection process. ACM—Association for Computing Machinery: Guide to Computing Literature; BASE—Bielefeld Academic Search Engine; NLM—National Library of Medicine: PubMed; ES—Elsevier Scopus.

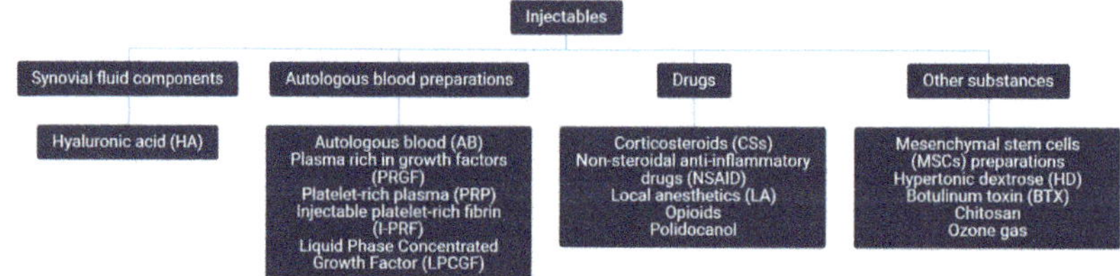

Figure 2. Classification of injectables (based on the included reports).

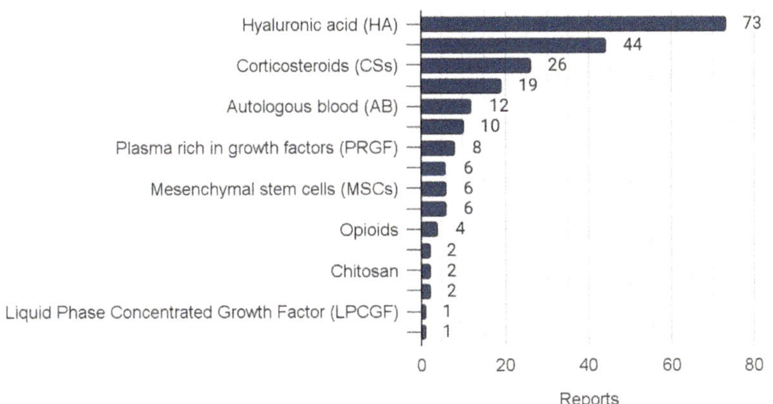

Figure 3. The number of reports on individual injectables.

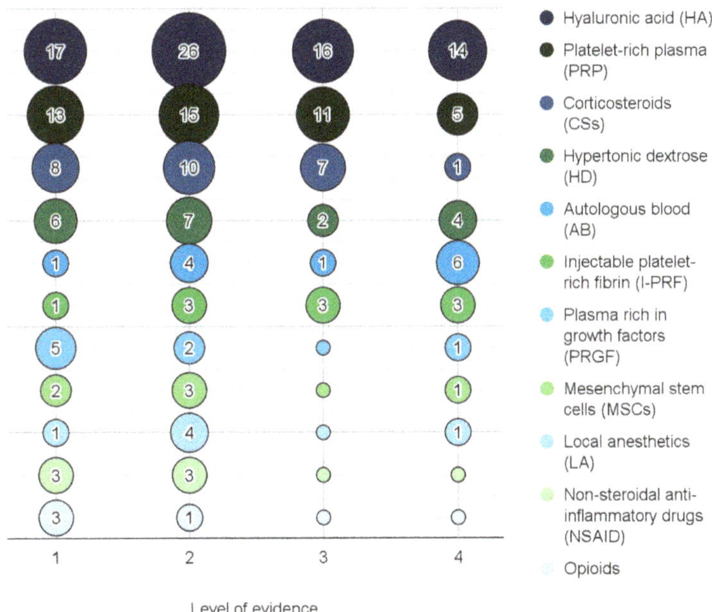

Figure 4. The number of reports by the level of evidence (horizontal axis) and injectables (vertical axis). Reports on injectables evaluated in less than 3 papers not included.

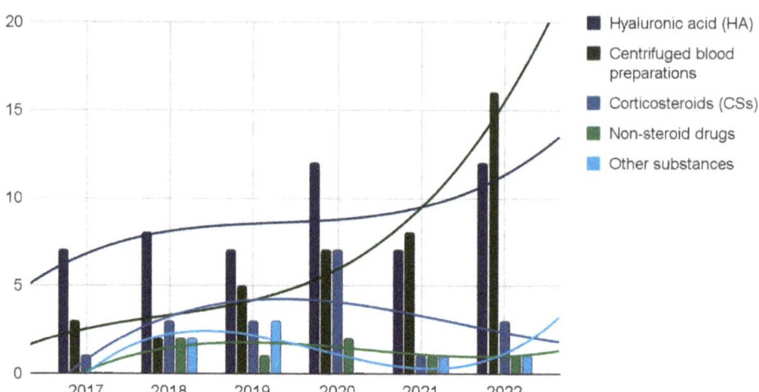

Figure 5. The number of primary research reports (level of evidence 2–4) on individual mandibular hypomobility treatment injectables with third-degree polynomial trendlines.

4. Discussion

TMDs that cause articular pain and mandibular mobility limitation are two main reasons for delivering intra-articular injections, except for HD and AB administrations which are performed to treat recurrent subluxation of the temporomandibular joint [22,24,37,40,44,53,55,57,61,65,66,69,72,73,90,94,96,97,110,111,131,149,160,165,168]. Currently, there is an intensive search for the gold standard of TMDs treatment, which is difficult due to the variety of etiologies and the specificity of individual dysfunctions. The main directions of research on the use of intra-articular injections in this indication are presented below.

4.1. Hyaluronic Acid (HA) Viscosupplementation

Improving the composition of the synovial fluid by supplementing its main ingredient, HA, is the most frequently described type of injection into the TMJs. 56 primary studies, including 26 randomized, summarized in 17 systematic reviews make this injectable substance the best studied. There is a steady upward trend in the number of primary studies on HA published in subsequent years. The primacy of HA from 2021 seems to be threatened by centrifuged blood products, but this group of substances is heterogeneous and cannot be compared to HA in terms of effectiveness as a whole [18,19,22,25,38,43,49,51,54,58–60,63,68, 70,71,74,75,78,80,80,82,86,88,89,91–93,95,98–102,105,106,110,113–117,120,121,124,126,127,129, 130,132–135,137,138,143,144,146,151–153,157,158,161,163,164,166,171].

4.2. Hypertonic Dextrose (HD) Prolotherapy

Unlike viscosupplementation, HD prolotherapy aims to reduce the range of motion of the mandible. The administration of HD as an irritant is one of the treatment methods for hypermobility in TMJs. So far, the concentration of HD has not been standardized and varies from 12.5% to 25%. Only the studies involving the administration of HD into the TMJs are included in this review, but the substance is frequently deposited peri-articularly in this indication. Of the two substances applied to TMJs (HD and autologous blood), HD injections are better documented. The 13 primary studies on intra-articular administration of HD have been summarized in 6 reviews [22,24,37,53,65,66,72,73,94,96,97,110,111,131,149,160,165].

4.3. Blood Preparations Autotransplantation

Blood preparations are a group of substances obtained from autologous peripheral blood including unprocessed blood and blood concentrates. Autologous blood (AB) is the second, next to HD, substance administered into the TMJs for the treatment of hypermobility. Reports since 2017 describing AB therapy are fewer and generally with a lower level of evidence than these regarding HD (four randomized, one non-randomized, and six uncontrolled) [37,40,44,55,57,61,65,66,69,90]. The only included systematic review on AB therapy administrated intra-articularly suggests the need for randomized trials [168].

Blood concentrates are obtained by centrifuging freshly taken venous blood and are delivered immediately after the preparation. Different protocols allow obtaining various concentrates without the red cell fraction. Some of the concentrates can be collected in liquid form and injected into TMDs. In the discussed years, the administration of preparations referred to as plasma rich in growth factors (PRGF), PRP, I-PRF and liquid phase concentrated growth factor (LPCGF) into the TMJs cavities was described. They differ in the centrifugation procedure, and thus in the composition and effectiveness in anti-inflammatory action and stimulation of tissue regeneration. The lack of a standardized centrifugation protocol for platelet-rich concentrates for injection into TMJs clearly illustrates the active development of a therapeutic standard. Of the 152 reports on blood concentrates included, 120 primary studies were published (including 53 randomized trials) as well as 32 systematic reviews. From 2021, primary research on the substances in question has been more numerous than on HA. [18,23,25,36,46,56,60,62,64,67,68,70,74,75,81,84,85,87,95,101,103,104,107,112–114,118,119, 122,123,123,128,133–137,141–143,145,147,148,152,158,161,162,167,169,170].

4.4. Mesenchymal Stem Cells (MSCSs) Autotransplantation

MSCs, obtained primarily from autogenous fat, are an attractive injectable due to their high potential to stimulate the regeneration of TMJ structures. Only four primary studies using MSCs for intracavitary administration are known, of which three were randomized.

4.5. Drugs Administration

Substances used as drugs for other indications, normally with other routes of administration, are included in this group. CSs are definitely the best studied among them. After HA and PRP, CSs were the third most frequently reported injectables group in 2017–2022 (8 systematic reviews, 10 randomized trials, and 8 other trials) [19,25,38,39,41, 45,47,47,49,51,54,62,75,86,121,125,127,134,135,137,140,156,164,166,171]. Nevertheless, since 2020, the number of primary studies on intra-articular injection of CSs has clearly decreased. Other papers describe the use of non-steroidal anti-inflammatory drugs (NSAIDs), local anesthetics (LAs), opioids, and polidocanol. These substances have been used exceptionally and so far there is no well-established knowledge about their effectiveness and safety [24,37,38,62,73,77,79,97,101,108,137,150,165].

4.6. Other Substances Injections

Unique studies on the administration of botulinum toxin (BTX), chitosan, and ozone gas provide potential directions for the future development of intra-articular injections. At present, however, these methods should be regarded as insufficiently researched [24,42,83,139,159].

4.7. Limitations

This systematic map was limited to injections into the temporomandibular joint cavities. Therefore, studies focusing on pericapsular injections, which are used in the treatment of mandibular hypermobility, were omitted. Therefore, this paper covers only a part of the articles on AB and HD injections.

A separate large group of interventions, not included in this review, is stand-alone arthrocentesis. They have been excluded as there was no intention to administer any substance intra-articularly. However, TMJs lavage relieves pain and increases mandibular mobility similarly to injections of, for example, HA or PRP, and future mapping of papers on this topic should be considered.

5. Conclusions

In the years 2017–2023, hyaluronic acid was the most common topic of scientific publications among injectables administered into temporomandibular joint cavities (26 randomized controlled trials and 30 other clinical studies). In the same period, there was a significant upward trend in the number of published primary studies focused on centrifuged blood preparations used in the treatment of TMDs. As of 2021, blood products administered into TMJs cavities have become a more popular topic for professional medical articles than hyaluronan. Nevertheless, it should be emphasized that this is a group of substances that differ in composition depending on the centrifugation protocol. The therapeutic efficacies of substances evaluated in at least three clinical trials were synthesized in systematic reviews.

Author Contributions: Conceptualization, M.C. and M.S.; methodology, M.C. and K.C.; software, M.C. and K.C.; validation, K.C. and M.S.; formal analysis, M.C. and K.C.; investigation, M.C. and A.B.; resources, K.C.; data curation, M.C., K.C. and A.B.; writing—original draft preparation, M.C., K.C., N.T., A.B. and M.S.; writing—review and editing, M.C., M.S. and D.C.; visualization, M.C. and K.C.; supervision, M.S. and D.C.; project administration, D.C. All authors have read and agreed to the published version of the manuscript.

Funding: This research received no external funding.

Institutional Review Board Statement: Not applicable.

Informed Consent Statement: Not applicable.

Data Availability Statement: All collected data are included in the content of this article.

Conflicts of Interest: The authors declare no conflict of interest.

Appendix A

Table A1. Search queries.

Engine	Inclusion Criteria
ACM	[[All: temporomandibular] OR [All: tmj] OR [All: tmjs]] AND [[All: injection] OR [All: injections] OR [All: puncture] OR [All: punctures] OR [All: administration] OR [All: administrations]] AND [E-Publication Date: (1 January 2017 TO 31 December 2023)]
BASE	(temporomandibular OR TMJ OR TMJs) AND (injection OR injections OR puncture OR punctures OR administration OR administrations) year: [2017 TO *]
NLM	(temporomandibular OR tmj OR tmjs) AND (injection OR injections OR puncture OR punctures OR administration OR administrations) AND ("1 January 2017" [Date—Publication]: "3000" [Date—Publication])
ES	TITLE-ABS-KEY ((temporomandibular OR tmj OR tmjs) AND (injection OR injections OR puncture OR punctures OR administration OR administrations)) AND PUBYEAR > 2016

Table A2. Included reports.

First Author	Publication Year	Level of Evidence	Injectables	DOI Number
Bayramoglu	2023	2	NSAID	10.1186/s12903-023-02852-z
Bhargava	2023	2	AB, HD, LA	10.1007/s12663-022-01738-x
Chęciński	2023	1	HA, HD	10.3390/jcm12041664
Chhapane	2023	2	AB, HD	10.1007/s12663-023-01848-0
Gupta	2023	2	NSAID	10.7759/cureus.34580
Hegab	2023	2	HA, PRP	10.1016/j.jormas.2022.11.016
Li	2023	4	HA	10.1007/s11282-022-00621-2
Vingender	2023	3	HA, PRP, I-PRF	10.1016/j.jcms.2023.01.017
Wu	2023	4	Chitosan	10.3390/jcm12041657
Abbadi	2022	2	PRP	10.7759/cureus.31396
Ansar	2022	3	PRP	10.25122/jml-2021-0240
Asadpour	2022	2	HA, PRP	10.1016/j.joms.2022.05.002
Bera	2022	4	I-PRF	10.1111/ors.12665
Mazzara Bou	2022	1	CSs	10.20986/recom.2022.1344/2022
Castaño-Joaqui	2022	4	HA	10.1016/j.jcms.2022.06.004

Table A2. *Cont.*

First Author	Publication Year	Level of Evidence	Injectables	DOI Number
Cen	2022	3	HA	10.1007/s00784-021-04241-8
Chęciński	2022	1	MSCs	10.3390/cells11172709
Chęciński	2022	1	HD, LA, MSCs, NSAID, ozone, opioids	10.3390/jcm11092305
Chęciński	2022	1	HA	10.3390/jcm11071901
Dasukil	2022	2	HA, PRP	10.1016/j.jcms.2022.10.002
Dharamsi	2022	2	CSs, HA	10.1007/s12663-022-01804-4
Ferreira	2022	3	HA	10.1080/08869634.2022.2141784
Ghoneim	2022	2	I-PRF	10.1016/j.jds.2021.07.027
Gutiérrez	2022	1	PRP, PRGF	10.1016/j.jormas.2021.12.006
Haggag	2022	2	HD	10.1016/j.jcms.2022.02.009
Hyder	2022	2	HA	10.1016/j.jormas.2022.05.007
Işık	2022	2	I-PRF	10.1016/j.jcms.2022.06.006
Jacob	2022	2	HA, PRP	10.1007/s12663-021-01519-y
Shah	2022	4	AB	10.4103/jiaomr.jiaomr_199_21
Leketas	2022	2	HA, PRGF	10.1080/08869634.2022.2081445
Liu	2022	3	PRP	10.1111/joor.13261
Macedo de Sousa	2022	3	CSs, HA, PRP	10.3390/life12111739
Manafikhi	2022	4	I-PRF	10.1186/s12891-022-05421-7
Massé	2022	1	CSs, HA, HD, PRP	N/A
Memiş	2022	4	HA	10.1016/j.jcms.2022.07.003
Memiş	2022	3	HD	10.5125/jkaoms.2022.48.1.33
Pandey	2022	4	AB, HD	10.4103/njms.njms_509_21
Rajput	2022	2	PRP	10.1007/s12663-020-01351-w
Ramakrishnan	2022	2	HA, PRP	10.4103/njms.NJMS_94_20
Sari	2022	3	BTX	10.1016/j.jormas.2022.04.019
Sharma	2022	4	AB	10.1007/s12663-021-01540-1
Sikora	2022	4	PRP	10.3390/ijerph192013299
Sikora	2022	4	PRP	10.3390/jcm11154281
Singh	2022	2	CSs	10.4103/njms.njms_291_21
Vaidyanathan	2022	4	Polidocanol	10.4103/ams.ams_138_22
Xie	2022	1	CSs, HA, PRP	10.1016/j.jebdp.2022.101720
AbdulRazzak	2021	3	CSs	10.1007/s10006-020-00901-3
Al-Hamed	2021	1	PRP, PRGF	10.1177/2380084420927326
Amer	2021	2	AB	10.21608/EJENTAS.2021.56244.1300
Chandra	2021	3	PRP	10.4103/jfmpc.jfmpc_1633_20
Cömert Kılıç	2021	2	HA	10.1016/j.jcms.2021.02.012
Dasukil	2021	4	HD, LA	10.1007/s12663-020-01328-9
Derwich	2021	1	CSs, HA, PRP	10.3390/ijms22147405
Ferreira	2021	4	HA	10.1080/07853890.2021.1897446
Goker	2021	1	HA	10.23812/21-2supp1-3
Harba	2021	3	HA, PRP	10.17219/dmp/127446
Karadayi	2021	2	I-PRF	10.1016/j.jcms.2021.01.018
Li	2021	3	PRP	10.1016/j.joms.2020.09.016
Liapaki	2021	1	HA, PRP	10.1016/j.ijom.2021.01.019
Liu	2021	1	NSAID, opioids	10.1111/joor.13105
Romero-Tapia	2021	3	CSs, HA	10.5005/JP-JOURNALS-10024-2890
Rossini	2021	4	HA	10.6061/clinics/2021/e2840
Sàbado-Bundó	2021	1	HA	10.1080/08869634.2021.1925029
Sarwar	2021	3	PRP	10.51253/pafmj.v71i4.5361
Sembronio	2021	2	MSCs	10.1016/j.joms.2021.01.038
Singh	2021	2	HA	10.4103/jpbs.JPBS_675_20
Singh	2021	2	PRP	10.1007/s12663-019-01320-y
Sit	2021	1	HD	10.1038/s41598-021-94119-2
Torul	2021	3	HA, I-PRF	10.1016/j.ijom.2021.03.004
Wang	2021	4	HA	10.1016/j.bjoms.2020.07.013
Zubair	2021	4	PRP	10.21276/apjhs.2021.8.2.2
Aamir	2020	4	AB	N/A

Table A2. *Cont.*

First Author	Publication Year	Level of Evidence	Injectables	DOI Number
Abrahamsson	2020	1	AB, HD	10.1007/s00784-019-03126-1
Ahmed	2020	2	PRGF	N/A
Albilia	2020	4	I-PRF	10.1080/08869634.2018.1516183
Bukhari	2020	3	AB	10.5455/JPMA.5002
Dolwick	2020	2	CSs	10.1016/j.joms.2020.02.022
Fayed	2020	2	HA, opioids	N/A
Hammoodi	2020	3	CSs, PRP	N/A
Hosgor	2020	3	HA	10.1016/j.jcms.2020.07.008
Jara Armijos	2020	1	HA, PRP, PRGF, NSAID	10.4321/s0213-12852020000100005
Li	2020	1	PRP	10.11607/ofph.2470
Liu	2020	1	CSs, HA, PRGF, opioids	10.1016/j.joms.2019.10.016
Marzook	2020	3	CSs, HA	10.1016/j.jormas.2019.05.003
Mohammed	2020	2	CSs, HA	10.37506/ijfmt.v14i2.2817
Pihut	2020	2	HA, PRP	10.3390/ijerph17134726
Santagata	2020	4	HA	10.3390/jfmk5010018
Sezavar	2020	2	PRP	10.29252/jrdms.5.3.7
Sikora	2020	4	HA	10.3390/jcm9061749
Singh	2020	3	CSs, HA	10.1007/s12070-019-01738-3
Macedo De Sousa	2020	2	CSs, HA, PRP	10.3390/medicina56030113
Stasko	2020	3	HA	10.4149/BLL_2020_056
Taşkesen	2020	3	HD	10.1080/08869634.2020.1861887
Yuce	2020	3	HA, I-PRF	10.1097/SCS.0000000000006545
Zarate	2020	2	HD, LA	10.1089/acm.2020.0207
Zigmantavičius	2020	1	HA, PRP	N/A
Abd	2019	2	PRP	N/A
Bergstrand	2019	2	HA	10.2334/josnusd.17-0423
Brignardello-Petersen	2019	1	I-PRF	10.1016/j.adaj.2019.01.015
Carboni	2019	2	MSCs	10.1097/SCS.0000000000004884
Chung	2019	1	PRP	10.1016/j.oooo.2018.09.003
De Riu	2019	2	MSCs	10.1016/j.jcms.2018.11.025
Gavin Clavero	2019	4	HA	10.1007/s10006-019-00789-8
Giacomello	2019	4	PRGF	N/A
Gokçe Kutuk	2019	2	CSs, HA, PRP	10.1097/SCS.0000000000005211
Henk	2019	4	CSs	10.5935/0946-5448.20190003
Isacsson	2019	2	CSs	10.1111/joor.12718
Khallaf	2019	4	PRP	N/A
Louw	2019	2	HD, LA	10.1016/j.mayocp.2018.07.023
Mahmmood	2019	4	MSCs	10.1097/SCS.0000000000004938
Sequeira	2019	4	HA	10.1007/s12663-018-1093-4
Su	2019	4	HA	10.11607/ofph.2044
Toameh	2019	2	HA, PRP	10.17219/dmp/109329
Yilmaz	2019	2	HA	10.1016/j.jcms.2019.07.030
Batifol	2018	4	BTX	10.1016/j.jormas.2018.06.002
Bousnaki	2018	1	HA, PRP	10.1016/j.ijom.2017.09.014
Brignardello-Petersen	2018	1	PRP	10.1016/j.adaj.2017.11.012
Cen	2018	2	HA	10.1111/odi.12760
Daif	2018	4	Ozone	10.7203/jo3t.2.2.2018.11132
Davoudi	2018	1	CSs	10.4317/medoral.21925
Ferreira	2018	1	HA	10.1016/j.jcms.2018.08.007
Fonseca	2018	4	HA	10.1155/2018/5392538
Fouda	2018	2	HD	10.1016/j.bjoms.2018.07.022
Ganti	2018	3	HA	10.5005/jp-journals-10024-2456
Gupta	2018	2	CSs, LA, PRP	10.4103/njms.NJMS_69_16
Haigler	2018	1	PRP, PRGF	10.1016/j.adaj.2018.07.025
Lin	2018	3	PRP	10.1097/MD.0000000000010477

Table A2. *Cont.*

First Author	Publication Year	Level of Evidence	Injectables	DOI Number
Liu	2018	1	CSs, HA	10.1016/j.joms.2017.10.028
Machon	2018	2	AB	10.1007/s10006-017-0666-6
Moldez	2018	1	CSs, HA	10.11607/ofph.1783
Mustafa	2018	2	HD	10.1097/SCS.0000000000004480
Nagori	2018	1	HD	10.1111/joor.12698
Srinivas	2018	2	HA	10.5958/0974-360X.2018.00643.1
Sun	2018	3	HA	10.12659/MSM.908821
Vingender	2018	3	CSs, HA	10.1556/650.2018.31138
Yang	2018	2	HA	10.1016/j.joms.2018.04.031
Yapici-Yavuz	2018	2	CSs, HA, NSAID	10.4317/medoral.22237
Yoshida	2018	4	AB	10.1016/j.bjoms.2017.08.009
Al-Delayme	2017	4	PRP	10.1007/s12663-016-0911-9
Bouloux	2017	2	CSs, HA	10.1016/j.joms.2016.08.006
Castaño-Joaqui	2017	1	HA	10.1016/j.maxilo.2016.11.002
Cezairli	2017	4	HD	10.1089/acm.2017.0068
Gorrela	2017	2	HA	10.1007/s12663-016-0955-x
Guarda-Nardini	2017	4	HA	10.1080/08869634.2016.1232788
Gurung	2017	2	HA	10.4103/njms.NJMS_84_16
Iturriaga	2017	1	HA	10.1016/j.ijom.2017.01.014
Ozdamar	2017	2	HA	10.1111/joor.12467
Patel	2017	4	AB	10.4317/jced.53812
Peng	2017	4	HA	10.1016/j.ijom.2017.02.1219
Pihut	2017	3	HA, PRP	N/A
Refai	2017	4	HD	10.1016/j.bjoms.2016.12.002
Yang	2017	4	LPCGF	10.1097/MD.0000000000006302

N/A—not applicable; AB—autologous blood; BTX—botulinum toxin; CSs—corticosteroids; HA—hyaluronic acid; HD—hypertonic dextrose; I-PRF—injectable platelet-rich fibrin; LA—local anesthetics; LPCGF—liquid-phase concentrated growth factor; MSCs—mesenchymal stem cells; NSAID—non-steroidal anti-inflammatory drug; PRGF—plasma rich in growth factors; PRP—platelet-rich plasma.

References

1. Iturriaga, V.; Bornhardt, T.; Velasquez, N. Temporomandibular Joint. *Dent. Clin. N. Am.* **2023**, *67*, 199–209. [CrossRef]
2. Serrano-Hernanz, G.; Kothari, S.; Castrillón, E.; Álvarez-Méndez, A.; Ardizone-García, I.; Svensson, P. Importance of Standardized Palpation of the Human Temporomandibular Joint. *J. Oral Facial Pain Headache* **2019**, *33*, 220–226. [CrossRef]
3. Osiewicz, M.A.; Manfredini, D.; Loster, B.W.; Van Selms, M.K.A.; Lobbezoo, F. Comparison of the Outcomes of Dynamic/Static Tests and Palpation Tests in TMD-Pain Patients. *J. Oral Rehabil.* **2018**, *45*, 185–190. [CrossRef]
4. Wilkie, G.; Al-Ani, Z. Temporomandibular Joint Anatomy, Function and Clinical Relevance. *Br. Dent. J.* **2022**, *233*, 539–546. [CrossRef]
5. Machado, E.; Machado, P.; Wandscher, V.F.; Marchionatti, A.M.E.; Zanatta, F.B.; Kaizer, O.B. A Systematic Review of Different Substance Injection and Dry Needling for Treatment of Temporomandibular Myofascial Pain. *Int. J. Oral Maxillofac. Surg.* **2018**, *47*, 1420–1432. [CrossRef]
6. Valesan, L.F.; Da-Cas, C.D.; Réus, J.C.; Denardin, A.C.S.; Garanhani, R.R.; Bonotto, D.; Januzzi, E.; De Souza, B.D.M. Prevalence of Temporomandibular Joint Disorders: A Systematic Review and Meta-Analysis. *Clin. Oral Investig.* **2021**, *25*, 441–453. [CrossRef]
7. Klasser, G.D.; Abt, E.; Weyant, R.J.; Greene, C.S. Temporomandibular Disorders: Current Status of Research, Education, Policies, and Its Impact on Clinicians in the United States of America. *Quintessence Int.* **2023**, *54*, 328–334. [CrossRef]
8. Alona, E.-P.; Ilana, E. One Year into the COVID-19 Pandemic—Temporomandibular Disorders and Bruxism: What We Have Learned and What We Can Do to Improve Our Manner of Treatment. *Dent. Med. Probl.* **2021**, *58*, 215–218. [CrossRef]
9. Haddad, C.; Sayegh, S.M.; El Zoghbi, A.; Lawand, G.; Nasr, L. The Prevalence and Predicting Factors of Temporomandibular Disorders in COVID-19 Infection: A Cross-Sectional Study. *Cureus* **2022**, *14*, e28167. [CrossRef]
10. Ginszt, M.; Zieliński, G.; Szkutnik, J.; Wójcicki, M.; Baszczowski, M.; Litko-Rola, M.; Rózyło-Kalinowska, I.; Majcher, P. The Effects of Wearing a Medical Mask on the Masticatory and Neck Muscle Activity in Healthy Young Women. *J. Clin. Med.* **2022**, *11*, 303. [CrossRef]
11. Gębska, M.; Dalewski, B.; Pałka, Ł.; Kołodziej, Ł.; Sobolewska, E. Chronotype Profile, Stress, Depression Level, and Temporomandibular Symptoms in Students with Type D Personality. *J. Clin. Med.* **2022**, *11*, 1886. [CrossRef]
12. Chatzopoulos, G.S.; Sanchez, M.; Cisneros, A.; Wolff, L.F. Prevalence of Temporomandibular Symptoms and Parafunctional Habits in a University Dental Clinic and Association with Gender, Age, and Missing Teeth. *CRANIO®* **2019**, *37*, 159–167. [CrossRef]

13.   Minakuchi, H.; Fujisawa, M.; Abe, Y.; Iida, T.; Oki, K.; Okura, K.; Tanabe, N.; Nishiyama, A. Managements of Sleep Bruxism in Adult: A Systematic Review. *Jpn. Dent. Sci. Rev.* **2022**, *58*, 124–136. [CrossRef]

14.   Andre, A.; Kang, J.; Dym, H. Pharmacologic Treatment for Temporomandibular and Temporomandibular Joint Disorders. *Oral Maxillofac. Surg. Clin. N. Am.* **2022**, *34*, 49–59. [CrossRef]

15.   Arribas-Pascual, M.; Hernández-Hernández, S.; Jiménez-Arranz, C.; Grande-Alonso, M.; Angulo-Díaz-Parreño, S.; La Touche, R.; Paris-Alemany, A. Effects of Physiotherapy on Pain and Mouth Opening in Temporomandibular Disorders: An Umbrella and Mapping Systematic Review with Meta-Meta-Analysis. *J. Clin. Med.* **2023**, *12*, 788. [CrossRef]

16.   Manrriquez, S.L.; Robles, K.; Pareek, K.; Besharati, A.; Enciso, R. Reduction of Headache Intensity and Frequency with Maxillary Stabilization Splint Therapy in Patients with Temporomandibular Disorders-Headache Comorbidity: A Systematic Review and Meta-Analysis. *J. Dent. Anesth. Pain Med.* **2021**, *21*, 183. [CrossRef]

17.   Pietruszka, P.; Chruścicka, I.; Duś-Ilnicka, I.; Paradowska-Stolarz, A. PRP and PRF—Subgroups and Divisions When Used in Dentistry. *J. Pers. Med.* **2021**, *11*, 944. [CrossRef]

18.   Liapaki, A.; Thamm, J.R.; Ha, S.; Monteiro, J.L.G.C.; McCain, J.P.; Troulis, M.J.; Guastaldi, F.P.S. Is There a Difference in Treatment Effect of Different Intra-Articular Drugs for Temporomandibular Joint Osteoarthritis? A Systematic Review of Randomized Controlled Trials. *Int. J. Oral Maxillofac. Surg.* **2021**, *50*, 1233–1243. [CrossRef]

19.   Liu, Y.; Wu, J.; Fei, W.; Cen, X.; Xiong, Y.; Wang, S.; Tang, Y.; Liang, X. Is There a Difference in Intra-Articular Injections of Corticosteroids, Hyaluronate, or Placebo for Temporomandibular Osteoarthritis? *J. Oral Maxillofac. Surg.* **2018**, *76*, 504–514. [CrossRef]

20.   Bhattacharjee, B.; Bera, R.N.; Verma, A.; Soni, R.; Bhatnagar, A. Efficacy of Arthrocentesis and Stabilization Splints in Treatment of Temporomandibular Joint Disc Displacement Disorder Without Reduction: A Systematic Review and Meta-Analysis. *J. Maxillofac. Oral Surg.* **2023**, *22*, 83–93. [CrossRef]

21.   Thorpe, A.R.D.S.; Haddad, Y.; Hsu, J. A Systematic Review and Meta-Analysis of Randomized Controlled Trials Comparing Arthrocentesis with Conservative Management for Painful Temporomandibular Joint Disorder. *Int. J. Oral Maxillofac. Surg.* **2023**, *52*, 889–896. [CrossRef] [PubMed]

22.   Chęciński, M.; Chęcińska, K.; Turosz, N.; Sikora, M.; Chlubek, D. Intra-Articular Injections into the Inferior versus Superior Compartment of the Temporomandibular Joint: A Systematic Review and Meta-Analysis. *J. Clin. Med.* **2023**, *12*, 1664. [CrossRef]

23.   Gutiérrez, I.Q.; Sábado-Bundó, H.; Gay-Escoda, C. Intraarticular Injections of Platelet Rich Plasma and Plasma Rich in Growth Factors with Arthrocenthesis or Arthroscopy in the Treatment of Temporomandibular Joint Disorders: A Systematic Review. *J. Stomatol. Oral Maxillofac. Surg.* **2022**, *123*, e327–e335. [CrossRef] [PubMed]

24.   Chęciński, M.; Chęcińska, K.; Nowak, Z.; Sikora, M.; Chlubek, D. Treatment of Mandibular Hypomobility by Injections into the Temporomandibular Joints: A Systematic Review of the Substances Used. *J. Clin. Med.* **2022**, *11*, 2305. [CrossRef] [PubMed]

25.   Xie, Y.; Zhao, K.; Ye, G.; Yao, X.; Yu, M.; Ouyang, H. Effectiveness of intra-articular injections of sodium hyaluronate, corticosteroids, platelet-rich plasma on temporomandibular joint osteoarthritis: A systematic review and network meta-analysis of randomized controlled trials. *J. Evid.-Based Dent. Pract.* **2022**, *22*, 101720. [CrossRef]

26.   Samson, D.; Schoelles, K.M. Chapter 2: Medical Tests Guidance (2) Developing the Topic and Structuring Systematic Reviews of Medical Tests: Utility of PICOTS, Analytic Frameworks, Decision Trees, and Other Frameworks. *J. Gen. Intern. Med.* **2012**, *27*, 11–19. [CrossRef]

27.   Methley, A.M.; Campbell, S.; Chew-Graham, C.; McNally, R.; Cheraghi-Sohi, S. PICO, PICOS and SPIDER: A Comparison Study of Specificity and Sensitivity in Three Search Tools for Qualitative Systematic Reviews. *BMC Health Serv. Res.* **2014**, *14*, 579. [CrossRef]

28.   Eriksen, M.B.; Frandsen, T.F. The Impact of Patient, Intervention, Comparison, Outcome (PICO) as a Search Strategy Tool on Literature Search Quality: A Systematic Review. *J. Med. Libr. Assoc.* **2018**, *106*, 420. [CrossRef]

29.   The ACM Guide to Computing Literature. Available online: https://libraries.acm.org/digital-library/acm-guide-to-computing-literature (accessed on 4 May 2023).

30.   BASE—Bielefeld Academic Search Engine | What Is BASE? Available online: https://www.base-search.net/about/en/index.php (accessed on 4 May 2023).

31.   About. Available online: https://pubmed.ncbi.nlm.nih.gov/about/ (accessed on 4 May 2023).

32.   Gusenbauer, M. Search Where You Will Find Most: Comparing the Disciplinary Coverage of 56 Bibliographic Databases. *Scientometrics* **2022**, *127*, 2683–2745. [CrossRef]

33.   Scopus Content. Available online: https://www.elsevier.com/solutions/scopus/how-scopus-works/content (accessed on 4 May 2023).

34.   Ouzzani, M.; Hammady, H.; Fedorowicz, Z.; Elmagarmid, A. Rayyan—A Web and Mobile App for Systematic Reviews. *Syst. Rev.* **2016**, *5*, 210. [CrossRef]

35.   OCEBM Levels of Evidence—Centre for Evidence-Based Medicine (CEBM), University of Oxford. Available online: https://www.cebm.ox.ac.uk/resources/levels-of-evidence/ocebm-levels-of-evidence (accessed on 1 May 2023).

36.   Rajput, A.; Bansal, V.; Dubey, P.; Kapoor, A. A Comparative Analysis of Intra-Articular Injection of Platelet-Rich Plasma and Arthrocentesis in Temporomandibular Joint Disorders. *J. Maxillofac. Oral Surg.* **2022**, *21*, 168–175. [CrossRef]

37. Bhargava, D.; Sivakumar, B.; Bhargava, P.G. A Comparative Preliminary Randomized Clinical Study to Evaluate Heavy Bupivacaine Dextrose Prolotherapy (HDP) and Autologous Blood Injection (ABI) for Symptomatic Temporomandibular Joint Hypermobility Disorder. *J. Maxillofac. Oral Surg.* **2023**, *22*, 110–118. [CrossRef]

38. Yapici-Yavuz, G.; Simsek-Kaya, G.; Ogul, H. A Comparison of the Effects of Methylprednisolone Acetate, Sodium Hyaluronate and Tenoxicam in the Treatment of Non-Reducing Disc Displacement of the Temporomandibular Joint. *Med. Oral Patol. Oral Cir. Bucal* **2018**, *23*, e351. [CrossRef]

39. Singh, S.; Prasad, R.; Punga, R.; Datta, R.; Singh, N. A Comparison of the Outcomes Following Intra-Articular Steroid Injection Alone or Arthrocentesis Alone in the Management of Internal Derangement of the Temporomandibular Joint. *Natl. J. Maxillofac. Surg.* **2022**, *13*, 80. [CrossRef] [PubMed]

40. Machon, V.; Levorova, J.; Hirjak, D.; Wisniewski, M.; Drahos, M.; Sidebottom, A.; Foltan, R. A Prospective Assessment of Outcomes Following the Use of Autologous Blood for the Management of Recurrent Temporomandibular Joint Dislocation. *Oral Maxillofac. Surg.* **2018**, *22*, 53–57. [CrossRef] [PubMed]

41. Dolwick, M.F.; Diaz, D.; Freburg-Hoffmeister, D.L.; Widmer, C.G. A Randomized, Double-Blind, Placebo-Controlled Study of the Efficacy of Steroid Supplementation After Temporomandibular Joint Arthrocentesis. *J. Oral Maxillofac. Surg.* **2020**, *78*, 1088–1099. [CrossRef] [PubMed]

42. Wu, C.-B.; Sun, H.-J.; Sun, N.-N.; Zhou, Q. Analysis of the Curative Effect of Temporomandibular Joint Disc Release and Fixation Combined with Chitosan Injection in the Treatment of Temporomandibular Joint Osteoarthrosis. *J. Clin. Med.* **2023**, *12*, 1657. [CrossRef]

43. Santagata, M.; De Luca, R.; Lo Giudice, G.; Troiano, A.; Lo Giudice, G.; Corvo, G.; Tartaro, G. Arthrocentesis and Sodium Hyaluronate Infiltration in Temporomandibular Disorders Treatment. Clinical and MRI Evaluation. *J. Funct. Morphol. Kinesiol.* **2020**, *5*, 18. [CrossRef]

44. Sharma, V.; Anchlia, S.; Sadhwani, B.S.; Bhatt, U.; Rajpoot, D. Arthrocentesis Followed by Autologous Blood Injection in the Treatment of Chronic Symptomatic Subluxation of Temporomandibular Joint. *J. Maxillofac. Oral Surg.* **2022**, *21*, 1218–1226. [CrossRef]

45. AbdulRazzak, N.J.; Sadiq, J.A.; Jiboon, A.T. Arthrocentesis versus Glucocorticosteroid Injection for Internal Derangement of Temporomandibular Joint. *Oral Maxillofac. Surg.* **2021**, *25*, 191–197. [CrossRef]

46. Abbadi, W.; Kara Beit, Z.; Al-Khanati, N.M. Arthrocentesis, Injectable Platelet-Rich Plasma and Combination of Both Protocols of Temporomandibular Joint Disorders Management: A Single-Blinded Randomized Clinical Trial. *Cureus* **2022**, *14*, e31396. [CrossRef]

47. Mazzara Bou, C.; González Sarrión, Ó. Artrocentesis de La Articulación Temporomandibular y Corticoides. Revisión de La Literatura. *Rev. Esp. Cir. Oral Maxilofac.* **2022**, *44*, 303. [CrossRef]

48. Mahmmood, V.H.; Shihab, S.M. Assessment of Therapeutic Effect of Intra-Articular Nanofat Injection for Temporomandibular Disorders. *J. Craniofac. Surg.* **2019**, *30*, 659–662. [CrossRef]

49. Mohammed, S.M.; Abusanna, M.M.H.; Daily, Z.A. Assessment the Efficacy of Arthrocentesis with Corticosteroid and Arthrocentesis with Sodium Hyaluronate in Treatment Temporomandibular Joint Disorders: A Comparative Study. *Indian J. Forensic Med. Toxicol.* **2020**, *14*, 361–366. [CrossRef]

50. Chęciński, M.; Chęcińska, K.; Turosz, N.; Kamińska, M.; Nowak, Z.; Sikora, M.; Chlubek, D. Autologous Stem Cells Transplants in the Treatment of Temporomandibular Joints Disorders: A Systematic Review and Meta-Analysis of Clinical Trials. *Cells* **2022**, *11*, 2709. [CrossRef]

51. Vingender, S.; Restár, L.; Csomó, K.B.; Schmidt, P.; Hermann, P.; Vaszilkó, M. Az állkapocsízületi károsodás kezelése szteroiddal, illetve hialuronsavval. *Orv. Hetil.* **2018**, *159*, 1475–1482. [CrossRef]

52. De Riu, G.; Vaira, L.A.; Carta, E.; Meloni, S.M.; Sembronio, S.; Robiony, M. Bone Marrow Nucleated Cell Concentrate Autograft in Temporomandibular Joint Degenerative Disorders: 1-Year Results of a Randomized Clinical Trial. *J. Cranio-Maxillofac. Surg.* **2019**, *47*, 1728–1738. [CrossRef]

53. Fouda, A.A. Change of Site of Intra-Articular Injection of Hypertonic Dextrose Resulted in Different Effects of Treatment. *Br. J. Oral Maxillofac. Surg.* **2018**, *56*, 715–718. [CrossRef] [PubMed]

54. Gokce Kutuk, S.; Gökçe, G.; Arslan, M.; Özkan, Y.; Kütük, M.; Kursat Arikan, O. Clinical and Radiological Comparison of Effects of Platelet-Rich Plasma, Hyaluronic Acid, and Corticosteroid Injections on Temporomandibular Joint Osteoarthritis. *J. Craniofac. Surg.* **2019**, *30*, 1144–1148. [CrossRef] [PubMed]

55. Patel, J.; Nilesh, K.; Parkar, M.; Vaghasiya, A. Clinical and Radiological Outcome of Arthrocentesis Followed by Autologous Blood Injection for Treatment of Chronic Recurrent Temporomandibular Joint Dislocation. *J. Clin. Exp. Dent.* **2017**, *9*, e962. [CrossRef] [PubMed]

56. Yang, J.-W.; Huang, Y.-C.; Wu, S.-L.; Ko, S.-Y.; Tsai, C.-C. Clinical Efficacy of a Centric Relation Occlusal Splint and Intra-Articular Liquid Phase Concentrated Growth Factor Injection for the Treatment of Temporomandibular Disorders. *Medicine* **2017**, *96*, e6302. [CrossRef]

57. Yoshida, H.; Nakatani, Y.-i.; Gamoh, S.; Shimizutani, K.; Morita, S. Clinical Outcome after 36 Months of Treatment with Injections of Autologous Blood for Recurrent Dislocation of the Temporomandibular Joint. *Br. J. Oral Maxillofac. Surg.* **2018**, *56*, 64–66. [CrossRef]

58. Sun, H.; Su, Y.; Song, N.; Li, C.; Shi, Z.; Li, L. Clinical Outcome of Sodium Hyaluronate Injection into the Superior and Inferior Joint Space for Osteoarthritis of the Temporomandibular Joint Evaluated by Cone-Beam Computed Tomography: A Retrospective Study of 51 Patients and 56 Joints. *Med. Sci. Monit.* **2018**, *24*, 5793–5801. [CrossRef]
59. Ferreira, J.R.; Nunes, M.A.; Salvado, F. Clinical Outcomes in TMD Patients after Arthrocentesis with Lysis, Lavage and Viscossu-plementation. *Ann. Med.* **2021**, *53*, S87–S88. [CrossRef]
60. Asadpour, N.; Shooshtari, Z.; Kazemian, M.; Gholami, M.; Vatanparast, N.; Samieirad, S. Combined Platelet-Rich Plasma and Hyaluronic Acid Can Reduce Pain in Patients Undergoing Arthrocentesis for Temporomandibular Joint Osteoarthritis. *J. Oral Maxillofac. Surg.* **2022**, *80*, 1474–1485. [CrossRef]
61. Shah, J.; Joshi, K.; Jha, S.; Mathumathi, A. Comparative Analysis of Autologous Blood Injection and Conservative Therapy for the Management of Chronic Temporomandibular Joint Dislocation. *J. Indian Acad. Oral Med. Radiol.* **2022**, *34*, 394. [CrossRef]
62. Gupta, S.; Sharma, A.; Purohit, J.; Goyal, R.; Malviya, Y.; Jain, S. Comparison between Intra-Articular Platelet-Rich Plasma Injection versus Hydrocortisone with Local Anesthetic Injections in Temporomandibular Disorders: A Double-Blind Study. *Natl. J. Maxillofac. Surg.* **2018**, *9*, 205. [CrossRef] [PubMed]
63. Ferreira, N.R.; Oliveira, A.T.; Sanz, C.K.; Guedes, F.R.; Rodrigues, M.J.; Grossmann, E.; DosSantos, M.F. Comparison between Two Viscosupplementation Protocols for Temporomandibular Joint Osteoarthritis. *CRANIO®* **2022**, 1–9. [CrossRef] [PubMed]
64. Sarwar, H.; Shah, I.; Khan, A.A.; Afzal, M.; Babar, A.; Baig, A.M. Comparison of arthrocentesis plus platelet rich plasma with arthrocentesis alone in the treatment of temporomandibular joint dysfunction. *Pak. Armed Forces Med. J.* **2021**, *71*, 1377–1381. [CrossRef]
65. Chhapane, A.; Wadde, K.; Sachdev, S.S.; Barai, S.; Landge, J.; Wadewale, M. Comparison of Autologous Blood Injection and Dextrose Prolotherapy in the Treatment of Chronic Recurrent Temporomandibular Dislocation: A Randomized Clinical Trial. *J. Maxillofac. Oral Surg.* **2023**. [CrossRef]
66. Pandey, S.; Baidya, M.; Srivastava, A.; Garg, H. Comparison of Autologous Blood Prolotherapy and 25% Dextrose Prolotherapy for the Treatment of Chronic Recurrent Temporomandibular Joint Dislocation on the Basis of Clinical Parameters: A Retrospective Study. *Natl. J. Maxillofac. Surg.* **2022**, *13*, 398. [CrossRef]
67. Li, F.-L.; Wu, C.-B.; Sun, H.-J.; Zhou, Q. Comparison of Autologous Platelet-Rich Plasma and Chitosan in the Treatment of Temporomandibular Joint Osteoarthritis: A Retrospective Cohort Study. *J. Oral Maxillofac. Surg.* **2021**, *79*, 324–332. [CrossRef]
68. Ramakrishnan, D.; Kandamani, J.; Nathan, K.S. Comparison of Intraarticular Injection of Platelet-Rich Plasma Following Arthrocentesis, with Sodium Hyaluronate and Conventional Arthrocentesis for Management of Internal Derangement of Temporomandibular Joint. *Natl. J. Maxillofac. Surg.* **2022**, *13*, 254. [CrossRef] [PubMed]
69. Bukhari, A.; Rahim, A. Comparison of Mean Decrease in Mouth Opening by Autologous Blood Injection in Superior Joint Space with and without Pericapsular Tissue in Treatment of Chronic Recurrent Temporomandibular Joint Dislocation in Mayo Hospital Lahore. *J. Pak. Med. Assoc.* **2020**, *70*, 1878–1882. [CrossRef] [PubMed]
70. Yuce, E.; Komerik, N. Comparison of the Efficacy of Intra-Articular Injection of Liquid Platelet-Rich Fibrin and Hyaluronic Acid After in Conjunction With Arthrocentesis for the Treatment of Internal Temporomandibular Joint Derangements. *J. Craniofac. Surg.* **2020**, *31*, 1870–1874. [CrossRef]
71. Yilmaz, O.; Korkmaz, Y.T.; Tuzuner, T. Comparison of Treatment Efficacy between Hyaluronic Acid and Arthrocentesis plus Hyaluronic Acid in Internal Derangements of Temporomandibular Joint. *J. Cranio-Maxillofac. Surg.* **2019**, *47*, 1720–1727. [CrossRef]
72. Haggag, M.A.; Al-Belasy, F.A.; Said Ahmed, W.M. Dextrose Prolotherapy for Pain and Dysfunction of the TMJ Reducible Disc Displacement: A Randomized, Double-Blind Clinical Study. *J. Cranio-Maxillofac. Surg.* **2022**, *50*, 426–431. [CrossRef]
73. Zarate, M.A.; Frusso, R.D.; Reeves, K.D.; Cheng, A.-L.; Rabago, D. Dextrose Prolotherapy Versus Lidocaine Injection for Temporomandibular Dysfunction: A Pragmatic Randomized Controlled Trial. *J. Altern. Complement. Med.* **2020**, *26*, 1064–1073. [CrossRef]
74. Leketas, M.; Dvylys, D.; Sakalys, D.; Simuntis, R. Different Intra-Articular Injection Substances Following Temporomandibular Joint Arthroscopy and Their Effect on Early Postoperative Period: A Randomized Clinical Trial. *CRANIO®* **2022**, 1–6. [CrossRef] [PubMed]
75. Macedo De Sousa, B.; López-Valverde, N.; López-Valverde, A.; Caramelo, F.; Flores Fraile, J.; Herrero Payo, J.; Rodrigues, M.J. Different Treatments in Patients with Temporomandibular Joint Disorders: A Comparative Randomized Study. *Medicina* **2020**, *56*, 113. [CrossRef]
76. Li, Z.; Zhou, J.; Yu, L.; He, S.; Li, F.; Lin, Y.; Xu, J.; Chen, S. Disc–Condyle Relationship Alterations Following Stabilization Splint Therapy or Arthrocentesis plus Hyaluronic Acid Injection in Patients with Anterior Disc Displacement: A Retrospective Cohort Study. *Oral Radiol.* **2023**, *39*, 198–206. [CrossRef] [PubMed]
77. Liu, S.; Hu, Y.; Zhang, X. Do Intra-articular Injections of Analgesics Improve Outcomes after Temporomandibular Joint Arthro-centesis?: A Systematic Review and Meta-analysis. *J. Oral Rehabil.* **2021**, *48*, 95–105. [CrossRef]
78. Cömert Kılıç, S. Does Glucosamine, Chondroitin Sulfate, and Methylsulfonylmethane Supplementation Improve the Outcome of Temporomandibular Joint Osteoarthritis Management with Arthrocentesis plus Intraarticular Hyaluronic Acid Injection. A Randomized Clinical Trial. *J. Cranio-Maxillofac. Surg.* **2021**, *49*, 711–718. [CrossRef] [PubMed]

79. Bayramoglu, Z.; Yavuz, G.Y.; Keskinruzgar, A.; Koparal, M.; Kaya, G.S. Does Intra-Articular Injection of Tenoxicam after Arthrocentesis Heal Outcomes of Temporomandibular Joint Osteoarthritis? A Randomized Clinical Trial. *BMC Oral Health* **2023**, *23*, 131. [CrossRef] [PubMed]

80. Rossini, R.; Grossmann, E.; Poluha, R.L.; Setogutti, Ê.T.; Dos Santos, M.F. Double-Needle Arthrocentesis with Viscosupplementation in Patients with Temporomandibular Joint Disc Displacement without Reduction. *Clinics* **2021**, *76*, e2840. [CrossRef]

81. Lin, S.-L.; Tsai, C.-C.; Wu, S.-L.; Ko, S.-Y.; Chiang, W.-F.; Yang, J.W. Effect of Arthrocentesis plus Platelet-Rich Plasma and Platelet-Rich Plasma Alone in the Treatment of Temporomandibular Joint Osteoarthritis: A Retrospective Matched Cohort Study (A STROBE-Compliant Article). *Medicine* **2018**, *97*, e0477. [CrossRef]

82. Iturriaga, V.; Bornhardt, T.; Manterola, C.; Brebi, P. Effect of Hyaluronic Acid on the Regulation of Inflammatory Mediators in Osteoarthritis of the Temporomandibular Joint: A Systematic Review. *Int. J. Oral Maxillofac. Surg.* **2017**, *46*, 590–595. [CrossRef]

83. Batifol, D.; Huart, A.; Finiels, P.J.; Nagot, N.; Jammet, P. Effect of Intra-Articular Botulinum Toxin Injections on Temporo-Mandibular Joint Pain. *J. Stomatol. Oral Maxillofac. Surg.* **2018**, *119*, 319–324. [CrossRef]

84. Li, F.; Wu, C.; Sun, H.; Zhou, Q. Effect of Platelet-Rich Plasma Injections on Pain Reduction in Patients with Temporomandibular Joint Osteoarthrosis: A Meta-Analysis of Randomized Controlled Trials. *J. Oral Facial Pain Headache* **2020**, *34*, 149–156. [CrossRef]

85. Liu, S.; Xu, L.; Fan, S.; Lu, S.; Jin, L.; Liu, L.; Yao, Y.; Cai, B. Effect of Platelet-rich Plasma Injection Combined with Individualised Comprehensive Physical Therapy on Temporomandibular Joint Osteoarthritis: A Prospective Cohort Study. *J. Oral Rehabil.* **2022**, *49*, 150–159. [CrossRef]

86. Moldez, M.; Camones, V.; Ramos, G.; Padilla, M.; Enciso, R. Effectiveness of Intra-Articular Injections of Sodium Hyaluronate or Corticosteroids for Intracapsular Temporomandibular Disorders: A Systematic Review and Meta-Analysis. *J. Oral Facial Pain Headache* **2018**, *32*, 53–66. [CrossRef] [PubMed]

87. Chung, P.-Y.; Lin, M.-T.; Chang, H.-P. Effectiveness of Platelet-Rich Plasma Injection in Patients with Temporomandibular Joint Osteoarthritis: A Systematic Review and Meta-Analysis of Randomized Controlled Trials. *Oral Surg. Oral Med. Oral Pathol. Oral Radiol.* **2019**, *127*, 106–116. [CrossRef] [PubMed]

88. Fonseca, R.M.D.F.B.; Januzzi, E.; Ferreira, L.A.; Grossmann, E.; Carvalho, A.C.P.; de Oliveira, P.G.; Vieira, É.L.M.; Teixeira, A.L.; Almeida-Leite, C.M. Effectiveness of Sequential Viscosupplementation in Temporomandibular Joint Internal Derangements and Symptomatology: A Case Series. *Pain Res. Manag.* **2018**, *2018*, 5392538. [CrossRef] [PubMed]

89. Gurung, T.; Singh, R.; Mohammad, S.; Pal, U.; Mahdi, A.; Kumar, M. Efficacy of Arthrocentesis versus Arthrocentesis with Sodium Hyaluronic Acid in Temporomandibular Joint Osteoarthritis: A Comparison. *Natl. J. Maxillofac. Surg.* **2017**, *8*, 41. [CrossRef]

90. Amer, I.; Kukereja, P.; Gaber, A. Efficacy of Autologous Blood Injection for Treatment of Chronic Recurrent Temporo-Mandibular Joint Dislocation. *Egypt. J. Ear Nose Throat Allied Sci.* **2021**, *22*, 1–6. [CrossRef]

91. Hyder, A.; Tawfik, B.E.; Elmohandes, W. Efficacy of Computer-Guided versus Conventional Sodium Hyaluronate Injection in Superior Joint Space in Treatment of Temporomandibular Joint (TMJ) Internal Derangement: Comparative Randomized Controlled Trial. *J. Stomatol. Oral Maxillofac. Surg.* **2022**, *123*, e321–e326. [CrossRef]

92. Srinivas, M.R.; James, D.; Muthusekhar, M.R. Efficacy of Hyaluronic Acid in the Treatment of Internal Derangement—Clinical Study. *Res. J. Pharm. Technol.* **2018**, *11*, 3483. [CrossRef]

93. Peng, C.Y.; Lu, M.Y. Efficacy of Hyaluronic Acid Injection in Superior Joint Space for the Treatment of Temporomandibular Disorder in Taiwan. *Int. J. Oral Maxillofac. Surg.* **2017**, *46*, 362. [CrossRef]

94. Sit, R.W.-S.; Reeves, K.D.; Zhong, C.C.; Wong, C.H.L.; Wang, B.; Chung, V.C.; Wong, S.Y.; Rabago, D. Efficacy of Hypertonic Dextrose Injection (Prolotherapy) in Temporomandibular Joint Dysfunction: A Systematic Review and Meta-Analysis. *Sci. Rep.* **2021**, *11*, 14638. [CrossRef]

95. Jacob, S.M.; Bandyopadhyay, T.K.; Chattopadhyay, P.K.; Parihar, V.S. Efficacy of Platelet-Rich Plasma Versus Hyaluronic Acid Following Arthrocentesis for Temporomandibular Joint Disc Disorders: A Randomized Controlled Trial. *J. Maxillofac. Oral Surg.* **2022**, *21*, 1199–1204. [CrossRef]

96. Taşkesen, F.; Cezairli, B. Efficacy of Prolotherapy and Arthrocentesis in Management of Temporomandibular Joint Hypermobility. *CRANIO®* **2020**, 1–9. [CrossRef]

97. Dasukil, S.; Shetty, S.K.; Arora, G.; Degala, S. Efficacy of Prolotherapy in Temporomandibular Joint Disorders: An Exploratory Study. *J. Maxillofac. Oral Surg.* **2021**, *20*, 115–120. [CrossRef] [PubMed]

98. Sequeira, J.; Rao, B.H.S.; Kedia, P.R. Efficacy of Sodium Hyaluronate for Temporomandibular Joint Disorder by Single-Puncture Arthrocentesis. *J. Maxillofac. Oral Surg.* **2019**, *18*, 88–92. [CrossRef]

99. Gorrela, H.; Prameela, J.; Srinivas, G.; Reddy, B.V.B.; Sudhir, M.; Arakeri, G. Efficacy of Temporomandibular Joint Arthrocentesis with Sodium Hyaluronate in the Management of Temporomandibular Joint Disorders: A Prospective Randomized Control Trial. *J. Maxillofac. Oral Surg.* **2017**, *16*, 479–484. [CrossRef] [PubMed]

100. Ferreira, N.; Masterson, D.; Lopes de Lima, R.; de Souza Moura, B.; Oliveira, A.T.; Kelly da Silva Fidalgo, T.; Carvalho, A.C.P.; DosSantos, M.F.; Grossmann, E. Efficacy of Viscosupplementation with Hyaluronic Acid in Temporomandibular Disorders: A Systematic Review. *J. Cranio-Maxillofac. Surg.* **2018**, *46*, 1943–1952. [CrossRef] [PubMed]

101. Jara Armijos, J.; Hidalgo Andrade, B.; Velásquez Ron, B. Eficacia Del Ácido Hialurónico En El Tratamiento de Los Trastornos Temporomandibulares. Revisión Sistemática. *Av. Odontoestomatol.* **2020**, *36*, 35–47. [CrossRef]

102. Castaño-Joaqui, O.G.; Muñoz-Guerra, M.F.; Campo, J.; Martínez-Bernardini, G.; Cano, J. Estado actual de la viscosuplementación con ácido hialurónico en el tratamiento de los trastornos temporomandibulares: Revisión sistemática. *Rev. Esp. Cir. Oral Maxilofac.* **2017**, *39*, 213–220. [CrossRef]

103. Bera, R.N.; Tiwari, P. Evaluating the Role of Intra Articular Injection of Platelet-rich Fibrin in the Management of Temporomandibular Joint Osteoarthritis: A STROBE Compliant Retrospective Study. *Oral Surg.* **2022**, *15*, 218–223. [CrossRef]

104. Singh, A.K.; Sharma, N.K.; Kumar, P.G.N.; Singh, S.; Mishra, N.; Bera, R.N. Evaluation of Arthrocentesis with and Without Platelet-Rich Plasma in the Management of Internal Derangement of Temporomandibular Joint: A Randomized Controlled Trial. *J. Maxillofac. Oral Surg.* **2021**, *20*, 252–257. [CrossRef]

105. Goker, F. Evaluation of Arthrocentesis with Hyaluronic Acid Injections for Management of Temporomandibular Disorders: A Systematic Review and Case Series. *J. Biol. Regul. Homeost. Agents* **2021**, *35*, 21–35. [CrossRef]

106. Kapadia, J.M.; Ganti, S.; Shriram, P.; Ansari, A.S.; Azad, A.; Dubey, A. Evaluation of Effect of Glucosamine-Chondroitin Sulfate, Tramadol, and Sodium Hyaluronic Acid on Expression of Cytokine Levels in Internal Derangement of Temporomandibular Joint. *J. Contemp. Dent. Pract.* **2018**, *19*, 1502–1506. [CrossRef]

107. Sezavar, M.; Shafaei Fard, S.; Sharifzadeh, H.; Pahlevan, R.; Badkoobeh, A. Evaluation of the Effect of Platelet-Rich Plasma on Temporomandibular Joint Disorders: A Split-Match Randomized Clinical Trial. *J. Res. Dent. Maxillofac. Sci.* **2020**, *5*, 7–14. [CrossRef]

108. Vaidyanathan, A.; Haidry, N.; Sinha, U.; Singh, A.; Salahudheen, A. Evaluation of the Effects of Polidocanol Injection in the Treatment of Temporomandibular Joint Hypermobility—A Prospective Study. *Ann. Maxillofac. Surg.* **2022**, *12*, 166. [CrossRef]

109. Memiş, S. Evaluation of the Effects of Prolotherapy on Condyles in Temporomandibular Joint Hypermobility Using Fractal Dimension Analysis. *J. Korean Assoc. Oral Maxillofac. Surg.* **2022**, *48*, 33–40. [CrossRef] [PubMed]

110. Memiş, S. Evaluation of the Effects of Temporomandibular Joint Arthrocentesis with Hyaluronic Acid Injection on Mandibular Condyles Using Fractal Dimension Analysis: A Retrospective Study. *J. Cranio-Maxillofac. Surg.* **2022**, *50*, 643–650. [CrossRef]

111. Mustafa, R.; Güngörmüş, M.; Mollaoğlu, N. Evaluation of the Efficacy of Different Concentrations of Dextrose Prolotherapy in Temporomandibular Joint Hypermobility Treatment. *J. Craniofac. Surg.* **2018**, *29*, e461–e465. [CrossRef]

112. Manafikhi, M.; Ataya, J.; Heshmeh, O. Evaluation of the Efficacy of Platelet Rich Fibrin (I-PRF) Intra-Articular Injections in the Management of Internal Derangements of Temporomandibular Joints—A Controlled Preliminary Prospective Clinical Study. *BMC Musculoskelet. Disord.* **2022**, *23*, 454. [CrossRef]

113. Vingender, S.; Dőri, F.; Schmidt, P.; Hermann, P.; Vaszilkó, M.T. Evaluation of the Efficiency of Hyaluronic Acid, PRP and I-PRF Intra-Articular Injections in the Treatment of Internal Derangement of the Temporomandibular Joint: A Prospective Study. *J. Cranio-Maxillofac. Surg.* **2023**, *51*, 1–6. [CrossRef]

114. Harba, A.; Harfoush, M. Evaluation of the Participation of Hyaluronic Acid with Platelet-Rich Plasma in the Treatment of Temporomandibular Joint Disorders. *Dent. Med. Probl.* **2021**, *58*, 81–88. [CrossRef]

115. Cen, X.; Liu, Y.; Wang, S.; Yang, X.; Shi, Z.; Liang, X. Glucosamine Oral Administration as an Adjunct to Hyaluronic Acid Injection in Treating Temporomandibular Joint Osteoarthritis. *Oral Dis.* **2018**, *24*, 404–411. [CrossRef]

116. Cen, X.; Pan, X.; Zhang, B.; Liu, C.; Huang, X.; Zhao, Z. Hyaluronan Injection versus Oral Glucosamine and Diclofenac in the Treatment of Temporomandibular Joint Osteoarthritis. *Clin. Oral Investig.* **2022**, *26*, 2703–2710. [CrossRef] [PubMed]

117. Stasko, J.; Statelova, D.; Janickova, M.; Mikuskova, K.; Bacinsky, M.; Sokol, J.; Frlickova, Z.; Hvizdos, D.; Malachovsky, I. Hyaluronic Acid Application vs Arthroscopy in Treatment of Internal Temporomandibular Joint Disorders. *Bratisl. Med. J.* **2020**, *121*, 352–357. [CrossRef]

118. Işık, G.; Kenç, S.; Özveri Koyuncu, B.; Günbay, S.; Günbay, T. Injectable Platelet-Rich Fibrin as Treatment for Temporomandibular Joint Osteoarthritis: A Randomized Controlled Clinical Trial. *J. Cranio-Maxillofac. Surg.* **2022**, *50*, 576–582. [CrossRef]

119. Brignardello-Petersen, R. Injection of Platelet-Rich Fibrin Probably Reduces Pain but Not Trismus When Used after Arthrocentesis or Arthroscopy in Patients with Temporomandibular Joint Osteoarthritis. *J. Am. Dent. Assoc.* **2019**, *150*, e57. [CrossRef]

120. Guarda-Nardini, L.; Cadorin, C.; Frizziero, A.; Masiero, S.; Manfredini, D. Interrelationship between Temporomandibular Joint Osteoarthritis (OA) and Cervical Spine Pain: Effects of Intra-Articular Injection with Hyaluronic Acid. *CRANIO®* **2017**, *35*, 276–282. [CrossRef] [PubMed]

121. Marzook, H.A.M.; Abdel Razek, A.A.; Yousef, E.A.; Attia, A.A.M.M. Intra-Articular Injection of a Mixture of Hyaluronic Acid and Corticosteroid versus Arthrocentesis in TMJ Internal Derangement. *J. Stomatol. Oral Maxillofac. Surg.* **2020**, *121*, 30–34. [CrossRef] [PubMed]

122. Dasukil, S.; Arora, G.; Boyina, K.K.; Jena, A.K.; Jose, A.; Das, S. Intra-Articular Injection of Hyaluronic Acid versus Platelet-Rich Plasma Following Single Puncture Arthrocentesis for the Management of Internal Derangement of TMJ: A Double-Blinded Randomised Controlled Trial. *J. Cranio-Maxillofac. Surg.* **2022**, *50*, 825–830. [CrossRef]

123. Brignardello-Petersen, R. Intra-Articular Injections of Platelet-Rich Plasma May Improve Pain Associated with Temporomandibular Disorders Compared with Arthrocentesis and Arthroscopy. *J. Am. Dent. Assoc.* **2018**, *149*, e52. [CrossRef]

124. Sàbado-Bundó, H.; Sánchez-Garcés, M.; Camps-Font, O.; Gay-Escoda, C. Intraarticular Injections of Hyaluronic Acid in Arthrocentesis and Arthroscopy as a Treatment of Temporomandibular Joint Disorders: A Systematic Review. *CRANIO®* **2021**, 1–10. [CrossRef]

125. Davoudi, A.; Khaki, H.; Mohammadi, I.; Daneshmand, M.; Tamizifar, A.; Bigdelou, M.; Ansaripoor, F. Is Arthrocentesis of Temporomandibular Joint with Corticosteroids Beneficial? A Systematic Review. *Med. Oral Patol. Oral Cir. Bucal* **2018**, *23*, e367. [CrossRef]

126. Hosgor, H. Is Arthrocentesis plus Hyaluronic Acid Superior to Arthrocentesis Alone in the Treatment of Disc Displacement without Reduction in Patients with Bruxism? *J. Cranio-Maxillofac. Surg.* **2020**, *48*, 1023–1027. [CrossRef]

127. Bouloux, G.F.; Chou, J.; Krishnan, D.; Aghaloo, T.; Kahenasa, N.; Smith, J.A.; Giannakopoulos, H. Is Hyaluronic Acid or Corticosteroid Superior to Lactated Ringer Solution in the Short-Term Reduction of Temporomandibular Joint Pain After Arthrocentesis? Part 1. *J. Oral Maxillofac. Surg.* **2017**, *75*, 52–62. [CrossRef] [PubMed]

128. Albilia, J.; Herrera-Vizcaíno, C.; Weisleder, H.; Choukroun, J.; Ghanaati, S. Liquid Platelet-Rich Fibrin Injections as a Treatment Adjunct for Painful Temporomandibular Joints: Preliminary Results. *CRANIO®* **2020**, *38*, 292–304. [CrossRef] [PubMed]

129. Castaño-Joaqui, O.G.; Maza Muela, C.; Casco Zavala, B.; Casares García, G.; Domínguez Gordillo, A.Á. Long Term Oral Health Related Quality of Life after TMJ Arthrocentesis with Hyaluronic Acid. A Retrospective Cohort Study. *J. Cranio-Maxillofac. Surg.* **2022**, *50*, 583–589. [CrossRef] [PubMed]

130. Bergstrand, S.; Ingstad, H.K.; Møystad, A.; Bjørnland, T. Long-Term Effectiveness of Arthrocentesis with and without Hyaluronic Acid Injection for Treatment of Temporomandibular Joint Osteoarthritis. *J. Oral Sci.* **2019**, *61*, 82–88. [CrossRef]

131. Refai, H. Long-Term Therapeutic Effects of Dextrose Prolotherapy in Patients with Hypermobility of the Temporomandibular Joint: A Single-Arm Study with 1–4 Years' Follow up. *Br. J. Oral Maxillofac. Surg.* **2017**, *55*, 465–470. [CrossRef]

132. Singh, N.; Dubey, S.; Bhanawat, N.; Rai, G.; Kumar, A.; Vatsa, R. Management of Internal Disc Derangement Using Normal Saline and Sodium Hyaluronate: A Comparative Study. *J. Pharm. Bioallied Sci.* **2021**, *13*, 207. [CrossRef]

133. Toameh, M.; Alkhouri, I.; Karman, M.A. Management of Patients with Disk Displacement without Reduction of the Temporomandibular Joint by Arthrocentesis Alone, plus Hyaluronic Acid or plus Platelet-Rich Plasma. *Dent. Med. Probl.* **2019**, *56*, 265–272. [CrossRef]

134. Derwich, M.; Mitus-Kenig, M.; Pawlowska, E. Mechanisms of Action and Efficacy of Hyaluronic Acid, Corticosteroids and Platelet-Rich Plasma in the Treatment of Temporomandibular Joint Osteoarthritis—A Systematic Review. *Int. J. Mol. Sci.* **2021**, *22*, 7405. [CrossRef]

135. Macedo de Sousa, B.; López-Valverde, A.; Caramelo, F.; Rodrigues, M.J.; López-Valverde, N. Medium-Term Effect of Treatment with Intra-Articular Injection of Sodium Hyaluronate, Betamethasone and Platelet-Rich Plasma in Patients with Temporomandibular Arthralgia: A Retrospective Cohort Study. *Life* **2022**, *12*, 1739. [CrossRef] [PubMed]

136. Chandra, L.; Goyal, M.; Srivastava, D. Minimally Invasive Intraarticular Platelet Rich Plasma Injection for Refractory Temporomandibular Joint Dysfunction Syndrome in Comparison to Arthrocentesis. *J. Fam. Med. Prim. Care* **2021**, *10*, 254. [CrossRef] [PubMed]

137. Liu, Y.; Wu, J.; Tang, Y.; Tang, Y.; Fei, W.; Liang, X. Multiple Treatment Meta-Analysis of Intra-Articular Injection for Temporomandibular Osteoarthritis. *J. Oral Maxillofac. Surg.* **2020**, *78*, 373.e1–373.e18. [CrossRef]

138. Yang, W.; Liu, W.; Miao, C.; Sun, H.; Li, L.; Li, C. Oral Glucosamine Hydrochloride Combined With Hyaluronate Sodium Intra-Articular Injection for Temporomandibular Joint Osteoarthritis: A Double-Blind Randomized Controlled Trial. *J. Oral Maxillofac. Surg.* **2018**, *76*, 2066–2073. [CrossRef] [PubMed]

139. Daif, E.T.; Basha, H.Y. Ozone Therapy as a treatment modality for temporo-mandibular joint dysfunction. *J. Ozone Ther.* **2018**, *2*, 3. [CrossRef]

140. Isacsson, G.; Schumann, M.; Nohlert, E.; Mejersjö, C.; Tegelberg, Å. Pain Relief Following a Single-dose Intra-articular Injection of Methylprednisolone in the Temporomandibular Joint Arthralgia—A Multicentre Randomised Controlled Trial. *J. Oral Rehabil.* **2019**, *46*, 5–13. [CrossRef] [PubMed]

141. Sikora, M.; Sielski, M.; Chęciński, M.; Chęcińska, K.; Czerwińska-Niezabitowska, B.; Chlubek, D. Patient-Reported Quality of Life versus Physical Examination in Treating Temporomandibular Disorders with Intra-Articular Platelet-Rich Plasma Injections: An Open-Label Clinical Trial. *Int. J. Environ. Res. Public. Health* **2022**, *19*, 13299. [CrossRef]

142. Al-Hamed, F.S.; Hijazi, A.; Gao, Q.; Badran, Z.; Tamimi, F. Platelet Concentrate Treatments for Temporomandibular Disorders: A Systematic Review and Meta-Analysis. *JDR Clin. Transl. Res.* **2021**, *6*, 174–183. [CrossRef]

143. Bousnaki, M.; Bakopoulou, A.; Koidis, P. Platelet-Rich Plasma for the Therapeutic Management of Temporomandibular Joint Disorders: A Systematic Review. *Int. J. Oral Maxillofac. Surg.* **2018**, *47*, 188–198. [CrossRef] [PubMed]

144. Su, N.; Wang, H.; van Wijk, A.; Visscher, C.; Lobbezoo, F.; Shi, Z.; van der Heijden, G. Prediction Models for Oral Health–Related Quality of Life in Patients with Temporomandibular Joint Osteoarthritis 1 and 6 Months After Arthrocentesis with Hyaluronic Acid Injections. *J. Oral Facial Pain Headache* **2019**, *33*, 54–66. [CrossRef]

145. Ansar, A.S.; Munna, K.; Iqbal, A.; Mohammad, F.; Naved, A.; Shamimul, H. Prognostic Criteria for the Management of Temporomandibular Disorders Using Arthrocentesis with Normal Saline and Arthrocentesis with Normal Saline and Platelet-Rich Plasma. *J. Med. Life* **2022**, *15*, 698–704. [CrossRef]

146. Gavin Clavero, M.A.; Simón Sanz, M.V.; Mur Til, A.; Blasco Palacio, J. Prospective Study to Evaluate the Influence of Joint Washing and the Use of Hyaluronic Acid on 111 Arthrocentesis. *Oral Maxillofac. Surg.* **2019**, *23*, 415–421. [CrossRef]

147. Karadayi, U.; Gursoytrak, B. Randomised Controlled Trial of Arthrocentesis with or without PRF for Internal Derangement of the TMJ. *J. Cranio-Maxillofac. Surg.* **2021**, *49*, 362–367. [CrossRef]

148. Sikora, M.; Sielski, M.; Chęciński, M.; Nowak, Z.; Czerwińska-Niezabitowska, B.; Chlubek, D. Repeated Intra-Articular Administration of Platelet-Rich Plasma (PRP) in Temporomandibular Disorders: A Clinical Case Series. *J. Clin. Med.* **2022**, *11*, 4281. [CrossRef]

149. Cezairli, B.; Sivrikaya, E.C.; Omezli, M.M.; Ayranci, F.; Seyhan Cezairli, N. Results of Combined, Single-Session Arthrocentesis and Dextrose Prolotherapy for Symptomatic Temporomandibular Joint Syndrome: A Case Series. *J. Altern. Complement. Med.* **2017**, *23*, 771–777. [CrossRef] [PubMed]

150. Gupta, A.; Ali, I.; Zeeshan, M.; Singh, S.; Kumar, A.; Adil, A. Role of Intra-Articular Piroxicam in the Temporomandibular Joint After Arthrocentesis for Anterior Disc Displacement Without Reduction. *Cureus* **2023**, *15*, e34580. [CrossRef]

151. Sikora, M.; Czerwińska-Niezabitowska, B.; Chęciński, M.A.; Sielski, M.; Chlubek, D. Short-Term Effects of Intra-Articular Hyaluronic Acid Administration in Patients with Temporomandibular Joint Disorders. *J. Clin. Med.* **2020**, *9*, 1749. [CrossRef] [PubMed]

152. Hegab, A.F.; Hameed, H.I.A.A.; Hassaneen, A.M.; Hyder, A. Synergistic Effect of Platelet Rich Plasma with Hyaluronic Acid Injection Following Arthrocentesis to Reduce Pain and Improve Function in TMJ Osteoarthritis. *J. Stomatol. Oral Maxillofac. Surg.* **2023**, *124*, 101340. [CrossRef] [PubMed]

153. Wang, X.W.; Fang, W.; Li, Y.J.; Long, X.; Cai, H.X. Synovial Fluid Levels of VEGF and FGF-2 before and after Intra-Articular Injection of Hyaluronic Acid in Patients with Temporomandibular Disorders: A Short-Term Study. *Br. J. Oral Maxillofac. Surg.* **2021**, *59*, 64–69. [CrossRef] [PubMed]

154. Carboni, A.; Amodeo, G.; Perugini, M.; Arangio, P.; Orsini, R.; Scopelliti, D. Temporomandibular Disorders Clinical and Anatomical Outcomes After Fat-Derived Stem Cells Injection. *J. Craniofac. Surg.* **2019**, *30*, 793–797. [CrossRef]

155. Sembronio, S.; Tel, A.; Tremolada, C.; Lazzarotto, A.; Isola, M.; Robiony, M. Temporomandibular Joint Arthrocentesis and Microfragmented Adipose Tissue Injection for the Treatment of Internal Derangement and Osteoarthritis: A Randomized Clinical Trial. *J. Oral Maxillofac. Surg.* **2021**, *79*, 1447–1456. [CrossRef]

156. Henk, K.; Mark, K. Temporomandibular Steroids in Patients with Tinnitus: Only on Indication. *Int. Tinnitus J.* **2019**, *23*, 10–17. [CrossRef]

157. Chęciński, M.; Sikora, M.; Chęcińska, K.; Nowak, Z.; Chlubek, D. The Administration of Hyaluronic Acid into the Temporomandibular Joints' Cavities Increases the Mandible's Mobility: A Systematic Review and Meta-Analysis. *J. Clin. Med.* **2022**, *11*, 1901. [CrossRef] [PubMed]

158. Pihut, M.; Gala, A. The Application of Intra-Articulr Injections for Management of the Consequences of Disc Displacement without Reduction. *Int. J. Environ. Res. Public. Health* **2020**, *17*, 4726. [CrossRef] [PubMed]

159. Sari, B.C.; Develi, T. The Effect of Intraarticular Botulinum Toxin-A Injection on Symptoms of Temporomandibular Joint Disorder. *J. Stomatol. Oral Maxillofac. Surg.* **2022**, *123*, e316–e320. [CrossRef] [PubMed]

160. Nagori, S.A.; Jose, A.; Gopalakrishnan, V.; Roy, I.D.; Chattopadhyay, P.K.; Roychoudhury, A. The Efficacy of Dextrose Prolotherapy over Placebo for Temporomandibular Joint Hypermobility: A Systematic Review and Meta-Analysis. *J. Oral Rehabil.* **2018**, *45*, 998–1006. [CrossRef]

161. Torul, D.; Cezairli, B.; Kahveci, K. The Efficacy of Intra-Articular Injectable Platelet-Rich Fibrin Application in the Management of Wilkes Stage III Temporomandibular Joint Internal Derangement. *Int. J. Oral Maxillofac. Surg.* **2021**, *50*, 1485–1490. [CrossRef]

162. Al-Delayme, R.M.A.; Alnuamy, S.H.; Hamid, F.T.; Azzamily, T.J.; Ismaeel, S.A.; Sammir, R.; Hadeel, M.; Nabeel, J.; Shwan, R.; Alfalahi, S.J.; et al. The Efficacy of Platelets Rich Plasma Injection in the Superior Joint Space of the Tempromandibular Joint Guided by Ultra Sound in Patients with Non-Reducing Disk Displacement. *J. Maxillofac. Oral Surg.* **2017**, *16*, 43–47. [CrossRef]

163. Ozdamar, S.M.; Alev, B.; Yarat, A. The Impact of Arthrocentesis with and without Hyaluronic Acid Injection in the Prognosis and Synovial Fluid Myeloperoxidase Levels of Patients with Painful Symptomatic Internal Derangement of Temporomandibular Joint: A Randomised Controlled Clinical Trial. *J. Oral Rehabil.* **2017**, *44*, 73–80. [CrossRef]

164. Romero-Tapia, P.; Sedano-Balbin, G.; Mayta-Tovalino, F.; Pérez-Vargas, F.; Marín, J. Therapeutic Effect of Sodium Hyaluronate and Corticosteroid Injections on Pain and Temporomandibular Joint Dysfunction: A Quasi-Experimental Study. *J. Contemp. Dent. Pract.* **2021**, *21*, 1084–1090. [CrossRef]

165. Louw, W.F.; Reeves, K.D.; Lam, S.K.H.; Cheng, A.-L.; Rabago, D. Treatment of Temporomandibular Dysfunction With Hypertonic Dextrose Injection (Prolotherapy): A Randomized Controlled Trial With Long-Term Partial Crossover. *Mayo Clin. Proc.* **2019**, *94*, 820–832. [CrossRef]

166. Singh, J.; Bhardwaj, B. Treatment of Temporomandibular Joint Arthritis with Triamcinolone Acetonide and Hyaluronic Acid Injection: An Observational Study. *Indian J. Otolaryngol. Head Neck Surg.* **2020**, *72*, 403–410. [CrossRef] [PubMed]

167. Ghoneim, N.I.; Mansour, N.A.; Elmaghraby, S.A.; Abdelsameaa, S.E. Treatment of Temporomandibular Joint Disc Displacement Using Arthrocentesis Combined with Injectable Platelet Rich Fibrin versus Arthrocentesis Alone. *J. Dent. Sci.* **2022**, *17*, 468–475. [CrossRef] [PubMed]

168. Abrahamsson, H.; Eriksson, L.; Abrahamsson, P.; Häggman-Henrikson, B. Treatment of Temporomandibular Joint Luxation: A Systematic Literature Review. *Clin. Oral Investig.* **2020**, *24*, 61–70. [CrossRef] [PubMed]

169. Zubair, H.; Hashmi, G.S.; Rahman, S.A.; Ahmed, S.S.; Ahmad, M. Ultrasonography-Guided Injections of Platelet-Rich Plasma in the Management of Internal Derangement of Temporomandibular Joint. *Asian Pac. J. Health Sci.* **2021**, *8*, 5–8. [CrossRef]

170. Haigler, M.C.; Abdulrehman, E.; Siddappa, S.; Kishore, R.; Padilla, M.; Enciso, R. Use of Platelet-Rich Plasma, Platelet-Rich Growth Factor with Arthrocentesis or Arthroscopy to Treat Temporomandibular Joint Osteoarthritis. *J. Am. Dent. Assoc.* **2018**, *149*, 940–952.e2. [CrossRef]

171. Dharamsi, R.; Nilesh, K.; Mouneshkumar, C.D.; Patil, P. Use of Sodium Hyaluronate and Triamcinolone Acetonide Following Arthrocentesis in Treatment of Internal Derangement of Temporomandibular Joint: A Prospective Randomized Comparative Study. *J. Maxillofac. Oral Surg.* **2022**. [CrossRef]

Systematic Review

Intra-Articular Injections into the Inferior versus Superior Compartment of the Temporomandibular Joint: A Systematic Review and Meta-Analysis

Maciej Chęciński [1], Kamila Chęcińska [2], Natalia Turosz [3], Maciej Sikora [4,5] and Dariusz Chlubek [5,*]

1   Department of Oral Surgery, Preventive Medicine Center, Komorowskiego 12, 30-106 Cracow, Poland
2   Department of Glass Technology and Amorphous Coatings, Faculty of Materials Science and Ceramics, AGH University of Science and Technology, Mickiewicza 30, 30-059 Cracow, Poland
3   Institute of Public Health, Jagiellonian University Medical College, Skawińska 8, 31-066 Cracow, Poland
4   Department of Maxillofacial Surgery, Hospital of the Ministry of Interior, Wojska Polskiego 51, 25-375 Kielce, Poland
5   Department of Biochemistry and Medical Chemistry, Pomeranian Medical University, Powstańców Wielkopolskich 72, 70-111 Szczecin, Poland
*   Correspondence: dchlubek@pum.edu.pl

Abstract: This systematic review and meta-analysis aimed to validate the hypothesis that intra-articular injections into the inferior temporomandibular joint compartment are more efficient than analogous superior compartment interventions. Publications reporting differences between the above-mentioned techniques in the domains of revealing articular pain, decreasing the Helkimo index, and abolishing mandibular mobility limitation were included. Medical databases covered by the Bielefeld Academic Search Engine, Google Scholar, PubMed, ResearchGate, and Scopus engines were searched. The risk of bias was assessed using dedicated Cochrane tools (RoB2, ROBINS-I). The results were visualized with tables, charts, and a funnel plot. Six reports describing five studies with a total of 342 patients were identified. Of these, four trials on a total of 337 patients were qualified for quantitative synthesis. Each eligible report was at moderate risk of bias. From 19% to 51% improvement in articular pain, 12–20% lower Helkimo index, and 5–17% higher maximum mouth opening were observed. The evidence was limited by the small number of eligible studies, discrepancies regarding the substances used, possible biases, and the differences in observation times and scheduled follow-up visits. Despite the above, the advantage of inferior over superior compartment temporomandibular joint intra-articular injections is unequivocal and encourages further research in this direction.

Keywords: temporomandibular joint; inferior TMJ compartment; discomandibular space; temporomandibular disorders; intra-articular injections; hyaluronic acid; arthrocentesis; hypertonic dextrose

Citation: Chęciński, M.; Chęcińska, K.; Turosz, N.; Sikora, M.; Chlubek, D. Intra-Articular Injections into the Inferior versus Superior Compartment of the Temporomandibular Joint: A Systematic Review and Meta-Analysis. *J. Clin. Med.* **2023**, *12*, 1664. https://doi.org/10.3390/jcm12041664

Academic Editors: Luís Eduardo Almeida and Eiji Tanaka

Received: 2 January 2023
Revised: 30 January 2023
Accepted: 16 February 2023
Published: 19 February 2023

1. Introduction

1.1. Background

The temporomandibular joints (TMJs) are located symmetrically on both sides of the head. Properly functioning, they allow the teeth to move relative to each other, thus biting and chewing [1,2]. The articular disc divides each TMJ cavity into superior (discotemporal) and inferior (discomandibular) compartments, which dictates a complicated pattern of possible movements of the articular head relative to the acetabulum. (Figure 1) [3,4]. Articular disc displacement with or without reduction is referred to as TMJ internal derangement [5]. This dysfunction causes disc clicking, hence inflammation manifested by articular pain and reduced mandibular mobility, resulting in deterioration in patient-reported quality of life [6,7]. The severity of the above, along with imaging tests, allows for classifying the internal derangement stage [8–12]. Depending on the severity of the temporomandibular disorders, treatment regimens consist of pharmacotherapy, physiotherapy, occlusal rearrangement (including splint therapy), intra-muscular injections, intra-articular injections, and arthroplasty [13–20].

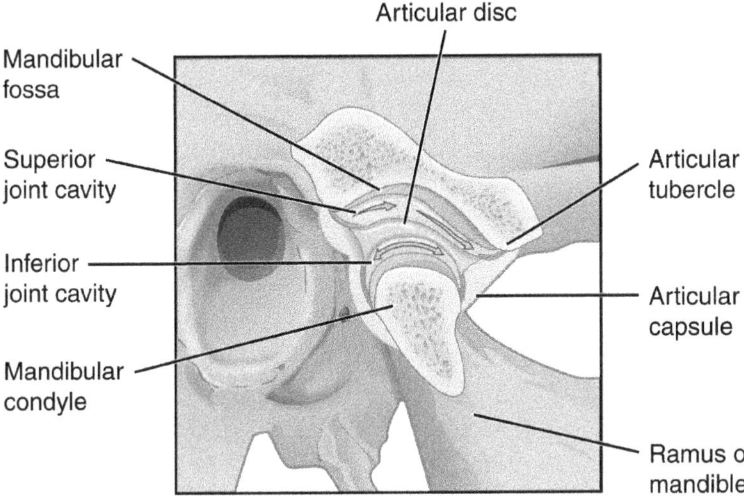

Figure 1. Temporomandibular joint. Modified. OpenStax College, CC BY 3.0 (creativecommons.org/licenses/by/3.0).

Efficient in immediate ailment relief, intra-articular injections also provide satisfactory effects in several months of observation [21,22]. They allow for TMJ cavity rinsing (arthrocentesis) and the administration of autogenous preparations (blood derivatives and cell transplants) or drugs [6,23–27]. Arthrocentesis can be used as a stand-alone technique or precede the injection of a selected substance [23,28,29]. The protocols of drug administration with or without prior lavage differ from each other in the injectable used [21,30–34]. In the group of pharmaceuticals, hyaluronic acid (HA), corticosteroids, hypertonic dextrose, and anesthetics are applicable [30–35]. The significance of the type of injectable used is still debated, prompting the assessment of the importance of other factors, such as the specific injection site [27,28,36].

The specific intra-articular location of depositing the injected substance is presumed to be determinative due to the different motor functions of individual TMJ compartments [37,38]. The full assessment of the complexity of these movements is still pending, but it is assumed that hinge or rotational movements take place below the articular disc, and translation or gliding movements above it [37–39]. Mastication capacity, resulting from a combination of the movements described above, translates into an overall assessment of the quality of life [7,40]. Based on subdiscal arthroscopy, the role of the inferior compartment for the proper functioning of the TMJ and the hitherto underestimated role of abnormalities below the articular disc in the etiopathogenesis of internal disarrangement are regarded to be greater than previously thought [39].

This all leads to careful selection of the rinsed TMJ compartment and the exact place of injectable deposition [41,42]. The superior compartment, bounded superiorly by the articular fossa on the temporal bone, is wider (approximately 1.2 mL) and, therefore, easier to access for a needle or an endoscope [12,37–39,43]. Puncturing into the inferior compartment partially surrounding the head of the mandible is technically more difficult due to the shape of this part of the joint cavity and its smaller volume (approximately 0.9 mL) [37–39,42]. Precise injection poses a challenge for clinicians due to the blind nature of this intra-articular intervention type [44–47]. This problem is being addressed with ever-improved puncture protocols and various imaging techniques (radiography, ultrasonography, and magnetic resonance imaging) [41,44–50].

1.2. Rationale

The more significant role of the inferior TMJ compartment for the initial phase of mouth opening encourages the consideration of inferior compartment intra-articular injections as potentially more efficient in relieving articular pain, reducing the overall severity of internal disarrangement expressed by the Helkimo index, and improving mandibular mobility [37–40,42]. The comparison of the effects of inferior or both compartments versus superior intra-articular injections was the subject of a systematic review with meta-analysis published in 2012 by Li et al. [42]. A greater improvement in both articular pain and maximum mouth opening domains with the interventions studied compared to standard upper compartment administrations was observed [42,49–52]. Nevertheless, both the meta-analysis result itself and later published comments argued for the need to support the initial conclusions with further clinical trials [42,53,54]. Heterogeneity in terms of technique (lower or both spaces) and more than a decade since publication further justify the need to determine the current state of knowledge about the validity of intra-articular injections for specific TMJ compartments [42].

1.3. Objective

This meta-analysis aims to validate the hypothesis that intra-articular injections into the inferior temporomandibular joint compartment are more efficient in relieving articular pain and abolishing mouth-opening limitation than analogous superior compartment interventions.

2. Materials and Methods

2.1. Eligibility Criteria

Eligibility was determined according to the PICO methodology, specifying inclusion and exclusion criteria for the problem, intervention, comparison, and outcomes (Table 1) [55–57]. Any types of publications containing data from original clinical trials were included without any time frame limit. The problem studied was the diagnosis of temporomandibular joint internal derangement in the Wilkes classification stages II to V [11]. Studies in which patients underwent TMJ inferior compartment arthrocentesis and/or intra-articular injections of self-derived preparations (e.g., PRP, PRGF, I-PRF, MSC) or drugs (e.g., hyaluronan, corticosteroids, hypertonic dextrose, anesthetics) and their combinations were qualified. Treatment involving arthroscopy or open joint surgery within the same procedure was rejected. Studies that did not include the treatment group for the sole inferior TMJ compartment only were also excluded. This decision was motivated by the inability to estimate the effectiveness of the individual components of a combined injection into both TMJ compartments [42,51,52]. As a reference, the same type of intervention in terms of substance, dosage regimen, and duration of treatment within the superior joint compartment was required. Comparisons conducted by injections into both joint spaces were excluded for the reasons mentioned above [42,51,52]. The quantitative evaluation of the effectiveness of therapy in the domains of articular pain severity, Helkimo index, and the range of mandibular mobility was taken into account [58,59]. Subjective pain assessment on a visual analog scale (VAS) or numeric rating scale (NRS) was accepted [19,59]. The range of mandibular mobility expressed as abduction, protrusive movement, and lateral movements were allowed, respecting this hierarchy. Papers reporting at least one of the outcomes mentioned above were accepted.

Table 1. Summarized eligibility criteria.

	Inclusion	Exclusion
Problem	TMJ internal derangement	Animal studies
Intervention	Arthrocentesis, intra-articular injection, or a combination thereof within the inferior TMJ compartment	Atroscopy or open joint surgery as part of the same procedure
Comparison	Same intervention for the superior TMJ compartment	Intervention in both TMJ compartments in one patient group
Outcomes	Articular pain, Helkimo index, mandibular mobility	-

TMJ—temporomandibular joint.

2.2. Information Sources

The Bielefeld Academic Search Engine (BASE), ClinicalTrials.gov, Google Scholar, PubMed, ResearchGate, and Scopus search engines were used to identify potentially eligible reports throughout medical databases [60]. All final searches were performed on 1 November 2022. Additionally, the references of each eligible publication were searched for further records.

2.3. Search Strategy

The following search strategy was used:

(inferior OR lower) AND (superior OR upper) AND (compartment OR space) AND temporomandibular AND (joint OR articulation) AND (arthrocentesis OR rinsing OR lavage OR injection OR administration OR viscosupplementation).

Individual search queries for each search engine are shown in Table A1, Appendix A.

2.4. Selection Process

Reports were selected in accordance with the Preferred Reporting Items for Systematic Reviews and Meta-Analyses (PRISMA) protocol using the Rayyan tool [61–63]. The convergence of blind assessments of two judges (M.C. and K.C.) was expressed by the value of Cohen's kappa coefficient. Reports considered potentially eligible by any of the judges during screening were promoted to the full-text assessment stage. A reference search of the included studies was conducted for further potentially eligible items.

2.5. Data Collection Process

Data were extracted from the content of reports without the use of automation tools. In the case of discrepancies between the values collected by two independent reviewers (M.C. and N.T.), joint verification was performed, and the decision was made through discussion.

2.6. Data Items

The following data items were collected to identify individual studies and to characterize the test and control groups: (1) first author of the report; (2) publication year; (3) study type; (4) diagnosis; (5) study and control group sizes; (6) sex and age structures; (7) injectable and dosage; (8) eligible outcome domains; (9) follow-up time. The study group was considered to be patients receiving injections into the inferior TMJ compartment, and the study group consisted of individuals injected within the superior TMJ compartment.

For the purposes of the synthesis, the values of the following variables were extracted from the study reports: (1) the intensity of articular pain; (2) Helkimo index (HI); (3) maximum mouth opening (MMO) [19,58,59]. The values of these variables before treatment initiation (initial), at intermediate visits, and after treatment completion (final) were collected for the study and control groups.

2.7. Study Risk of Bias Assessment

The risk of bias within the studies was assessed using the revised Cochrane risk-of-bias tool for randomized trials (RoB2) and the tool for assessing the risk of bias in non-randomized studies of interventions (ROBINS-I) for randomized and non-randomized trials, respectively [64,65]. In the case of reports covering a larger number of patient groups or other interventions, only the data useful in this meta-analysis were considered in assessing the risk of bias.

2.8. Effect Measures

To assess the efficiency of inferior versus superior compartment TMJ treatment in pain, Helkimo Index, and MMO domains, the percentage decrease in the corresponding variable values was compared each time in the study and control groups according to the formula:

$$e = e_i - e_s = v_{if} \div v_{i0} \times 100\% - v_{sf} \div v_{s0} \times 100\%,$$

where e_i and es are the treatment efficiencies in the inferior and superior TMJ compartment groups, respectively, and v stands for final (v_{if}, v_{sf}) and initial (v_{i0}, v_{s0}) variable values, with an analogous compartment designation.

2.9. Synthesis Methods

Studies with no greater than a moderate risk of bias were allowed for syntheses. In the absence of information on the values of the necessary variables, a given study was not taken into account in a given synthesis. Pain severity in VAS or NRS was converted proportionally to values in the range 0–10, with only the VAS values used when using both scales in one study [19,59]. With the different methods of measuring the extent of the mandibular abduction within one study (e.g., maximum mouth opening, maximum unassisted opening, maximum pain-free opening, etc.), only the one with the highest values was used for synthesis. For data visualization purposes, it was assumed that a month consists of 4 weeks. The results of individual studies were presented in tabular form, and the results of syntheses were presented graphically in charts. The exploration of possible reasons for the heterogeneity of the studies was carried out using the meta-regression method. Each of the syntheses was carried out under the condition of completeness of data in a given domain, which excluded the need to assess the risk of reporting bias. For all assessments, a significance level of 0.05 was adopted. For the meta-assessment of the publication bias presence, the results were visualized in a funnel plot.

3. Results

3.1. Study Selection

Searching with six engines led to the identification of 97 records, of which 54 duplicates indicated by the Rayyan tool were manually removed [61,62]. Thus, 43 entries qualified for the screening stage and assessment of the titles' and abstracts' content, resulting in the exclusion of a further 36 records inconsistent with the problem or intervention sought. The compliance of abstract qualifications according to two judges, expressed by Cohen's kappa coefficient, was 0.91. Seven reports qualified for full-text evaluation, three of which described the same trial by Long [50,66]. Among them, a record with errors in the title and author list fields pointing to the same full text as the correct one was excluded [50]. The same was done with a conference abstract from 2008 with consistent characteristics of patient groups as the scientific article from 2009 [50,66]. However, the formal reason for rejecting the latter was the lack of numerical values of the variables in any of the eligible outcome domains. One of the articles qualifying for the synthesis was a systematic review containing the required data from two studies meeting the inclusion criteria [42,49,50]. The report from the first one was found in the course of this selection process, regardless of its inclusion in the review paper [50]. The second study was originally published in the Chinese-language journal "Guoji kouqiang yixue zazhi", from which the source report was retrieved and included in the synthesis [49]. Ultimately, the meta-analysis was based on six reports describing five eligible studies comparing the effectiveness of injection therapies to the lower versus the upper TMJ compartment [41,42,48–50,67]. These studies were conducted on a total of 342 patients [41,48–50,67]. The reference search yielded no further results. The main steps of the selection process were visualized as a PRISMA-compliant diagram (Figure 2) [68].

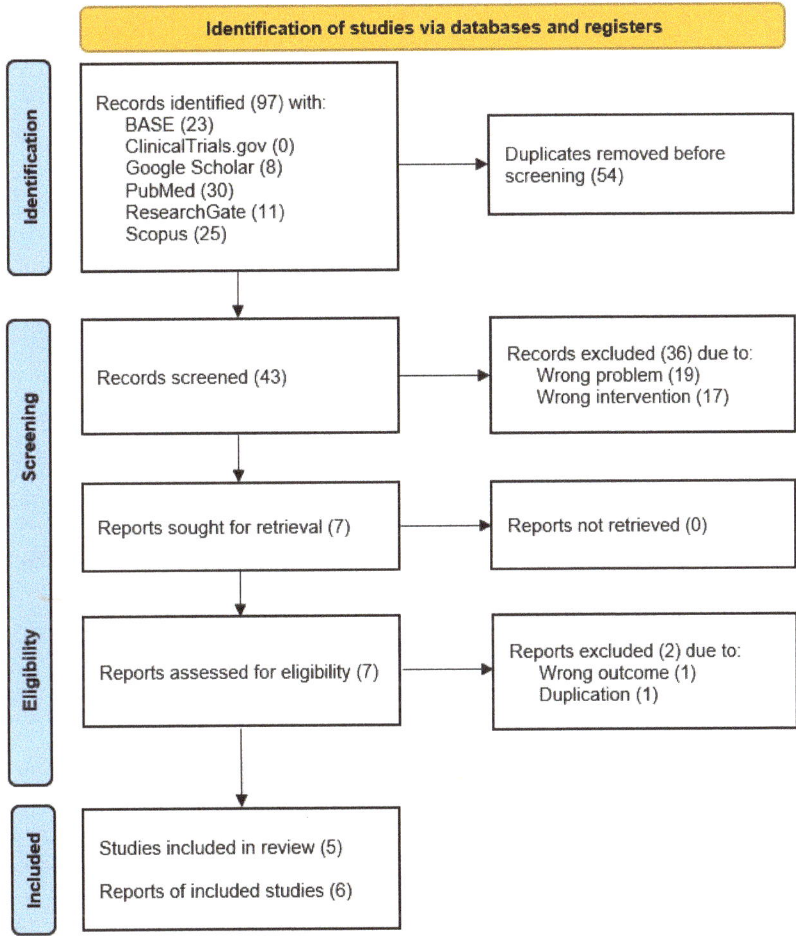

Figure 2. PRISMA flow diagram.

3.2. Study Characteristics

All qualifying studies were conducted on the basis of groups diagnosed with TMJs internal derangement. Only in the study by Fouda et al. did the stage of the disease in patients allow for the reduction of the displaced disc; in others, the blockage was irreducible. The size of the study and control groups ranged from 1 to 73 subjects, which was taken into account in the meta-analysis. The numerous missing data did not allow for precise age determination and, in particular, the age differentiation between the study and control groups. In general, the interventions consisted of intra-articular injection of 1 to 2 mL of fluid, although the substances used differed depending on the study design. The number of administrations ranged from 1 to 4, and in the case of several interventions, the intervals between them were 1–2 weeks. The amplitude of mandibular abduction was determined in four out of five studies, and both articular pain and Helkimo index values were presented in two out of five reports. The maximum observation time for at least some members of the study group ranged from 2–3 days to 9 months. The characteristics of the individual studies are presented in Table 2.

Table 2. Study characteristics.

First Author	Publication Year	Study Type	Diagnosis	Group Sizes	Age, Mean	Intervention	Number of Instances, Interval	Eligible Outcomes	Follow-Up Visits
Fouda [48]	2018	RT	DDwR	18—inferior, 18—superior	In range 18–42, N/S	1.5 mL of 25% hypertonic dextrose + 0.2 mL anesthetic	4, 1 week	Pain, MMO	2 weeks and 3 months
Li [41]	2014	RT	DDwoR	68—inferior, 73—superior	N/S, in range 31.4–34.1	1 mL HA	4, 2 weeks	MMO, HI	3 and 9 months
Liu [49]	2010	RT	DDwoR	28—inferior, 28—superior	14–48, 25.7	Arthrocentesis + 1–2 mL HA + 2 months of splint therapy	1, N/A	MMO	2 months
Long [50]	2009	RT	DDwoR	54—inferior, 50—superior	N/S, in range 25.6–30.6	1 mL anesthetic + 1 mL HA	3, 2 weeks	Pain, HI, MMO	3 and 6 months
Ozawa [67]	1996	nRT	DDwoR	4—inferior, 1—superior	17–37, 22.2	Pumping arthrocentesis	1, N/A	MMO	2–3 days

RT—randomized trial; nRT—non-randomized trial; DDwR—disk displacement with reduction; DDwoR—disk displacement without reduction; N/S—not specified; HA—Hyaluronic Acid; N/A—not applicable; MMO—maximum mouth opening; HI—Helkimo Index.

3.3. Risk of Bias in Studies

The overall risk of bias for reporting changes in pain, Helkimo's index, and MMO domains for the patient groups receiving lower and upper TMJ compartment injections was moderate in all included studies but one (Table 3). The report by Ozawa et al. was included in the review due to an excerpt of the results that were considered eligible according to the PICOS criteria. Concerns about the selectively included data from this paper led to an assessment of the high risk of bias, thus excluding the study from syntheses.

Table 3. Risk of bias assessment.

First Author	Randomization Process	Confounding	Selection of Participants in the Study	Classification of Interventions	Deviations from Intended Interventions	Missing Data	Measurement of Outcomes	Selection of the Reported Result	Overall Bias
Fouda [48]	Low	N/A	N/A	N/A	Low	Low	Unclear	Low	Moderate
Li [41]	Unclear	N/A	N/A	N/A	Low	Moderate	Unclear	Low	Moderate
Liu [49]	Unclear	N/A	N/A	N/A	Low	Low	Moderate	Low	Moderate
Long [50]	Low	N/A	N/A	N/A	Low	Low	Unclear	Low	Moderate
Ozawa [67]	N/A	High	High	Low	Low	Low	Unclear	Low	High

Columns 2–9 list the domains for which the bias was assessed. N/A—not applicable.

3.4. Results of Individual Studies

3.4.1. Articular Pain

In both studies that included articular pain in the VAS as a separate variable, improvement was observed for both superior and inferior TMJ injections. The best fits were obtained in the study by Fouda et al. for logarithmic curves, and in the report by Long et al., trends followed second-degree polynomials. The differences between the results in the study group and the control group were statistically significant in the reports of both teams of authors (Table 4, Figure 3).

Table 4. Articular pain.

First Author	TMJ Compartment	Initial Pain	Pain after 2 Weeks	Pain after 3 Months	Pain after 6 Months
Fouda [48]	Superior	3.7 ± 2.7 (100%; 18 pts)	3.4 ± 3.0 (91.9%; 18 pts)	2.9 ± 3.1 (78.4%; 18 pts)	N/S
	Inferior	6.6 ± 2.5 (100%; 18 pts)	2.8 ± 2.8 (42.4%; 18 pts)	1.8 ±2.1 (27.3%; 18 pts)	N/S
	Difference	0% ($p < 0.05$)	−49.5% ($p < 0.05$)	−51.1% ($p < 0.05$)	N/S
Long [50]	Superior	6.2 ± 0.2 (100%; 50 pts)	N/S	4.1 ± 1.9 (66.1%; 50 pts)	3.3 ± 2.3 (53.2%; 50 pts)
	Inferior	6.0 ± 0.2 (100%; 54 pts)	N/S	2.8 ± 1.7 (46.7%; 54 pts)	1.1 ± 1.3 (18.3%; 54 pts)
	Difference	0% ($p > 0.05$)	N/S	−19.4% ($p < 0.05$)	−34.9% ($p < 0.05$)

TMJ—temporomandibular joint; pain—articular pain in 0–10 VAS; pts—patients; N/S—not specified.

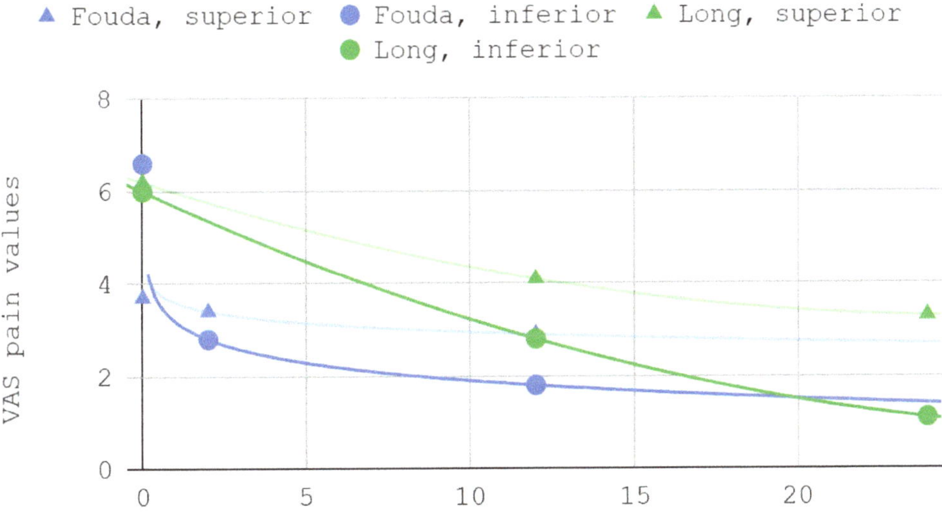

Figure 3. Articular pain in VAS over time.

3.4.2. Helkimo Index

The severity of TMJ dysfunction expressed by the Helkimo index subsided as a result of treatment for both superior and inferior compartment injections. A good fit was obtained in each series using a second-degree polynomial trend line. Efficiency differences in favor of the group injected in the lower part of the joint were statistically significant during both follow-up visits in the course of observation of the team of Li et al. In a study by Long et al. numerically greater differences were observed, but statistically significant only on the second of the two follow-up visits (Table 5, Figure 4).

Table 5. Helkimo index.

First Author	TMJ Compartment	Initial Helkimo Index	Helkimo Index after 3 Months	Helkimo Index after 6 Months	Helkimo Index after 9 Months
Li [41]	Superior	7.4 ± 3.1 (100%; 73 pts)	2.2 ± 2.0 (29.6%; 65 pts)	N/S	1.7 ± 2.2 (24.5%; 44 pts)
	Inferior	7.7 ± 3.3 (100%; 68 pts)	1.3 ± 1.4 (17.8%; 61 pts)	N/S	1.3 ± 1.9 (15.8%; 30 pts)
	Difference	0% ($p > 0.05$)	−11.8% ($p < 0.05$)	N/S	−8.7% ($p < 0.05$)
Long [50]	Superior	6.1 ± 3.3 (100%; 50 pts)	3.8 ± 2.8 (62.3%; 50 pts)	3.0 ± 2.4 (49.2%; 50 pts)	N/S
	Inferior	6.6 ± 2.3 (100%; 54 pts)	2.8 ± 2.0 (42.4%; 54 pts)	1.2 ± 1.5 (18.2%; 54 pts)	N/S
	Difference	0% ($p > 0.05$)	−19.9% ($p > 0.05$)	−31.0% ($p < 0.05$)	N/S

TMJ—temporomandibular joint; pts—patients; N/S—not specified.

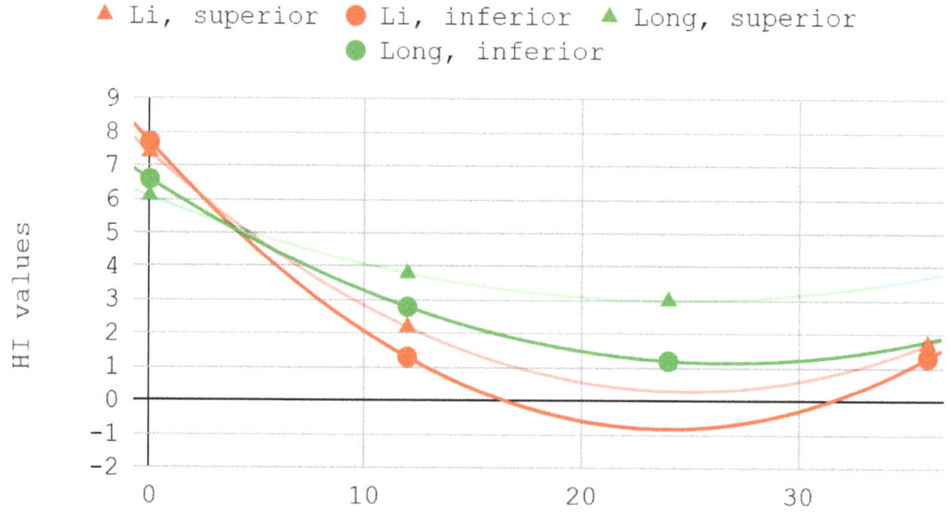

Figure 4. Helkimo index (HI) over time.

3.4.3. Maximum Mouth Opening

The mobility of the mandible, expressed by the values of its abduction, was measured in each of the discussed reports. Injection treatment was effective in this domain each time, regardless of the injected TMJ compartment. The results of Ozawa et al. for acute cases are numerically presented but not illustrated due to the small size of the patient groups. MMO values from reports by Li et al. and Long et al. showed a second-degree polynomial trend. The other series differed from the pattern of standard (linear, polynomial, logarithmic) fits. For a general view, linear trend lines have been used in these cases. Of the seven measurements during the follow-up periods described in various reports, three differences between the inferior versus superior groups were not statistically significant, including the only difference in favor of the superior group (Table 6, Figures 5 and 6).

Table 6. Maximum mouth opening.

First Author	TMJ Compartment	Initial MMO	MMO after 2–3 Days	MMO after 2 Weeks	MMO after 2 Months	MMO after 3 Months	MMO after 6 Months	MMO after 9 Months
Fouda [48]	Superior	35.6 ± 5.5 (100%; 18 pts)	N/S	37.1 ± 4.4 (104.2%; 18 pts)	N/S	36.0 ± 4.2 (101.1%; 18 pts)	N/S	N/S
	Inferior	34.6 ± 2.4 (100%; 18 pts)	N/S	36.6 ± 1.4 (105.8%; 18 pts)	N/S	36.8 ± 1.2 (106.4%; 18 pts)	N/S	N/S
	Difference	0% ($p > 0.05$)	N/S	1.6% ($p < 0.05$)	N/S	5.3% ($p < 0.05$)	N/S	N/S
Li [41]	Superior	31.1 ± 7.9 (100%; 73 pts)	N/S	N/S	N/S	37.6 ± 6.5 (120.9%; 65 pts)	N/S	41.5 ± 6.4 (133.4%; 44 pts)
	Inferior	30.0 ± 6.8 (100%; 68 pts)	N/S	N/S	N/S	37.9 ± 5.9 (126.3%; 61 pts)	N/S	39.6 ± 5.8 (132.0%; 30 pts)
	Difference	0% ($p > 0.05$)	N/S	N/S	N/S	5.4% ($p > 0.05$)	N/S	−1.4% ($p > 0.05$)
Liu [49]	Superior	32.4 ± 2.3 (100%, 28 pts)	N/S	N/S	43.6 ± 5.1 (134.6%; 28 pts)	N/S	N/S	N/S
	Inferior	36.8 ± 1.4 (100%; 28 pts)	N/S	N/S	55.9 ± 2.9 (151.9%; 28 pts)	N/S	N/S	N/S
	Difference	0% ($p > 0.05$)	N/S	N/S	17.3% ($p < 0.05$)	N/S	N/S	N/S
Long [50]	Superior	30.8 ± 4.9 (100%; 50 pts)	N/S	N/S	N/S	35.3 ± 4.7 (114.6%; 50 pts)	36.4 ± 5.0 (118.2%; 50 pts)	N/S
	Inferior	29.0 ± 4.7 (100%; 54 pts)	N/S	N/S	N/S	36.9 ± 4.6 (127.2%; 54 pts)	39.4 ± 4.4 (135.9%; 54 pts)	N/S
	Difference	0% ($p > 0.05$)	N/S	N/S	N/S	12.6% ($p > 0.05$)	17.7% ($p < 0.05$)	N/S
Ozawa [67]	Superior	34.0 ± 0.0 (100%, 1 pts)	40.0 (117.6%; 1 pts)	N/S	N/S	N/S	N/S	N/S
	Inferior	22.5 ± 4.3 (100%; 4 pts)	39.3 ± 2.9 (174.7%; 4 pts)	N/S	N/S	N/S	N/S	N/S
	Difference	0% (N/A)	57.1% (N/A)	N/S	N/S	N/S	N/S	N/S

TMJ—temporomandibular joint; MMO—maximum mouth opening in millimeters; pts—patients; N/S—not specified; N/A—not applicable.

3.5. Results of Syntheses

The effectiveness of lower compartment treatment compared to control varied significantly between studies. Articular pain difference in the report of Long et al. increased gradually, in contrast to the high constant difference observed after treatment by Fouda et al. The pattern of differences in the effectiveness of both techniques suggests a possible loss of superiority of injection into the lower compartment over time in relation to the Helkimo index. However, during the reported follow-up period, this superiority was still present. The results achieved in the study groups relative to the control regarding MMO suggest an increasing advantage of treatment oriented to the lower TMJ compartment. Nevertheless, it can be suspected that this advantage peaks between 3 and 9 months of observation and decreases further. For all but one study, second-degree polynomial fits of inferior versus superior compartment efficiencies were found to be the most appropriate. Exceptionally, for Fouda et al.'s results, the natural-based logarithmic trendlines were presented as the most accurate fits. The difference in efficiencies between superior and inferior compartment TMJ injections after 2–3 months was assessed in each of the synthesized studies. Therefore, this data series was used to illustrate the convergence of results presented in individual reports. The outline of the funnel on the horizontal axis is the mean difference in performance minus and increased by the standard deviation. The height of the funnel on the vertical axis is determined by the total number of 337 patients treated with injections into the upper and lower TMJ compartments. The outlier outside the funnel contour coexists with outlying trend curves in the MMO domain for the study by Liu et al. Despite the datapoint outlier in the funnel plot, the result of Li et al. is consistent with that of Fouda et al. (Figures 7–10).

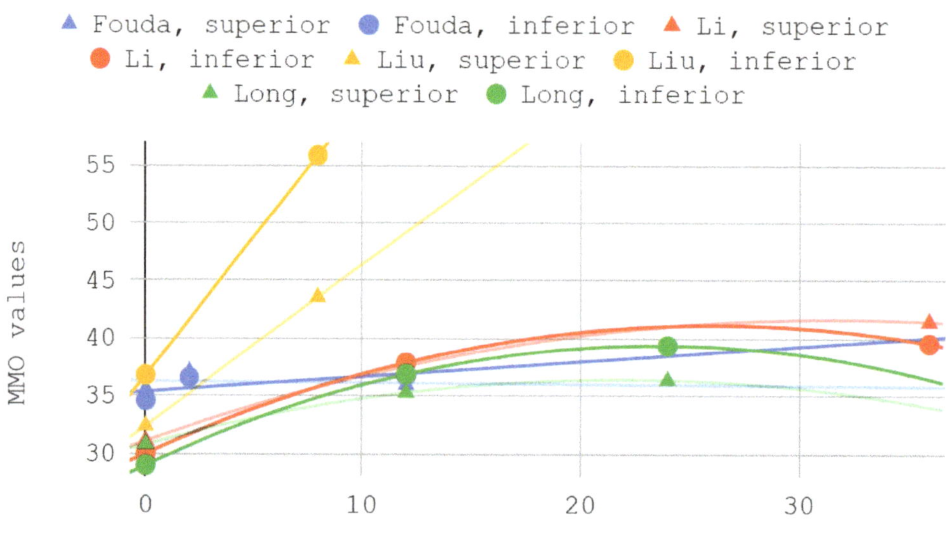

Figure 5. Maximum mouth opening (MMO) over time—whole chart.

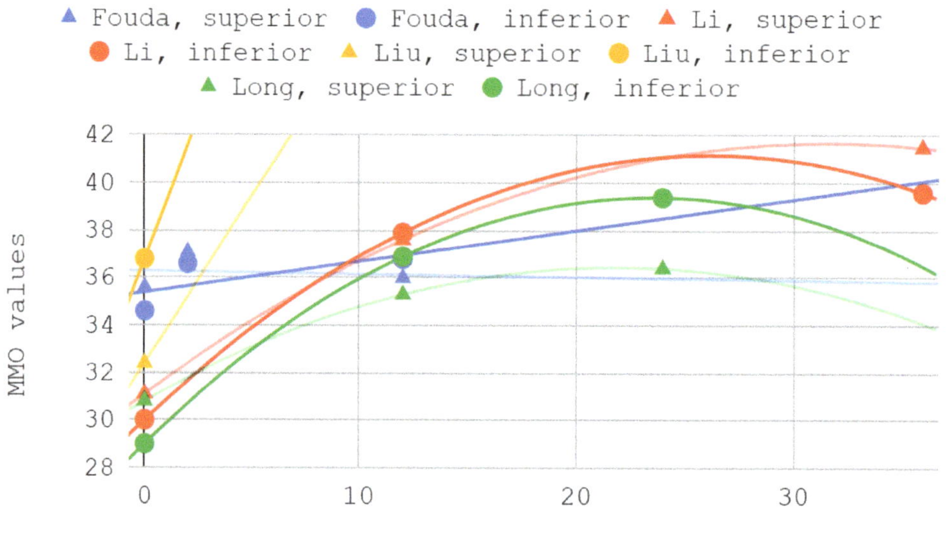

Figure 6. Maximum mouth opening (MMO) over time—chart fragment close-up.

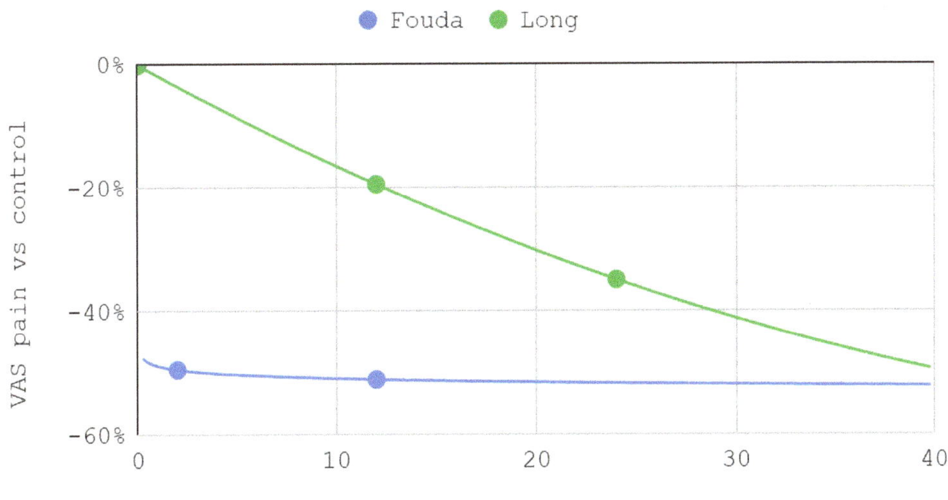

Figure 7. The efficiency of inferior versus superior compartment TMJ treatment in VAS articular pain domain over time.

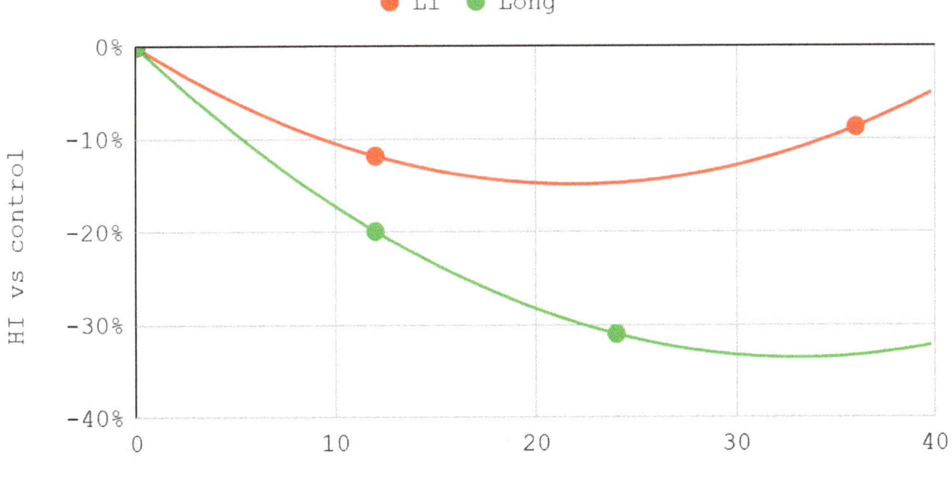

Figure 8. The efficiency of inferior versus superior compartment TMJ treatment in Helkimo index (HI) domain over time.

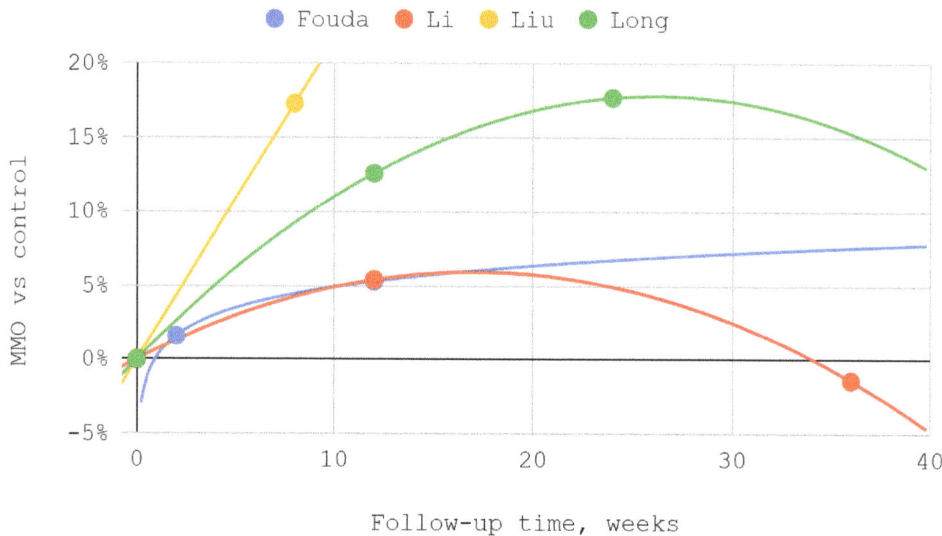

Figure 9. The efficiency of inferior versus superior compartment TMJ treatment in maximum mouth opening (MMO) domain over time.

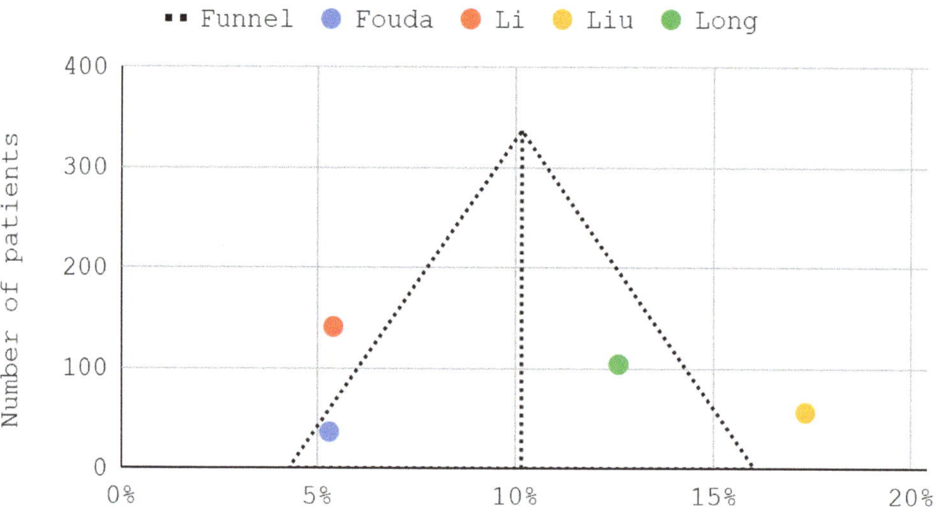

Figure 10. A funnel plot of the efficiency of inferior versus superior compartment TMJ treatment in maximum mouth opening (MMO) domain after 2–3 months.

4. Discussion

4.1. Interpretation of the Results

In all qualifying studies, a greater improvement in VAS articular pain, HI, and MMO domains occurred due to inferior than superior compartment TMJ intra-articular injections [41,48–50,67]. However, the lack of homogeneity in terms of the injectables used throughout the analyzed studies must be emphasized [41,48–50,67]. The reports by Li et al.

and Long et al. showed similar trends for HI and MMO rates [41,49]. The improvement in MMO was higher for injections into the inferior compartment of TMJ than into the superior compartment by 5.4% in the report of Li et al., and by 12.6% in the study of Long et al. [41,49]. Regression analysis showed that the MMO improvements progressed over time in smaller and smaller increments [41,49]. The described trends did not differ from the results of other studies observed for viscosupplementation of HA [21,41,49,69,70]. In both analyzed studies, patients received intra-articularly equal doses of 1 mL HA [41,49]. The decrease in the Helkimo index after 3 months was 11.8% higher for injections into the inferior compartment of the TMJ than for the superior compartment in the Li et al. study and 19.9% higher in the Long et al. study [41,49]. Both Li at al. and Long et al. administered 1 mL of HA intra-articularly using a congruent technique [41,49]. However, Li et al. administered one dose more, which may explain the better results obtained [41,49].

Of note are the results of Liu et al., who also reported a significant increase in MMO administering HA to the TMJ space [49]. After two months, MMO increased by 51.9% for injections into the inferior compartment of TMJ, while it increased by 34.6% for the superior compartment [49]. In the course of the discussed trial, apart from the injections, the patients also underwent splint therapy [49]. In a recent systematic review, the first administration of HA was proven to be more effective than subsequent administrations [21]. These partially explain the outlier result of single HA administration preceded by arthrocentesis and combined with splint therapy in the study by Liu et al. [21,24,49,71].

Fouda et al. noted the lowest increase in MMO (1.1–6.4%) among analyzed studies [48]. This may be due to the diagnosis of disk displacement with reduction, in contrast to the other reports where disk displacement without reduction was treated [41,48–50,67]. The weaker results in the report of Fouda et al. may be also explained by the administration of 25% hypertonic dextrose (after receiving 2% mepivacaine locally) in contrast to the HA used in other trials [41,48–50]. Recent systematic reviews showed that the use of HA results in a greater increase in MMO and often greater improvement in VAS than administering hypertonic dextrose [21,28]. According to these data, in patients who received HA, the final values of pain were from 14 to 62% of the initial intensity [21]. The reduction in joint pain levels among patients treated with dextrose ranged from 33 to 76% of initial pain [28]. In the report of Fouda et al., articular pain on the VAS scale was also measured, yielding better results within the inferior compartment of TMJ where a 72.7% decrease in pain was noted (51.1% for the superior TMJ compartment) [48]. These values fit into the range specified in the review mentioned above [28]. The results of the Fouda et al. trial show that in both domains, improvement occurred immediately after the intervention, with no significant gain in effectiveness over follow-up [48]. This is consistent with the results of other prolotherapy reports [28,48,72–74]. In the material collected in this review, it was observed that the longer the follow-up time, the greater the improvement in MMO [41,48–50]. Li et al. and Long et al. monitored patients for 3 to 6 months longer than Fouda et al., which makes it possible that the highest MMO values for the prolotherapy group were not recorded [41,48,50].

4.2. Injection Technique

Making sure that the drug is administered to the correct compartment of the temporomandibular joint is crucial for taking advantage of this aspect [41,44–50,67]. Ahlqvist et al. used an X-ray source under the patient's head to correctly locate the bevel of the needle tip [45]. The bevel should be positioned next to the condyle facing its surface in case of inferior compartment injections, and below the posterior slope of the articular tubercle when administering to the upper compartment [45]. Yeung et al. showed the possibility of using an intraoperative navigation system (Stryker Leibinger, Freiburg, Germany) which, based on magnetic resonance imaging (MRI) of the TMJ, enables injection into the correct compartment of the joint [46]. Recent studies presented the use of ultrasonography when performing intra-articular TMJ injections [44,47]. Cha et al. reported that ultrasound-guided injections, especially into the lower compartment, are performed with significantly higher accuracy than those based solely on anatomical landmarks [44]. Januzzi et al. additionally used cone beam computed tomography (CBCT) to confront the precise location of osseous structures with facial access points [47]. The lack of universal acceptance of a specific method of puncture into the inferior TMJ compartment inspires further anatomical and clinical research in this direction [44–47]. Injecting both compartments as part of one intervention, preferably with a single injection, seems to be a desirable solution [18,51,52].

4.3. Limitations

A similar systematic review was published more than a decade ago by Li et al. [42]. These authors then identified two studies comparing inferior compartment injections and another two comparing both-compartment injections to upper compartment TMJ injections [42,49–52]. On the basis of the meta-analysis, the superiority of both techniques over administrations to the upper TMJ compartment was demonstrated [42,49–52]. Limitations of the work of Li et al. resulting from the small number of heterogeneous source studies are still a current problem [42]. Despite no mixing of inferior compartment injection with both-compartment injection studies within the systematic review reported here, the heterogeneity resulting from the variety of administered substances remained [41,48–50,67]. Literature reviews indicate that the type of substance used may have a significant impact on the effectiveness of injection therapy [27–29,75]. Therefore, the small number of studies identified, discrepancies regarding the substances used and intra-articular injection techniques, uncertainties regarding possible biases, and the differences in observation times and scheduled follow-up visits between individual trials are the limitations of the evidence included in this review [41,48–50].

5. Conclusions

The advantage of inferior over superior compartment intra-articular injections within the temporomandibular joint is unequivocal. However, the evidence is limited to four studies (337 patients in total), each with a moderate risk of bias. It can be conservatively assumed that after 2–3 months of follow-up, the numerical benefits of the lower compartment intra-articular injections are close to the ranges shown in this meta-analysis. These are from 19% to 51% improvement in articular pain, from 12% to 20% lower Helkimo index, and from 5% to 17% higher maximum mouth opening compared to supradiscal interventions. The inhomogeneity of the analyzed studies in terms of the substances used does not allow for the determination of average efficiency gain due to the application of the technique in question. Further clinical trials followed by a meta-analysis taking into account subgroups depending on the injectables seem to be justified.

Author Contributions: Conceptualization, M.C., M.S. and D.C.; methodology, M.C. and M.S.; software, M.C., K.C. and N.T.; validation, K.C. and N.T.; formal analysis, M.C. and N.T.; investigation, M.C., K.C. and N.T.; resources, M.C. and K.C.; data curation, M.C., K.C. and N.T.; writing—original draft preparation, M.C., K.C. and N.T.; writing—review and editing, M.C., N.T., M.S. and D.C.; visualization, M.C. and N.T.; supervision, D.C.; project administration, M.C. and M.S. All authors have read and agreed to the published version of the manuscript.

Funding: This research received no external funding.

Institutional Review Board Statement: Not applicable.

Informed Consent Statement: Not applicable.

Data Availability Statement: The protocol of this systematic review is available in the Prospective Register of Systematic Reviews (PROSPERO) under number CRD42022381667. All collected data are available within the body of this article.

Conflicts of Interest: The authors declare no conflict of interest.

Appendix A

Table A1. Search queries.

Search Engine	Search Query
BASE	(inferior OR lower) AND (superior OR upper) AND (compartment OR space) AND temporomandibular AND (joint OR articulation) AND (arthrocentesis OR rinsing OR lavage OR injection OR administration OR viscosupplementation)
ClinicalTrials.gov	(arthrocentesis OR rinsing OR lavage OR injection OR administration OR viscosupplementation) | (inferior OR lower) AND (superior OR upper) AND (compartment OR space) AND temporomandibular AND (joint OR articulation)
Google Scholar	allintitle: (inferior OR lower) AND (superior OR upper) AND (compartment OR space) AND temporomandibular AND (joint OR articulation) AND (arthrocentesis OR rinsing OR lavage OR injection OR administration OR viscosupplementation)
PubMed	(inferior OR lower) AND (superior OR upper) AND (compartment OR space) AND temporomandibular AND (joint OR articulation) AND (arthrocentesis OR rinsing OR lavage OR injection OR administration OR viscosupplementation)
ResearchGate	(inferior OR lower) AND (superior OR upper) AND (compartment OR space) AND temporomandibular AND (joint OR articulation) AND (arthrocentesis OR rinsing OR lavage OR injection OR administration OR viscosupplementation)
Scopus	TITLE-ABS-KEY ((inferior OR lower) AND (superior OR upper) AND (compartment OR space) AND temporomandibular AND (joint OR articulation) AND (arthrocentesis OR rinsing OR lavage OR injection OR administration OR viscosupplementation))

References

1. Speksnijder, C.M.; Mutsaers, N.E.A.; Walji, S. Functioning of the Masticatory System in Patients with an Alloplastic Total Temporomandibular Joint Prostheses Compared with Healthy Individuals: A Pilot Study. *Life* **2022**, *12*, 2073. [CrossRef] [PubMed]
2. Tobe, S.; Ishiyama, H.; Nishiyama, A.; Miyazono, K.; Kimura, H.; Fueki, K. Effects of Jaw-Opening Exercises with/without Pain for Temporomandibular Disorders: A Pilot Randomized Controlled Trial. *Int. J. Environ. Res. Public Health* **2022**, *19*, 16840. [CrossRef] [PubMed]
3. Habibi, H.A.; Ozturk, M.; Caliskan, E.; Turan, M. Quantitative assessment of temporomandibular disc and masseter muscle with shear wave elastography. *Oral Radiol.* **2022**, *38*, 49–56. [CrossRef]
4. Walker, T.F.; Broadwell, B.K.; Noujeim, M.E. MRI assessment of temporomandibular disc position among various mandibular positions: A pilot study. *CRANIO®* **2017**, *35*, 10–14. [CrossRef] [PubMed]
5. Yildiz, S.; Balel, Y.; Tumer, M.K. Evaluation of prevalence of temporomandibular disorders based on DC/TMD Axis I diagnosis in Turkish population and correlation with Axis II. *J. Stomatol. Oral Maxillofac. Surg.* **2022**, *124*, 101303. [CrossRef] [PubMed]
6. Sikora, M.; Sielski, M.; Chęciński, M.; Nowak, Z.; Czerwińska-Niezabitowska, B.; Chlubek, D. Repeated Intra-Articular Administration of Platelet-Rich Plasma (PRP) in Temporomandibular Disorders: A Clinical Case Series. *J. Clin. Med.* **2022**, *11*, 4281. [CrossRef]
7. Bitiniene, D.; Zamaliauskiene, R.; Kubilius, R.; Leketas, M.; Gailius, T.; Smirnovaite, K. Quality of life in patients with temporomandibular disorders. A systematic review. *Stomatologija* **2018**, *20*, 3–9. [PubMed]
8. Schiffman, E.; Ohrbach, R.; Truelove, E.; Look, J.; Anderson, G.; Goulet, J.-P.; List, T.; Svensson, P.; Gonzalez, Y.; Lobbezoo, F.; et al. Diagnostic Criteria for Temporomandibular Disorders (DC/TMD) for Clinical and Research Applications: Recommendations of the International RDC/TMD Consortium Network* and Orofacial Pain Special Interest Group†. *J. Oral Facial Pain Headache* **2014**, *28*, 6–27. [CrossRef]
9. Pigg, M.; Law, A.; Nixdorf, D.; Renton, T.; Sharav, Y.; Svensson, P.; Ernberg, M.; Peck, C.; Alstergren, P.; Kaspo, G.; et al. International Classification of Orofacial Pain, 1st edition (ICOP). *Cephalalgia* **2020**, *40*, 129–221. [CrossRef]
10. Iwaszenko, S.; Munk, J.; Baron, S.; Smoliński, A. New Method for Analysis of the Temporomandibular Joint Using Cone Beam Computed Tomography. *Sensors* **2021**, *21*, 3070. [CrossRef]
11. Smolka, W.; Yanai, C.; Smolka, K.; Iizuka, T. Efficiency of arthroscopic lysis and lavage for internal derangement of the temporomandibular joint correlated with Wilkes classification. *Oral Surg. Oral Med. Oral Pathol. Oral Radiol. Endodontol.* **2008**, *106*, 317–323. [CrossRef] [PubMed]
12. Machoy, M.; Szyszka-Sommerfeld, L.; Rahnama, M.; Koprowski, R.; Wilczyński, S.; Woźniak, K. Diagnosis of Temporomandibular Disorders Using Thermovision Imaging. *Pain Res. Manag.* **2020**, *2020*, 1–8. [CrossRef]
13. Turosz, N.; Chęcińska, K.; Chęciński, M.; Kamińska, M.; Nowak, Z.; Sikora, M.; Chlubek, D. A Scoping Review of the Use of Pioglitazone in the Treatment of Temporo-Mandibular Joint Arthritis. *Int. J. Environ. Res. Public Health* **2022**, *19*, 16518. [CrossRef] [PubMed]

14. Nitecka-Buchta, A.; Walczynska-Dragon, K.; Kempa, W.M.; Baron, S. Platelet-Rich Plasma Intramuscular Injections—Antinociceptive Therapy in Myofascial Pain Within Masseter Muscles in Temporomandibular Disorders Patients: A Pilot Study. *Front. Neurol.* **2019**, *10*, 250. [CrossRef]

15. Nowak, Z.; Chęciński, M.; Nitecka-Buchta, A.; Bulanda, S.; Ilczuk-Rypuła, D.; Postek-Stefańska, L.; Baron, S. Intramuscular Injections and Dry Needling within Masticatory Muscles in Management of Myofascial Pain. Systematic Review of Clinical Trials. *Int. J. Environ. Res. Public Health* **2021**, *18*, 9552. [CrossRef]

16. Kulesa-Mrowiecka, M.; Piech, J.; Gaździk, T.S. The Effectiveness of Physical Therapy in Patients with Generalized Joint Hypermobility and Concurrent Temporomandibular Disorders—A Cross-Sectional Study. *J. Clin. Med.* **2021**, *10*, 3808. [CrossRef] [PubMed]

17. Dowgierd, K.; Pokrowiecki, R.; Kulesa Mrowiecka, M.; Dowgierd, M.; Woś, J.; Szymor, P.; Kozakiewicz, M.; Lipowicz, A.; Roman, M.; Myśliwiec, A. Protocol for Multi-Stage Treatment of Temporomandibular Joint Ankylosis in Children and Adolescents. *J. Clin. Med.* **2022**, *11*, 428. [CrossRef] [PubMed]

18. Sun, H.; Su, Y.; Song, N.; Li, C.; Shi, Z.; Li, L. Clinical Outcome of Sodium Hyaluronate Injection into the Superior and Inferior Joint Space for Osteoarthritis of the Temporomandibular Joint Evaluated by Cone-Beam Computed Tomography: A Retrospective Study of 51 Patients and 56 Joints. *Med. Sci. Monit.* **2018**, *24*, 5793–5801. [CrossRef]

19. Byra, J.; Kulesa-Mrowiecka, M.; Pihut, M. Physiotherapy in hypomobility of temporomandibular joints. *Folia Med. Cracov.* **2020**, *60*, 123–134.

20. Sikora, M.; Chęciński, M.; Nowak, Z.; Chlubek, D. Variants and Modifications of the Retroauricular Approach Using in Temporomandibular Joint Surgery: A Systematic Review. *J. Clin. Med.* **2021**, *10*, 2049. [CrossRef]

21. Chęciński, M.; Sikora, M.; Chęcińska, K.; Nowak, Z.; Chlubek, D. The Administration of Hyaluronic Acid into the Temporomandibular Joints' Cavities Increases the Mandible's Mobility: A Systematic Review and Meta-Analysis. *J. Clin. Med.* **2022**, *11*, 1901. [CrossRef]

22. Chung, P.-Y.; Lin, M.-T.; Chang, H.-P. Effectiveness of platelet-rich plasma injection in patients with temporomandibular joint osteoarthritis: A systematic review and meta-analysis of randomized controlled trials. *Oral Surg. Oral Med. Oral Pathol. Oral Radiol.* **2019**, *127*, 106–116. [CrossRef]

23. Guarda-Nardini, L.; De Almeida, A.; Manfredini, D. Arthrocentesis of the Temporomandibular Joint: Systematic Review and Clinical Implications of Research Findings. *J. Oral Facial Pain Headache* **2021**, *35*, 17–29. [CrossRef]

24. Şentürk, M.F.; Yazıcı, T.; Gülşen, U. Techniques and modifications for TMJ arthrocentesis: A literature review. *CRANIO®* **2017**, *36*, 332–340. [CrossRef] [PubMed]

25. Manafikhi, M.; Ataya, J.; Heshmeh, O. Evaluation of the efficacy of platelet rich fibrin (I-PRF) intra-articular injections in the management of internal derangements of temporomandibular joints—A controlled preliminary prospective clinical study. *BMC Musculoskelet. Disord.* **2022**, *23*, 454. [CrossRef] [PubMed]

26. Chęciński, M.; Chęcińska, K.; Turosz, N.; Kamińska, M.; Nowak, Z.; Sikora, M.; Chlubek, D. Autologous Stem Cells Transplants in the Treatment of Temporomandibular Joints Disorders: A Systematic Review and Meta-Analysis of Clinical Trials. *Cells* **2022**, *11*, 2709. [CrossRef] [PubMed]

27. Derwich, M.; Mitus-Kenig, M.; Pawlowska, E. Mechanisms of Action and Efficacy of Hyaluronic Acid, Corticosteroids and Platelet-Rich Plasma in the Treatment of Temporomandibular Joint Osteoarthritis—A Systematic Review. *Int. J. Mol. Sci.* **2021**, *22*, 7405. [CrossRef]

28. Chęciński, M.; Chęcińska, K.; Nowak, Z.; Sikora, M.; Chlubek, D. Treatment of Mandibular Hypomobility by Injections into the Temporomandibular Joints: A Systematic Review of the Substances Used. *J. Clin. Med.* **2022**, *11*, 2305. [CrossRef]

29. Cömert Kiliç, S.; Güngörmüş, M.; Sümbüllü, M.A. Is Arthrocentesis Plus Platelet-Rich Plasma Superior to Arthrocentesis Alone in the Treatment of Temporomandibular Joint Osteoarthritis? A Randomized Clinical Trial. *J. Oral Maxillofac. Surg.* **2015**, *73*, 1473–1483. [CrossRef]

30. Sikora, M.; Czerwińska-Niezabitowska, B.; Chęciński, M.A.; Sielski, M.; Chlubek, D. Short-Term Effects of Intra-Articular Hyaluronic Acid Administration in Patients with Temporomandibular Joint Disorders. *J. Clin. Med.* **2020**, *9*, 1749. [CrossRef]

31. Liapaki, A.; Thamm, J.R.; Ha, S.; Monteiro, J.L.G.C.; McCain, J.P.; Troulis, M.J.; Guastaldi, F.P.S. Is there a difference in treatment effect of different intra-articular drugs for temporomandibular joint osteoarthritis? A systematic review of randomized controlled trials. *Int. J. Oral Maxillofac. Surg.* **2021**, *50*, 1233–1243. [CrossRef]

32. Sit, R.W.-S.; Reeves, K.D.; Zhong, C.C.; Wong, C.H.L.; Wang, B.; Chung, V.C.; Wong, S.Y.; Rabago, D. Efficacy of hypertonic dextrose injection (prolotherapy) in temporomandibular joint dysfunction: A systematic review and meta-analysis. *Sci. Rep.* **2021**, *11*, 14638. [CrossRef] [PubMed]

33. Sipahi, A.; Satilmis, T.; Basa, S. Comparative study in patients with symptomatic internal derangements of the temporomandibular joint: Analgesic outcomes of arthrocentesis with or without intra-articular morphine and tramadol. *Br. J. Oral Maxillofac. Surg.* **2015**, *53*, 316–320. [CrossRef]

34. Yapici-Yavuz, G.; Simsek-Kaya, G.; Ogul, H. A comparison of the effects of Methylprednisolone Acetate, Sodium Hyaluronate and Tenoxicam in the treatment of non-reducing disc displacement of the temporomandibular joint. *Med. Oral* **2018**, *23*, e351–e358. [CrossRef] [PubMed]

35. Kałużyński, K.; Trybek, G.; Smektała, T.; Masiuk, M.; Myśliwiec, L.; Sporniak-Tutak, K. Effect of methylprednisolone, hyaluronic acid and pioglitazone on histological remodeling of temporomandibular joint cartilage in rabbits affected by drug-induced osteoarthritis. *Postep. Hig. Med. Dosw (Online)* **2016**, *70*, 74–79. [CrossRef]
36. Xie, Y.; Zhao, K.; Ye, G.; Yao, X.; Yu, M.; Ouyang, H. Effectiveness of Intra-Articular Injections of Sodium Hyaluronate, Corticosteroids, Platelet-Rich Plasma on Temporomandibular Joint Osteoarthritis: A Systematic Review and Network Meta-Analysis of Randomized Controlled Trials. *J. Evid.-Based Dent. Pract.* **2022**, *22*, 101720. [CrossRef]
37. Whyte, A.; Boeddinghaus, R.; Bartley, A.; Vijeyaendra, R. Imaging of the temporomandibular joint. *Clin. Radiol.* **2021**, *76*, e21–e35. [CrossRef]
38. Alomar, X.; Medrano, J.; Cabratosa, J.; Clavero, J.A.; Lorente, M.; Serra, I.; Monill, J.M.; Salvador, A. Anatomy of the Temporomandibular Joint. *Semin. Ultrasound CT MRI* **2007**, *28*, 170–183. [CrossRef] [PubMed]
39. González-García, R.; Moreno-Sánchez, M.; Moreno-García, C.; Román-Romero, L.; Monje, F. Arthroscopy of the Inferior Compartment of the Temporomandibular Joint: A New Perspective. *J. Maxillofac. Oral Surg.* **2018**, *17*, 228–232. [CrossRef]
40. Sikora, M.; Sielski, M.; Chęciński, M.; Chęcińska, K.; Czerwińska-Niezabitowska, B.; Chlubek, D. Patient-Reported Quality of Life versus Physical Examination in Treating Temporomandibular Disorders with Intra-Articular Platelet-Rich Plasma Injections: An Open-Label Clinical Trial. *Int. J. Environ. Res. Public Health* **2022**, *19*, 13299. [CrossRef]
41. Li, C.; Long, X.; Deng, M.; Li, J.; Cai, H.; Meng, Q. Osteoarthritic Changes After Superior and Inferior Joint Space Injection of Hyaluronic Acid for the Treatment of Temporomandibular Joint Osteoarthritis with Anterior Disc Displacement Without Reduction: A Cone-Beam Computed Tomographic Evaluation. *J. Oral Maxillofac. Surg.* **2015**, *73*, 232–244. [CrossRef] [PubMed]
42. Li, C.; Zhang, Y.; Lv, J.; Shi, Z. Inferior or Double Joint Spaces Injection Versus Superior Joint Space Injection for Temporomandibular Disorders: A Systematic Review and Meta-Analysis. *J. Oral Maxillofac. Surg.* **2012**, *70*, 37–44. [CrossRef] [PubMed]
43. Pihut, M.; Szuta, M.; Ferendiuk, E.; Zeńczak-Więckiewicz, D. Evaluation of Pain Regression in Patients with Temporomandibular Dysfunction Treated by Intra-Articular Platelet-Rich Plasma Injections: A Preliminary Report. *BioMed Res. Int.* **2014**, *2014*, 132369. [CrossRef] [PubMed]
44. Cha, Y.H.; Park, J.K.; Yang, H.M.; Kim, S.H. Ultrasound-guided versus blind temporomandibular joint injections: A pilot cadaveric evaluation. *Int. J. Oral Maxillofac. Surg.* **2019**, *48*, 540–545. [CrossRef]
45. Ahlqvist, J.; Legrell, P.E. A technique for the accurate administration of corticosteroids in the temporomandibular joint. *Dentomaxillofac. Radiol.* **1993**, *22*, 211–213. [CrossRef] [PubMed]
46. Yeung, R.W.K.; Xia, J.J.; Samman, N. Image-Guided Minimally Invasive Surgical Access to the Temporomandibular Joint: A Preliminary Report. *J. Oral Maxillofac. Surg.* **2006**, *64*, 1546–1552. [CrossRef] [PubMed]
47. Januzzi, E.; Cunha, T.C.A.; Silva, G.; Souza, B.D.M.; Duarte, A.S.B.; Zanini, M.R.S.; Andrade, A.M.; Pedrosa, A.R.; Custódio, A.L.N.; Castro, M.A.A. Viscosupplementation in the upper and lower compartments of the temporomandibular joint checked by ultrasonography in an ex vivo and in vivo study. *Sci. Rep.* **2022**, *12*, 17976. [CrossRef]
48. Fouda, A.A. Change of site of intra-articular injection of hypertonic dextrose resulted in different effects of treatment. *Br. J. Oral Maxillofac. Surg.* **2018**, *56*, 715–718. [CrossRef]
49. Liu, J.J.; Mu, H.; Zhang, D.S.; Long, X.; Wang, Q.H.; Fu, X.L. Lower joint cavity treatment of temporomandibular joint with anterior disc displacement without reduction. *Int. J. Stomatol.* **2010**, *31*, 30. [CrossRef]
50. Long, X.; Chen, G.; an Cheng, A.H.; Cheng, Y.; Deng, M.; Cai, H.; Meng, Q. A Randomized Controlled Trial of Superior and Inferior Temporomandibular Joint Space Injection with Hyaluronic Acid in Treatment of Anterior Disc Displacement Without Reduction. *J. Oral Maxillofac. Surg.* **2009**, *67*, 357–361. [CrossRef]
51. Cheng, X.L. An outcome analysis of two methods of intra-capsular injection of prednisolone for temporomandibular disorders. *Chin. J. Prim. Med. Pharm.* **2005**, *12*, 63–64.
52. Liu, C.M.; Shi, Z.D.; Yi, X.Z.; Guo, C.Z.; Zhang, X. Observation of the therapeutic effects on TMD by injection HA into both superior and inferior temporomandibular joint cavities. *J. Dent. Prev. Treat* **2003**, *11*, 14–16.
53. Ebrahim, S. Methodological Limitations of a Systematic Review Evaluating Inferior or Double Joint Spaces Injection Versus Superior Joint Space Injection for Temporomandibular Disorders. *J. Oral Maxillofac. Surg.* **2012**, *70*, 504–505. [CrossRef] [PubMed]
54. Li, C.; Shi, Z. Reply to Dr Shanil Ebrahim on Inferior or Double Joint Spaces Injection Versus Superior Joint Space Injection for Temporomandibular Disorders: A Systematic Review and Meta-Analysis. *J. Oral Maxillofac. Surg.* **2012**, *70*, 505–506. [CrossRef]
55. Turfah, A.; Liu, H.; Stewart, L.A.; Kang, T.; Weng, C. Extending PICO with Observation Normalization for Evidence Computing. In *Studies in Health Technology and Informatics*; Otero, P., Scott, P., Martin, S.Z., Huesing, E., Eds.; IOS Press: Amsterdam, The Netherlands, 2022; ISBN 978-1-64368-264-8.
56. Methley, A.M.; Campbell, S.; Chew-Graham, C.; McNally, R.; Cheraghi-Sohi, S. PICO, PICOS and SPIDER: A comparison study of specificity and sensitivity in three search tools for qualitative systematic reviews. *BMC Heal. Serv. Res.* **2014**, *14*, 1–10. [CrossRef] [PubMed]
57. Eriksen, M.B.; Frandsen, T.F. The impact of patient, intervention, comparison, outcome (PICO) as a search strategy tool on literature search quality: A systematic review. *J. Med Libr. Assoc.* **2018**, *106*, 420–431. [CrossRef]
58. Leamari, V.M.; de Rodrigues, A.F.; Camino Junior, R.; Luz, J.G.C. Correlations between the Helkimo indices and the maximal mandibular excursion capacities of patients with temporomandibular joint disorders. *J. Bodyw. Mov. Ther.* **2019**, *23*, 148–152. [CrossRef]

59. Emshoff, R.; Emshoff, I.; Bertram, S. Estimation of clinically important change for visual analog scales measuring chronic temporomandibular disorder pain. *J. Orofac. Pain* **2010**, *24*, 262–269.

60. Gusenbauer, M. Search where you will find most: Comparing the disciplinary coverage of 56 bibliographic databases. *Scientometrics* **2022**, *127*, 2683–2745. [CrossRef]

61. Valizadeh, A.; Moassefi, M.; Nakhostin-Ansari, A.; Hosseini Asl, S.H.; Saghab Torbati, M.; Aghajani, R.; Maleki Ghorbani, Z.; Faghani, S. Abstract screening using the automated tool Rayyan: Results of effectiveness in three diagnostic test accuracy systematic reviews. *BMC Med. Res. Methodol.* **2022**, *22*, 160. [CrossRef]

62. Ouzzani, M.; Hammady, H.; Fedorowicz, Z.; Elmagarmid, A. Rayyan—A web and mobile app for systematic reviews. *Syst. Rev.* **2016**, *5*, 210. [CrossRef]

63. Guimarães, N.S.; Ferreira, A.J.; Silva, R.D.C.R.; de Paula, A.A.; Lisboa, C.S.; Magno, L.; Ichiara, M.Y.; Barreto, M.L. Deduplicating records in systematic reviews: There are free, accurate automated ways to do so. *J. Clin. Epidemiology* **2022**, *152*, 110–115. [CrossRef]

64. Sterne, J.A.C.; Savović, J.; Page, M.J.; Elbers, R.G.; Blencowe, N.S.; Boutron, I.; Cates, C.J.; Cheng, H.-Y.; Corbett, M.S.; Eldridge, S.M.; et al. RoB 2: A revised tool for assessing risk of bias in randomised trials. *BMJ* **2019**, *366*, l4898. [CrossRef]

65. Sterne, J.A.; Hernán, M.A.; Reeves, B.C.; Savović, J.; Berkman, N.D.; Viswanathan, M.; Henry, D.; Altman, D.G.; Ansari, M.T.; Boutron, I.; et al. ROBINS-I: A tool for assessing risk of bias in non-randomised studies of interventions. *BMJ* **2016**, *355*, i4919. [CrossRef] [PubMed]

66. Long, X. Comparative Study of Inferior Versus Superior Joint Space Injection of Sodium Hyaluronate in Patients with Anterior Disc Displacement Without Reduction of the Temporomandibular Joint: A Randomized Controlled Trial. *J. Oral Maxillofac. Surg.* **2008**, *66*, 65–66. [CrossRef]

67. Ozawa, M.; Okaue, M.; Kaneko, K.; Hasegawa, M.; Matsunaga, S.; Matsumoto, M.; Hori, M.; Kudo, I.; Takagi, M. Clinical Assessment of the Pumping Technique in Treating TMJ Arthrosis with Closed Lock. *J. Nihon Univ. Sch. Dent.* **1996**, *38*, 1–10. [CrossRef] [PubMed]

68. Page, M.J.; McKenzie, J.E.; Bossuyt, P.M.; Boutron, I.; Hoffmann, T.C.; Mulrow, C.D.; Shamseer, L.; Tetzlaff, J.M.; Akl, E.A.; Brennan, S.E.; et al. The PRISMA 2020 statement: An updated guideline for reporting systematic reviews. *BMJ* **2021**, *372*, n71. [CrossRef]

69. Yang, W.; Liu, W.; Miao, C.; Sun, H.; Li, L.; Li, C. Oral Glucosamine Hydrochloride Combined with Hyaluronate Sodium Intra-Articular Injection for Temporomandibular Joint Osteoarthritis: A Double-Blind Randomized Controlled Trial. *J. Oral Maxillofac. Surg.* **2018**, *76*, 2066–2073. [CrossRef]

70. Fonseca, R.M.D.F.B.; Januzzi, E.; Ferreira, L.A.; Grossmann, E.; Carvalho, A.C.P.; de Oliveira, P.G.; Vieira, É.L.M.; Teixeira, A.L.; Almeida-Leite, C.M. Effectiveness of Sequential Viscosupplementation in Temporomandibular Joint Internal Derangements and Symptomatology: A Case Series. *Pain Res. Manag.* **2018**, *2018*, 1–9. [CrossRef]

71. Wiechens, B.; Paschereit, S.; Hampe, T.; Wassmann, T.; Gersdorff, N.; Bürgers, R. Changes in Maximum Mandibular Mobility Due to Splint Therapy in Patients with Temporomandibular Disorders. *Healthcare* **2022**, *10*, 1070. [CrossRef]

72. Zarate, M.A.; Frusso, R.D.; Reeves, K.D.; Cheng, A.-L.; Rabago, D. Dextrose Prolotherapy Versus Lidocaine Injection for Temporomandibular Dysfunction: A Pragmatic Randomized Controlled Trial. *J. Altern. Complement. Med.* **2020**, *26*, 1064–1073. [CrossRef] [PubMed]

73. Louw, W.F.; Reeves, K.D.; Lam, S.K.H.; Cheng, A.-L.; Rabago, D. Treatment of Temporomandibular Dysfunction with Hypertonic Dextrose Injection (Prolotherapy): A Randomized Controlled Trial with Long-term Partial Crossover. *Mayo Clin. Proc.* **2019**, *94*, 820–832. [CrossRef] [PubMed]

74. Lam, S.K.H.; Reeves, K.D.; Rabago, D. Dextrose Prolotherapy for Chronic Temporomandibular Pain and Dysfunction: Results of a Pilot-Level Randomized Controlled Study. *Arch. Phys. Med. Rehabil.* **2016**, *97*, e139. [CrossRef]

75. Zotti, F.; Albanese, M.; Rodella, L.F.; Nocini, P.F. Platelet-Rich Plasma in Treatment of Temporomandibular Joint Dysfunctions: Narrative Review. *Int. J. Mol. Sci.* **2019**, *20*, 277. [CrossRef]

Journal of
Clinical Medicine

Article

Short-Term Effects of 3D-Printed Occlusal Splints and Conventional Splints on Sleep Bruxism Activity: EMG–ECG Night Recordings of a Sample of Young Adults

Andrea Bargellini [1,2], Elena Mannari [3], Giovanni Cugliari [4], Andrea Deregibus [1,2], Tommaso Castroflorio [1,2], Leila Es Sebar [5], Gianpaolo Serino [6,7], Andrea Roggia [3] and Nicola Scotti [8,*]

[1] Department of Surgical Sciences, Specialization School of Orthodontics, Dental School, University of Torino, 10126 Torino, Italy; bargelli@ipsnet.it (A.B.); andrea.deregibus@unito.it (A.D.); tommaso.castroflorio@unito.it (T.C.)

[2] Department of Surgical Sciences, Gnathology Unit, Dental School, University of Torino, 10126 Torino, Italy

[3] Department of Surgical Sciences, Dental School, University of Torino, 10126 Torino, Italy; elena.mannari@edu.unito.it (E.M.); andrea.roggia@gmail.com (A.R.)

[4] Department of Medical Sciences, University of Torino, 10126 Torino, Italy; giovanni.cugliari@unito.it

[5] Department of Applied Science and Technology, Politecnico di Torino, 10129 Turin, Italy; leila.essebar@polito.it

[6] Department of Mechanical and Aerospace Engineering, Politecnico di Torino, 10129 Turin, Italy; gianpaolo.serino@polito.it

[7] PolitoBioMedLab, Politecnico di Torino, 10129 Turin, Italy

[8] Department of Surgical Sciences, Restorative Dentistry Unit, Dental School, University of Torino, 10126 Torino, Italy

* Correspondence: nicola.scotti@unito.it

Citation: Bargellini, A.; Mannari, E.; Cugliari, G.; Deregibus, A.; Castroflorio, T.; Es Sebar, L.; Serino, G.; Roggia, A.; Scotti, N. Short-Term Effects of 3D-Printed Occlusal Splints and Conventional Splints on Sleep Bruxism Activity: EMG–ECG Night Recordings of a Sample of Young Adults. *J. Clin. Med.* **2024**, *13*, 776. https://doi.org/10.3390/jcm13030776

Academic Editor: Luís Eduardo Almeida

Received: 30 November 2023
Revised: 22 January 2024
Accepted: 23 January 2024
Published: 29 January 2024

Abstract: (1) **Background**: This study aims to compare the effects of 3D-printed splints and conventional manufactured splints on sleep bruxism (SB) EMG activity. (2) **Methods**: Twenty-six patients (19 M, 7 F, 25.8 ± 2.6 years) were randomly allocated to a study group (3D splints) and a control group (conventional manufactured splints) and followed for a period of three months with night EMG–ECG recordings. Samples of the involved materials were analyzed for nanoindentation. The outcomes of interest considered were the overall SB index, the total amount of surface masseter muscle activity (sMMA), and general and SB-related phasic and tonic contractions. A statistical evaluation was performed with a confidence interval (CI) between 2.5% and 97.5%. (3) **Results**: Differences between groups with OAs were observed for general tonic contraction ($p = 0.0009$), while differences between recording times were observed for general phasic contractions ($p = 0.002$) and general tonic contractions ($p = 0.00001$). Differences between recording times were observed for the total amount of sMMA ($p = 0.01$), for general phasic contractions ($p = 0.0001$), and for general tonic contractions ($p = 0.000009$) during night recordings without OAs. (4) **Conclusions**: Three-dimensional splints seem to have a higher impact on SB-related electromyographic activity but not on the overall sleep bruxism index. The more regular surfaces offered by 3D splints could be related to phasic contraction stabilization.

Keywords: sleep bruxism; 3D; occlusal splint; custom appliance

1. Introduction

According to the American Academy of Sleep Disorders (AASM), sleep bruxism (SB) is defined as a sleep-related movement disorder characterized by simple, often stereotyped movements occurring during sleep [1]. SB is not a movement disorder or a sleep disorder in otherwise healthy individuals, and it is characterized by involuntary phasic (rhythmic) or tonic (sustained) motor activity in the masticatory muscles (e.g., the masseter or the temporalis) during sleep [2,3]. Considerations of the physiological mechanisms beneath SB have shown that they are characterized by tachycardia, often followed by bradycardia, and can occur with or without EEG desynchronization [4]. These events are similar to the physiological sequences involved in rhythmic masticatory muscle activity in SB (RMMA/SB), which consist of the following activities: an increase in sympathetic activity before the RMMA/SB onset (-4 to -8 min) [5], followed by an increase in EEG activity (cortical arousal) (-4 s; tachycardia occurs 1 s before RMMA/SB [6,7], followed by an increase in the respiratory amplitude concomitant with RMMA/SB onset [8]. Further, SB is associated with significant increases in both systolic and diastolic blood pressure (BP) [9].

RMMA may lead to an imbalance of the stomatognathic system, involving the temporomandibular joints (TMJs) and their structures in the long term [10–13]. Rubin et al. and Wieckiewicz et al. reported that it is still questionable whether sleep bruxism is related to clinical signs of temporomandibular disorders (TMDs) [14,15]: in the literature, there have been authors reporting that there is a statistically significant correlation between TMDs and SB [16], as well as authors who have reported that this correlation is not statistically significant [17]. This may be due to different diagnostic criteria and study designs. Also, Topaloglu-Ak et al. determined that there is a significant association between negative sleeping habits and SB, TMDs, and dental caries [17].

Recent studies have also considered that the serotonin neurotransmission pathway may be involved in SB pathogenesis [18]. Serotonin (5-hydroxytryptamine, 5-HT) is a monoamine neurotransmitter of the central nervous system that is synthesized from tryptophan obtained from dietary sources [19]. During the synthesis, tryptophan is converted to 5-hydroxytryptophan (5-HTP) [20] through biopterin-dependent monooxygenation catalyzed via tryptophan hydroxylases 1 and 2 (TPH1 and TPH2), and then 5-HTP is decarboxylated via aromatic l-amino acid decarboxylase (DDC) to 5-HT [21]. However, the cause of the decreased serotonin levels in patients with severe SB is not known yet.

SB has been treated with different therapeutic approaches, such as oral appliance therapy (OAT) with stabilization splints, cognitive behavioral therapy (CBT), biofeedback therapy (BFT), and pharmacological therapy [22,23]. To date, the approach to SB has mainly focused on reducing SB's detrimental effects on the stomatognathic system [24–26], and oral appliances (OAs) seem to be the standard reference [27]. Three-dimensional technology has been spreading in all dentistry fields since the early 1980s [28,29], and to date, a digital workflow is commonly applied from prosthodontics to orthodontics [30–32]. This technique has been applied even for OAs in the gnathological field with good results of accuracy and precision [33]. Over time, different materials have been involved in OA manufacturing in place of acrylic resin, such as a light-cured composite that, in preliminary studies, was preferred in terms of comfort by patients [34]. It is reasonable to question whether different approaches to OA manufacturing procedures, as well as different material selections, may thus influence the mechanical characteristics of such devices and whether they may consequently affect a neuromuscular response of the stomatognathic system. Moreover, the use of a simplified digital workflow tends to reduce the number of possible errors and distortions compared to traditional manufacturing, leading to a more precise casting in the laboratory and, therefore, to fewer occlusal adjustments on the dentist chair [35].

The aim of this randomized clinical trial was to answer the following questions:

-    Do different OA fabrication techniques (i.e., 3D or traditional) influence SB?
-    Do different OA materials influence perceived comfort during usage?
-    Is there a difference in nanomechanical properties between the materials used to produce the two devices?

2. Materials and Methods

2.1. Study Design

The data reported in this investigation were gathered from a 2-arm parallel-group randomized controlled trial comparing SB index, sleep-time masticatory muscle activity (sMMA), and physiological and SB-related electromyographic contractions (i.e., tonic and phasic) between a group of subjects wearing a traditionally manufactured Michigan maxillary splint (control) and a group of patients wearing a 3D-printed Michigan maxillary splint (study). This trial was conducted following the Consolidated Standards of Reporting Trials (CONSORT) extension for pragmatic clinical trials [36]. Ethical approval was obtained from the Committee of the Research Department of the University of Torino, Italy (ethical approval #0089207). This trial was registered at ISRCTN.com (ISRCTN: ISRCTN91976427). All subjects provided their informed consent and signed an informed consent form before their enrollment in the study, and they were aware of the possibility of withdrawing from the study at any time.

2.2. Participants

To conduct this study, a total amount of 40 patients (26 M and 14 F; mean age: 26 ± 2.5 years) were randomly selected from a pool of patients at the Gnathology Unit of the University of Torino (Italy) with validated SB diagnosis conducted using a dedicated EMG–ECG holter (Bruxoff®, OT Bioelettronica, Torino, Italy) [37–40]. The patients who enrolled in this study had to fulfill the following inclusion criteria: (1) permanent dentition, (2) the absence of medications, (3) the absence of previous and/or active SB treatments, (4) the absence of periodontal disease evaluated according to the simplified oral hygiene index [41], (5) no medical history of neurological, mental, or sleep disorders, (6) an instrumental diagnosis of SB with at least 2 or more episodes per hour of sleep, and (7) no other underlying sleep disorders (such as obstructive sleep apnea or neurological sleep disorders) [42]. The exclusion criteria comprised the following: (1) active periodontal disease, (2) extended implant rehabilitations (>2 implants per arch), (3) active dental interventions (i.e., extractions, untreated caries, and prosthetic interventions with modifications of the occlusal plane), (4) missing teeth (>2 per arch), and (5) active SB treatments with OAs. After dropouts' removal, data recorded from 26 patients (19 M and 7 F; mean age: 25.8 ± 2.6 years) were analyzed for the study as follows: 12 patients in the study group (7 M and 5 F; mean age: 26.7 ± 2.9 years) and 14 in the control group (12 M and 2 F; mean age: 25.4 ± 2.5 years). All data were collected and stored at the Orthodontics Unit of the Dental School of the University of Torino.

2.3. Randomization and Blinding

The randomization procedure consisted of a two-stage procedure in which subjects who entered the trial were first grouped into strata according to clinical features that could have influenced the outcomes: age and sex. No statistically significant differences in age and sex were shown after a Wilcoxon test between the groups. The patients' assignment to the study or control group was randomly conducted using a computer-generated randomized table of numbers created by a statistician (G.C.). Individual and sequentially numbered index cards with random assignments were prepared, folded, and placed in sealed, opaque envelopes. The patients were blinded to the treatment. The clinicians' blinding was not possible.

2.4. Outcome Measures

The outcomes of interest considered were the overall SB index, the total amount of surface masseter muscle activity (sMMA), and general and SB-related phasic and tonic contractions. A statistical evaluation was performed with a confidence interval (CI) between 2.5% and 97.5%.

2.5. Sample Size Calculation

The sample size calculation was based on data relating to the SB index (number of SB events per hour of sleep) estimated during screening in the full sample (40 subjects). Dropouts were included in the statistical analysis as an intention-to-treat (ITT) analysis. The mean $\pm$ SD of the SB index was 5.75 ± 3.96 SB episodes per hour of sleep. A 50% increase/decrease in the SB index was considered clinically relevant with an assumed SD of 3.96 and 80% power at the 5% significance level, which achieved the required sample size of 12 subjects per group.

2.6. Statistical Analysis

The normality assumption of the data was evaluated with the Shapiro–Wilk test [43]; the homoscedasticity and autocorrelation of the variables were assessed using the Breusch–Pagan and Durbin–Watson tests [44,45]. A two-way analysis of variance (ANOVA) was performed to estimate the differences in the δ (Tn–T0) means between groups stratified by splint use (yes/no) during Bruxoff® recordings. All analyses were adjusted for sex, age, and the duration of the test. Descriptive values were shown with a consideration of the main indicators of distribution and variability. The level of significance was set to $p < 0.05$. Statistical analyses were conducted using the R statistical package (version 3.5.3, R Core Team, Foundation for Statistical Computing, Vienna, Austria) [46].

2.7. Sensitivity Analysis

Descriptive values were shown with a consideration of the main indicators of distribution and variability. The level of significance was set to $p < 0.05$.

2.8. Experimental Section

Digital impressions were recorded for both groups with an iTero Element 5D (Align Technology, Inc., San Jose, CA, USA), and stereo-lithographic (STL) models were crafted with a computer-aided design and computer-aided manufacturing (CAD-CAM) 3D printer, a SolFlex 650 (VOCO® GmbH, Cuxhaven, Germany).

The OAs used in the study were realized as follows:

(1) Study group: Splints were manufactured thanks to the 3D printer, a SolFlex 650 (VOCO® GmbH, Cuxhaven, Germany), with an additive technique. The light-cured V-Print Splint resin (VOCO® GmbH, Cuxhaven, Germany) was used for this purpose, classified as class IIa medical disposal [47] (medium risk, non-active disposal), biocompatible, tasteless, transparent, and with high resistance to abrasion (reported flexural strength: 75 Mpa). A 0.20 mm offset value was used for the additive technique, as suggested by Lo Giudice et al. [48]. After printing, all debris was removed with an ultrasonic bath of isopropyl alcohol, and the plates were dried with compressed air. Fifteen minutes after the final contact with isopropyl alcohol, the plates were post-cured on both sides with an OtoFlash-Polymerization Unit (VOCO® GmbH, Cuxhaven, Germany): 3.5 min per side with a frequency of 10 flashes per second for a total amount of 2000 flashes. At the end of the process, all the plates were finished and polished by an expert dental technician of the Dental School, University of Torino, Torino (Italy).

(2) Control group: Splints were manufactured with a 1.25 mm thermoformed retentive base of polyethylenterephthalat–glycol copolyester (PET-G) (DURAN® Scheu-Dental, Am Burgberg, Germany) (reported flexural strength: 69 Mpa) rebased with self-curing resin (Forestacry®, Forestadent Bernhard Förster GmbH Westliche, Pforzheim, Germany), classified as class I medical disposal [47] (non-critical and non-active disposal), biocompatible, tasteless, transparent, and with high resistance to abrasion (reported flexural strength: $\geq$50 Mpa). The plates were fabricated, finished, and polished by an external dental laboratory (GLOI Laboratorio Odontotecnico, Biella, Italy).

Both OAs were designed as Michigan maxillary splints with an increase in the occlusal vertical dimension (OVD) of at least 2 mm in the molar area (Maestro 3D software (www.maestro3d.com, accessed on 27 November 2023), AGE Solutions S.r.l, Pontedera, PI, Italy). Occlusion was checked with an 8 μm Bausch articulating paper (Bausch®, Nashua, NH 03062 USA) in order to let arches occlude evenly and uniformly.

All patients were instructed by expert clinicians (T.C. and E.M.) on the use and maintenance of their splints, with the recommendation to wear them every night for three months.

SB monitoring was performed for all patients with the Bruxoff® device [37–40] (OT Bioelettronica, Torino, Italy) as follows: at OA delivery (T0), after 1 month since delivery (T1), and after three months since delivery (T2). Only during the T0, T1, and T2 stages were the patients asked to perform two different night recordings: one night wearing the splint and the following night without wearing it.

All the patients were instructed by expert clinicians (T.C., A.B., and E.M.) on the use of the Bruxoff® device. The data were analyzed with dedicated software (Bruxmeter (version 2.0.2.4)® OT Bioelettronica, Torino, Italy) by an expert clinician blinded to the study (A.B.).

At the T1 stage, the patients were asked to fill in a modified version of the Oral Health Impact Profile (OHIP-14) [49] to investigate their perceived comfort with different OAs (Table 1). The data were inserted into an Excel® table by a researcher not blinded to the study (E.M.) for statistical analysis (Microsoft Corporation, Redmond, WA, USA).

Table 1. The modified version of the Oral Health Impact Profile (OHIP-14) questionnaire.

1.	Did you feel discomfort with your teeth when your splint was delivered?
2.	If yes, how much did it hurt on a scale from 1 to 10?
3.	I yes, which teeth hurt the most?
4.	Could you wear your splint all night long?
5.	If not, why?
6.	Did you find an improvement in splint fit within one month of delivery?
7.	How many nights did you wear your splint within one month of delivery?
8.	If less than three nights a week, could you tell us why?
9.	How much did your teeth hurt on a scale from 1 to 10 one month after splint delivery?
10.	Did you find an improvement in your lifestyle one month after splint delivery (i.e., head and facial muscle tenderness, headache, or dental sensitivity)?
11.	Did you perceive a reduction in your parafunction one month after splint delivery?

2.9. Nanoindentation Protocol

Square-designed specimens (1 cm² and 1 mm thick) were prepared by following the same procedure used to fabricate the OAs under investigation. The surfaces of the two specimens were properly polished and rinsed with isopropyl alcohol. The data were recorded by expert clinicians (N.S. and A.R.) and analyzed by an expert engineer in the field (G.S).

Nanoindentation tests were carried out, imposing a maximum indentation depth of 2000 nm. Forty-two indentations per specimen were performed in order to investigate the mechanical properties of an area of 16 mm². The nanoindentation protocol was characterized by three phases. During the first step, the nanoindenter tip was brought in contact with the specimen surface. After the contact detection, the penetration rate of the indenter was set to 15 mN/s, controlling the load imposed by the indenter on the surface sample. When the set penetration depth was reached (2000 nm), the load was held constant for a period of 5 s. Finally, the indenter tip was retracted at the same rate as the loading phase.

The nanoindentation curves were analyzed using the Oliver and Pharr method [50] and following the ISO 14577-1:2015 [51] in order to obtain the nanoindentation modulus (E_{IT}).

A one-way analysis of variance was performed with a level of confidence equal to 0.01 to verify the difference in the nanoindentation moduli (E_{IT}) obtained for the two materials.

3. Results

A CONSORT diagram with subjects' flow through the trial is shown in Figure 1. A total amount of 6287 min of sleep data were recorded. The mean sleep duration was 8.01 ± 1.2 h in the study group and 8.03 ± 1.4 h in the control group, with no differences between the groups. The mean SB index of the overall population was 5.4 ± 1.2 per hour of sleep. The patients in the study group showed 4.76 ± 1.8 SB episodes per night, and the control group showed 6.05 ± 2 SB episodes per night during three months of observation. The average values of outcome variables during the recording time (T0–T2) are reported in Table 2. The ANOVA results are reported in Tables 3 and 4. At the end of the observation time of three months, the SB index was affected by neither traditional nor 3D-fabricated occlusal splints. Differences between the groups with OAs were observed for general tonic contraction ($p = 0.0009$), while differences between the recording times were observed for general phasic contractions ($p = 0.002$) and general tonic contractions ($p = 0.00001$). Differences between the recording times were observed for the total amount of sMMA ($p = 0.01$), for general phasic contractions ($p = 0.0001$), and for general tonic contractions ($p = 0.000009$) during night recordings without OAs.

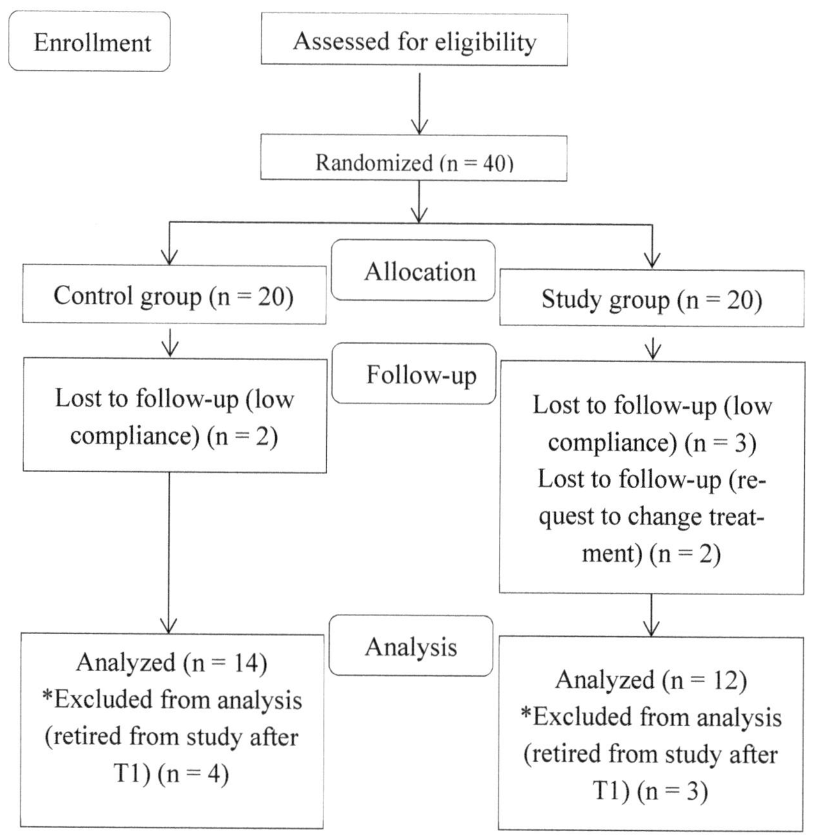

Figure 1. Consolidated Standards of Reporting Trials (CONSORT) diagram showing the flow of subjects in the study. Please note that * stands for retired patients from the study after T1.

Nanoindentation

Representative nanoindentation curves are reported in Figure 2. To reach the same indentation depth, the applied load on the molded specimen was higher compared to the printed specimen. Furthermore, the statistical analysis performed on the dataset of the nanoindentation modulus showed a significant difference between the elastic properties of the two materials ($p = 0.0002$).

Table 2. Average values (standard deviation, SD) of outcome variables over each recording night per group with and without splints.

Outcome Variable	Group	Recording Nights (T0–T2)	Average Values (SD) with Splint	Average Values (SD) without Splint
SB index	3D splint	T0	4.8 (1.6)	5.1 (1.3)
		T1	6.1 (3)	7.3 (3.8)
		T2	3 (2.1)	5.4 (5.4)
	Control splint	T0	6 (2)	5.7 (2)
		T1	4.3 (2.2)	6.2 (2.4)
		T2	5.3 (3.5)	5.5 (3.5)
Total sMMA contractions	3D splint	T0	76.3 (50)	69.7 (43.7)
		T1	221.4 (134.4)	217 (147.6)
		T2	255.6 (117.2)	265.3 (156.6)
	Control splint	T0	84 (41.5)	87 (45.6)
		T1	150 (92)	212.8 (213.6)
		T2	129.1 (83)	155 (71)
Phasic sMMA contractions	3D splint	T0	20.8 (18.3)	26.7 (13.4)
		T1	60.4 (42.9)	61.3 (45.2)
		T2	60.2 (20.6)	60.4 (36)
	Control splint	T0	26.9 (14)	20.3 (11.5)
		T1	46.5 (37.5)	54.1 (62)
		T2	85.8 (144.4)	44.7 (26)
Phasic sMMA contractions, SB-related	3D splint	T0	6.9 (4.2)	6.1 (3.7)
		T1	10.2 (8.2)	11.8 (9.1)
		T2	4.3 (4.8)	6.7 (4.9)
	Control splint	T0	12.8 (6.1)	11.3 (4.7)
		T1	8.2 (7.6)	8 (5.9)
		T2	16 (32.9)	7 (5.6)
Tonic sMMA contractions	3D splint	T0	22.4 (20.8)	23.5 (19.6)
		T1	69.2 (51.6)	67.3 (37.3)
		T2	91.6 (51.3)	79 (50.6)
	Control splint	T0	20.2 (15.1)	22.6 (16.1)
		T1	42.7 (29.4)	64.7 (70.8)
		T2	31 (31.8)	38.2 (19.9)
Tonic sMMA contractions, SB-related	3D splint	T0	8.1 (5.1)	8.5 (4.3)
		T1	12 (7)	17.5 (11.5)
		T2	7.7 (5.5)	9.6 (8.4)
	Control splint	T0	10.4 (8.2)	9.8 (10.4)
		T1	9.6 (9)	8.7 (7.8)
		T2	19.3 (48.5)	8.5 (4.6)

SB: sleep bruxism. sMMA: surface masseter muscle activity. T0: baseline. T1: 1 month. T2: 3 months.

Box plot and color maps of the nanoindentation modulus are shown in Figures 3 and 4, respectively. The spatial distribution and the dispersion of the elastic properties of the two materials were very different, indicating a significant difference between the two fabrication processes. The 3D-printed specimen was found to be non-homogeneous compared to the molded specimen. Indeed, the two color maps shown in Figure 4 corroborate the dispersion reported in the box plot of Figure 3. On the other hand, the molded specimen was found to be significantly stiffer than the printed one. As a matter of fact, while the printed specimen was found to be very homogeneous in terms of elastic properties, with a little spot where the elastic properties increased, the molded sample showed the opposite feature, with the spatial distribution of the elastic properties characterized by a predominance of higher values of the nanoindentation modulus compared to the lower ones measured in a smaller spot of the analyzed area.

Table 3. Results of the two-way ANOVA test for night recordings with OAs.

Outcome Variable	Estimate	F Value	Pr (>F)
SB index	Group	0.918	0.340
	Time	1.659	0.196
Total sMMA contractions	Group	3.139	0.0797
	Time	0.613	0.5438
Phasic sMMA contractions	Group	0.276	0.600464
	Time	6.335	0.002646 **
Phasic sMMA contractions, SB-related	Group	2.965	0.0885
	Time	0.093	0.9114
Tonic sMMA contractions	Group	11.659	0.000952 ***
	Time	12.928	0.00001 ***
Tonic sMMA contractions, SB-related	Group	0.710	0.4015
	Time	0.465	0.6296

SB: sleep bruxism. sMMA: surface masseter muscle activity. *, significant at $p < 0.05$. **, *** number of asteriscs is indicative for statistical significancy

Table 4. Results of the two-way ANOVA test for night recordings without OAs.

Outcome Variable	Estimate	F Value	Pr (>F)
SB index	Group	0.113	0.737
	Time	1.885	0.158
Total sMMA contractions	Group	0.738	0.39247
	Time	4.501	0.01365 *
Phasic sMMA contractions	Group	0.145	0.70456
	Time	9.551	0.00017 ***
Phasic sMMA contractions, SB-related	Group	0.863	0.35535
	Time	2.158	0.12136
Tonic sMMA contractions	Group	1.802	0.18281
	Time	13.176	0.000009 ***
Tonic sMMA contractions, SB-related	Group	1.929	0.16823
	Time	2.350	0.10104

SB: sleep bruxism. MMA: surface masseter muscle activity. *, significant at $p < 0.05$. *** number of asteriscs is indicative for statistical significancy

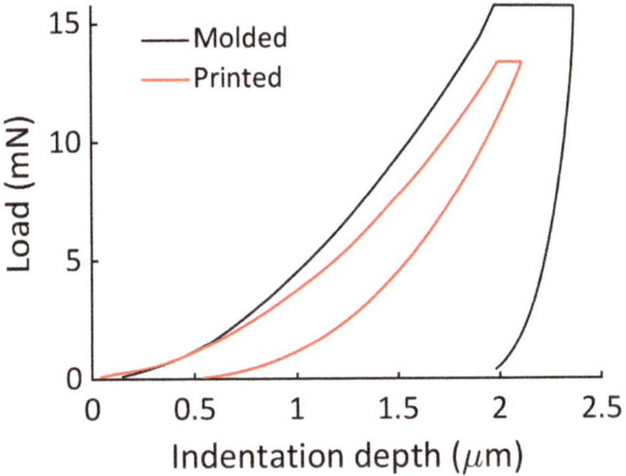

Figure 2. Representative nanoindentation curves obtained for molded and printed specimens.

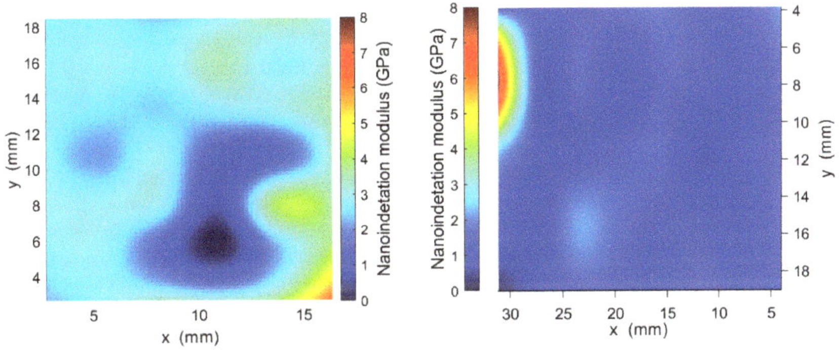

Figure 3. Color maps of nanoindentation moduli for molded (**on the left**) and printed (**on the right**) specimens.

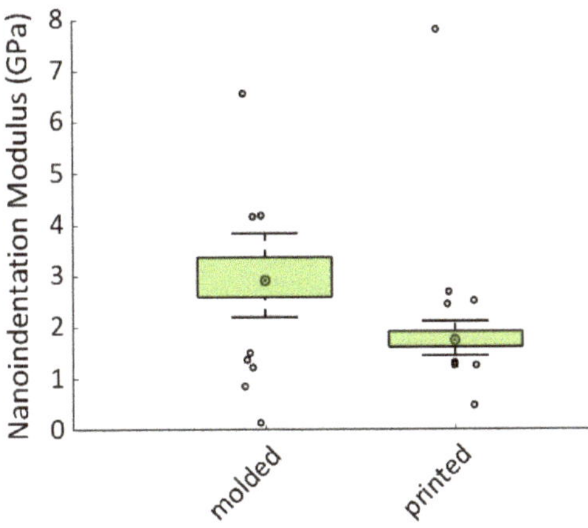

Figure 4. Dispersion of the nanoindentation modulus values reported through a box plot.

4. Discussion

The present study aimed to investigate possible effects on the SB index of 3D-printed splints. Although some effects were observed on general EMG signals for three months, no influences on SB activity were found. Three-dimensional splints seem to better control general phasic contractions over time [52]; indeed, from the first to the third month of observation, the patients with 3D splints tended to maintain a stable trend, while the patients wearing traditionally manufactured splints showed a constant increase from T0 to T2. Thus, 3D splints could lead to an easier and faster adjustment in patients showing a high ratio of phasic contractions; on the other hand, traditionally manufactured splints should be avoided in these cases since their use increases these specific EMG signals in the first three months after delivery. Given general tonic contractions [5], statistically significant differences between the groups and the observation times were detected: despite a general increase for both groups after one month from delivery, tonic contractions tended to be lower for the patients wearing traditionally manufactured splints, and this trend reduced after three months, while it drastically increased for the patients wearing 3D splints. Traditionally manufactured splints should, then, be indicated for patients with higher levels of tonic contractions, while 3D splints should be avoided since their use tends to increase these EMG signals during the first three months after delivery. These effects should be related to the different flexural strengths of the involved materials: 3D splints fabricated with multiple layers of the same material (V-Print Splint resin (VOCO® GmbH)) present a higher flexural strength (75 Mpa) than traditional splints fabricated with a thermoformed layer of polyethylenterephthalat–glycol copolyester (PET-G) (DURAN® Scheu-Dental, Am Burgberg 20, 58642 Iserlohn, Germany) (reported flexural strength: 69 Mpa) rebased with self-cured resin (Forestacry®, Forestadent Bernhard Förster GmbH Westliche, Pforzheim, Germany) with lower flexural strength (>50 Mpa). It could be inferred that layers of a single hard resin should positively influence general phasic (i.e., brief and rhythmic) contractions, leading to a stabilization trend within one month after use, but they negatively affect tonic (i.e., sustained) contractions, leading to a detrimental and significant increase in their level, while thermoformed layers of PET-G covered with more elastic resins tend to positively affect general tonic contractions, but by contrast, they negatively influence general phasic contractions, leading to their constant increase for three months. The sMMA activity shows significant variations over time when OAs are not in use during night recordings. This effect could be explained by different behaviors of the EMG signals composing the overall sMMA index: tonic contractions maintained an increasing trend for the study group, while they decreased in the control group; phasic contractions remained stable in the study group, while they decreased in the control group. Generally, phasic contractions seem to be more influenced by the use and the interruption of splints with low flexural strength, while both tonic and phasic contractions tend to maintain an imprint from splints with high flexural strength. When in use, traditional splints reduce tonic contractions and increase phasic contractions; when not in use, both tonic and phasic contractions are reduced. When in use, 3D splints stabilize general phasic contractions and increase tonic contractions; when their use is interrupted, the effect on motor units seems to be imprinted since the effect on general phasic and tonic contractions remains the same: general phasic contractions remain stable and tonic contractions increase. Hence, patients with higher levels of general phasic contractions will take more time to adapt to their device if it is fabricated traditionally with resins with low flexural strength (>50 Mpa). A similar effect for traditionally manufactured splints was reported by Matsumoto et al. [53] in patients wearing occlusal splints at intermittent times; they reported significant reductions in nocturnal EMG events and duration immediately after splint delivery and after one month later when compared to a pool of patients wearing occlusal splints for 29 nights continuously. Patients wearing occlusal splints for an entire month experienced an immediate reduction after splint delivery and after one week, without any reduction at 2, 3, and 4 weeks. Unfortunately, no other indications of EMG signals were reported (i.e., phasic or tonic EMG signals); thus, no other conclusions can be added. These findings could be helpful for future studies related to the cost benefit for patients: since 3D splints entail higher manufacturing costs, the use of traditional splints for patients showing higher tonic contractions could be helpful both for clinical use and for making their application more affordable for patients.

To answer the second question related to the possible influences of different materials on perceived comfort, only self-reported improvement in patients' quality of life and a perceived reduction in the night parafunction (questions 10 and 11 reported in Table 1) showed significant results: the patients wearing 3D-printed splints reported an improvement in their lifestyle in 64% of cases and referred to a perceived reduction of their parafunction in 63% of cases. These data suggest higher adaptation to 3D splints than to traditional ones if we consider the short-term effects on phasic and tonic contractions reported above. Anyway, these results reflect a perceived sensation and not the real trend of the instrumental data. As reported by Lobbezoo et al. [54], questionnaires are only able to indicate possible SB.

The mechanical properties of 3D-printed occlusal splints have been an area of active research in recent years, as the performance and durability of these splints depend on their material properties [55–57]. The present study results suggested that 3D-printed OAs had a lower nanoindentation modulus than the molded ones, suggesting lower mechanical properties. In the recent literature, several studies have investigated the mechanical properties of 3D-printed occlusal splints. A study by Cheah et al. [58] evaluated the mechanical properties of occlusal splints printed with two different 3D printing technologies, stereolithography (SLA) and digital light processing (DLP). The study found that the SLA-printed splints had higher flexural strength and wear resistance compared to the DLP-printed splints. Another study by Hwang et al. [59] investigated the mechanical properties of 3D-printed occlusal splints made from a biocompatible resin. The study found that the splints had a high degree of accuracy and precision, and they demonstrated good mechanical properties, including flexural strength and surface hardness. While most studies have suggested that 3D-printed occlusal splints have better mechanical properties than molded ones, there are some studies that have reported different findings, in accordance with the present study results. Li et al. [60] compared the mechanical properties of 3D-printed and traditionally polymerized occlusal splint materials and found that the traditionally polymerized splints had higher flexural strength values. Another study by Nayyer et al. [61] compared the wear resistance of 3D-printed and milled splints and found that the milled splints had better wear resistance. Paradowska–Starz et al. [62] pointed out that the long-term use of these resins must be taken into consideration, as well as aging's influence on their structure: in their study, these authors stated that artificial aging has a deeply negative impact on 3D-printed splints, involving both the compressive modulus and tension of the material, and polishing was suggested to increase resin's resistance to aging.

Overall, the literature suggests that 3D-printed occlusal splints can have good mechanical properties, but the performance can vary, depending on the printing technology and material used. It is important to note that the materials and printing/molding techniques used in each study may have differed, and these factors can have a significant impact on the mechanical properties of the resulting splints. Therefore, to answer the third question of this study, it is difficult to generalize about the strength of 3D-printed versus molded occlusal splints without considering the specific materials and techniques used in each case.

The uniformity of elasticity and hardness on the surface that meets the antagonistic teeth, on the other hand, should be considered when considering an OA. As shown in the color maps of Figure 3, the 3D-printed OAs seem to have a more uniform nanoindentation modulus than the molded ones. The better phasic contraction adjustment observed in patients wearing 3D-printed OAs during the study period may, therefore, be explained by the mechanical property pattern, which appears to be more regular over the device surface and results in a better and more consistent response to occlusal contact and sliding.

Outside the findings of this study, it is undoubted that the approach to SB should be carefully investigated with validated disposals to avoid overtreatments. The SB index, anyway, should be considered in association with all EMG activities to complete diagnoses and provide correct treatments. As stated by Lobbezoo et al., bruxism-related masticatory muscle activity should be assessed in its continuum, thus not only focusing on the raw number of bruxism events to correlate with clinical consequences [62]. It is not the number of bruxism events per se that represents a risk factor but, rather, the general level of EMG activity, which was found to be higher in temporomandibular disorder cases than in controls [63]. Such an approach is in line with the work of Greene et al. since the use of more technological and sophisticated investigation systems tends to be used to force treatments in non-symptomatic patients [64,65]; therefore, the use of occlusal splints in patients affected by SB and/or with a high number of EMG activities should be considered for symptomatic TMD patients.

Limitations

There were two main limitations to the current study. First, the high number of dropouts led to a low number of participants. Second, the short period of observation reduced the power of this study. Another minor limitation could be related to the comparison between only two materials.

5. Conclusions

The overall SB index seems not to be affected in the first three months using different OAs fabricated with traditional or CAD-CAM techniques, while their short-term effects can be observed on general EMG signals. Perceived improvements in patients' quality of life should be carefully considered since they could not evenly report the real physiological status observed with an instrumental investigation. The more regular surfaces achieved with 3D splints could be related to phasic contraction stabilization. Further studies conducted over a longer period with a higher number of participants are requested to improve the results obtained in this study.

Author Contributions: Conceptualization, T.C. and N.S.; methodology, G.S.; investigation, E.M.; data curation, G.C.; writing—original draft, A.B.; writing—review and editing, A.R. and L.E.S.; supervision, A.D. All authors have read and agreed to the published version of the manuscript.

Funding: Research funding was provided by the University of Torino, and materials and technical support for the 3D production of occlusal splints were provided by VOCO GmbH.

Institutional Review Board Statement: This trial was conducted following the Consolidated Standards of Reporting Trials (CONSORT) extension for pragmatic clinical trials [14]. Ethical approval was obtained from the Committee of the Research Department of the University of Torino, Italy (ethical approval #0089207). This trial was registered at ISRCTN.com (ISRCTN: ISRCTN91976427).

Informed Consent Statement: Informed consent was obtained from all the subjects involved in this study.

Data Availability Statement: Data is unavailable due to privacy restrictions.

Acknowledgments: The authors would like to thank GLOI dental laboratory (GLOI Laboratorio Odontotecnico, Via Bengasi, 5, 13900 Biella, Italy) for its cooperation with the University of Torino, VOCO GmbH for its contribution to this study, providing materials and technical support for the 3D production of occlusal splints involved in this study, and AGE Solutions S.r.l. for providing the Maestro 3D software used to design the splints.

Conflicts of Interest: The authors declare no conflicts of interest.

References

1. Sateia, M.J. *International Classification of Sleep Disorders*, 3rd ed.; Sleep Related Bruxism; American Academy of Sleep Medicine: Darien, IL, USA, 2014.
2. Lobbezoo, F.; Ahlberg, J.; Raphael, K.G.; Wetselaar, P.; Glaros, A.G.; Kato, T.; Santiago, V.; Winocur, E.; De Laat, A.; De Leeuw, R.; et al. International consensus on the assessment of bruxism: Report of a work in progress. *J. Oral. Rehabil.* **2018**, *45*, 837–844. [CrossRef]

3. Lavigne, G.J.; Khoury, S.; Abe, S.; Yamaguchi, T.; Raphael, K. Bruxism physiology and pathology: An overview for clinicians. *J. Oral. Rehabil.* **2008**, *35*, 476–494. [CrossRef]
4. Sforza, E.; Jouny, C.; Ibanez, V. Cardiac activation during arousal in humans: Further evidence for hierarchy in the arousal response. *Clin. Neurophysiol.* **2000**, *111*, 1611–1619. [CrossRef]
5. Huynh, N.; Kato, T.; Rompré, P.H.; Okura, K.; Saber, M.; Lanfranchi, P.A.; Montplaisir, J.Y.; Lavigne, G.J. Sleep bruxism is associated to micro-arousals and an increase in cardiac sympathetic activity. *J. Sleep. Res.* **2006**, *15*, 339–346. [CrossRef]
6. Macaluso, G.M.; Guerra, P.; Di Giovanni, G.; Boselli, M.; Parrino, L.; Terzano, M.G. Sleep bruxism is a disorder related to periodic arousals during sleep. *J. Dent. Res.* **1998**, *77*, 565–573. [CrossRef]
7. Kato, T.; Rompre, P.; Montplaisir, J.Y.; Sessle, B.J.; Lavigne, G.J. Sleep bruxism: An oromotor activity secondary to micro-arousal. *J. Dent. Res.* **2001**, *80*, 1940–1944. [CrossRef] [PubMed]
8. Khoury, S.; Rouleau, G.A.; Rompre, P.H.; Mayer, P.; Montplaisir, J.; Lavigne, G.J. A significant increase in breathing amplitude precedes sleep bruxism. *Chest* **2008**, *134*, 332–337. [CrossRef] [PubMed]
9. Nashed, A.; Lanfranchi, P.; Rompré, P.; Carra, M.C.; Mayer, P.; Colombo, R.; Huynh, N.; Lavigne, G. Sleep bruxism is associated with a rise in arterial blood pressure. *Sleep* **2012**, *35*, 529–536. [CrossRef] [PubMed]
10. Yazıcıoğlu, İ.; Çiftçi, V. Evaluation of signs and symptoms of temporomandibular disorders and incisal relationships among 7–10-year-old Turkish children with sleep bruxism: A cross-sectional study. *Cranio* **2021**, 1–7. [CrossRef] [PubMed]
11. Smardz, J.; Martynowicz, H.; Michalek-Zrabkowska, M.; Wojakowska, A.; Mazur, G.; Winocur, E.; Wieckiewicz, M. Sleep bruxism and occurrence of temporomandibular disorders-related pain: A polysomnographic study. *Front. Neurol.* **2019**, *10*, 168. [CrossRef] [PubMed]
12. Andrade de Alencar, N.; Nolasco Fernandes, A.B.; Gomes de Souza, M.M.; Luiz, R.R.; Fonseca-Gonçalves, A.; Maia, L.C. Lifestyle and oral facial disorders associated with sleep bruxism in children. *Cranio* **2017**, *35*, 168–174. [CrossRef]
13. Marpaung, C.; van Selms, M.K.; Lobbezoo, F. Prevalence and risk indicators of pain-related temporomandibular disorders among Indonesian children and adolescents. *Commun. Dent. Oral. Epidemiol.* **2018**, *46*, 400–406. [CrossRef]
14. Rubin, P.F.; Erez, A.; Peretz, B.; Birenboim-Wilensky, R.; Winocur, E. Prevalence of bruxism and temporomandibular disorders among orphans in southeast Uganda: A gender and age comparison. *Cranio* **2018**, *36*, 243–249. [CrossRef]
15. Wieckiewicz, M.; Smardz, J.; Martynowicz, H.; Wojakowska, A.; Mazur, G.; Winocur, E. Distribution of temporomandibular disorders among sleep bruxers and non-bruxers—A polysomnographic study. *J. Oral. Rehabil.* **2020**, *47*, 820–826. [CrossRef]
16. Lei, J.; Fu, J.; Yap, A.U.J.; Fu, K.Y. Temporomandibular disorders symptoms in Asian adolescents and their association with sleep quality and psychological distress. *Cranio* **2016**, *34*, 242–249. [CrossRef] [PubMed]
17. Topaloglu-Ak, A.; Kurtulmus, H.; Basa, S.; Sabuncuoglu, O. Can sleeping habits be associated with sleep bruxism, temporo-mandibular disorders and dental caries among children? *Dent. Med. Probl.* **2022**, *59*, 517–522. [CrossRef] [PubMed]
18. Cheifetz, A.T.; Osganian, S.K.; Allred, E.M.; Needleman, H.L. Prevalence of bruxism and associated correlates in children as reported by parents. *J. Dent. Child.* **2005**, *72*, 67–73.
19. Wieckiewicz, M.; Bogunia-Kubik, K.; Mazur, G.; Danel, D.; Smardz, J.; Wojakowska, A.; Poreba, R.; Dratwa, M.; Chaszczewska-Markowska, M.; Winocur, E.; et al. Genetic basis of sleep bruxism and sleep apnea-response to a medical puzzle. *Sci. Rep.* **2020**, *10*, 7497. [CrossRef] [PubMed]
20. Strasser, B.; Gostner, J.M.; Fuchs, D. Mood, food, and cognition: Role of tryptophan and serotonin. *Curr. Opin. Clin. Nutr. Metab. Care* **2016**, *19*, 55–61. [CrossRef] [PubMed]
21. Roberts, K.M.; Fitzpatrick, P.F. Mechanisms of tryptophan and tyrosine hydroxylase. *IUBMB Life* **2013**, *65*, 350–357. [CrossRef] [PubMed]
22. Jenkins, T.A.; Nguyen, J.C.; Polglaze, K.E.; Bertrand, P.P. Influence of tryptophan and serotonin on mood and cognition with a possible role of the gut-brain axis. *Nutrients* **2016**, *8*, 56. [CrossRef] [PubMed]
23. Minakuchi, H.; Fujisawa, M.; Abe, Y.; Iida, T.; Oki, K.; Okura, K.; Tanabe, N.; Nishiyama, A. Managements of sleep bruxism in adult: A systematic review. *Jpn. Dent. Sci. Rev.* **2022**, *58*, 124–136. [CrossRef]
24. Cerón, L.; Pacheco, M.; Delgado Gaete, A.; Bravo Torres, W.; Astudillo Rubio, D. Therapies for sleep bruxism in dentistry: A critical evaluation of systematic reviews. *Dent. Med. Probl.* **2023**, *60*, 335–344. [CrossRef]
25. Manfredini, D.; Ahlberg, J.; Winocur, E.; Lobbezoo, F. Management of sleep bruxism in adults: A qualitative systematic literature review. *J. Oral. Rehabil.* **2015**, *42*, 862–874. [CrossRef]
26. Melo, G.; Duarte, J.; Pauletto, P.; Porporatti, A.L.; Stuginski-Barbosa, J.; Winocur, E.; Flores-Mir, C.; De Luca Canto, G. Bruxism: An umbrella review of systematic reviews. *J. Oral. Rehabil.* **2019**, *46*, 666–690. [CrossRef] [PubMed]
27. Yap, A.U.; Chua, A.P. Sleep bruxism: Current knowledge and contemporary management. *J. Conserv. Dent.* **2016**, *19*, 383–389. [CrossRef]
28. Mörmann, W.H.; Brandestini, M. *State of the Art of CADS/CAM Restorations: 20 Years of CEREC*; Mörmann, W.H., Ed.; Quintessence: London, UK, 2006; pp. 1–8.
29. Pillai, S.; Upadhyay, A.; Khayambashi, P.; Farooq, I.; Sabri, H.; Tarar, M.; Lee, K.T.; Harb, I.; Zhou, S.; Wang, Y.; et al. Dental 3D-Printing: Transferring Art from the Laboratories to the Clinics. *Polymers* **2021**, *13*, 157. [CrossRef] [PubMed]
30. Duret, F.; Preston, J.D. CAD/CAM imaging in dentistry. *Curr. Opin. Dent.* **1991**, *1*, 150–154.
31. Blatz, M.B.; Conejo, J. The Current State of Chairside Digital Dentistry and Materials. *Dent. Clin. N. Am.* **2019**, *63*, 175–197. [CrossRef]

32. Cunha, T.M.A.D.; Barbosa, I.D.S.; Palma, K.K. Orthodontic digital workflow: Devices and clinical applications. *Dental Press. J. Orthod.* **2021**, *26*, e21spe6. [CrossRef]

33. Joda, T.; Zarone, F.; Ferrari, M. The complete digital workflow in fixed prosthodontics: A systematic review. *BMC Oral. Health* **2017**, *17*, 124. [CrossRef] [PubMed]

34. Marcel, R.; Reinhard, H.; Andreas, K. Accuracy of CAD/CAM-fabricated bite splints: Milling vs. 3D printing. *Clin. Oral. Investig.* **2020**, *24*, 4607–4615. [CrossRef]

35. Leib, A.M. Patient preference for light-cured composite bite splint compared to heat-cured acrylic bite splint. *J. Periodontol.* **2001**, *72*, 1108–1112. [CrossRef]

36. Patzelt, S.B.M.; Krügel, M.; Wesemann, C.; Pieralli, S.; Nold, J.; Spies, B.C.; Vach, K.; Kohal, R.J. In Vitro Time Efficiency, Fit, and Wear of Conventionally- versus Digitally-Fabricated Occlusal Splints. *Materials* **2022**, *15*, 1085. [CrossRef] [PubMed]

37. Zwarenstein, M.; Treweek, S.; Gagnier, J.J.; Altman, D.G.; Tunis, S.; Haynes, B.; Oxman, A.D.; Moher, D. Improving the reporting of pragmatic trials: An extension of the CONSORT statement. *BMJ* **2008**, *337*, a2390. [CrossRef] [PubMed]

38. Casett, E.; Réus, J.C.; Stuginski-Barbosa, J.; Porporatti, A.L.; Carra, M.C.; Peres, M.A.; de Luca Canto, G.; Manfredini, D. Validity of different tools to assess sleep bruxism: A meta-analysis. *J. Oral. Rehabil.* **2017**, *44*, 722–734. [CrossRef]

39. Castroflorio, T.; Mesin, L.; Tartaglia, G.M.; Sforza, C.; Farina, D. Use of electromyographic and electrocardiographic signals to detect sleep bruxism episodes in a natural environment. *IEEE J. Biomed. Health Inform.* **2013**, *17*, 994–1001. [CrossRef]

40. Castroflorio, T.; Deregibus, A.; Bargellini, A.; Debernardi, C.; Manfredini, D. Detection of sleep bruxism: Comparison between an electromyographic and electrocardiographic portable holter and polysomnography. *J. Oral. Rehabil.* **2014**, *41*, 163–169. [CrossRef]

41. Deregibus, A.; Castroflorio, T.; Bargellini, A.; Debernardi, C. Reliability of a portable device for the detection of sleep bruxism. *Clin. Oral. Investig.* **2014**, *18*, 2037–2043. [CrossRef]

42. Greene, J.C.; Vermillion, J.R. The simplified oral hygiene index. *J. Am. Dent. Assoc.* **1964**, *68*, 7–13. [CrossRef]

43. Mayer, P.; Heinzer, R.; Lavigne, G. Sleep Bruxism in Respiratory Medicine Practice. *Chest* **2016**, *149*, 262–271. [CrossRef] [PubMed]

44. Vetter, T.R. Fundamentals of Research Data and Variables: The Devil Is in the Details. *Anesth. Analg.* **2017**, *125*, 1375–1380. [CrossRef] [PubMed]

45. Breusch, T.S.; Pagan, A.R. A Simple Test for Heteroscedasticity and Random Coefficient Variation. *Econometrica* **1979**, *47*, 1287. [CrossRef]

46. Durbin, J.; Watson, G.S. Testing for Serial Correlation in Least Squares Regression. III. *Biometrika* **1971**, *58*, 1–19. [CrossRef]

47. R Core Team R. *A Language and Environment for Statistical Computing*; R Foundation for Statistical Computing: Vienna, Austria, 2013; ISBN 3-900051-07-0.

48. Regulation (EC) No 1394/2007 of the European Parliament and of the Council of 13 November 2007 on Advanced Therapy Medicinal Products and Amending Directive 2001/83/EC and Regulation (EC) No. 726/2004. 2019. Available online: https://ec.europa.eu/health//sites/health/files/files/eudralex/vol1/reg2007139/reg20071394en.pdf (accessed on 8 April 2022).

49. Lo Giudice, A.; Ronsivalle, V.; Pedullà, E.; Rugeri, M.; Leonardi, R. Digitally programmed (CAD) offset values for prototyped occlusal splints (CAM): Assessment of appliance-fitting using surface-based superimposition and deviation analysis. *Int. J. Comput. Dent.* **2021**, *24*, 53–63. [PubMed]

50. Robinson, P.G.; Gibson, B.; Khan, F.A.; Birnbaum, W. A comparison of OHIP 14 and OIDP as interviews and questionnaires. *Commun. Dent. Health.* **2001**, *18*, 144–149.

51. *ISO 14577-1:2015*; Metallic materials. Instrumented indentation test for hardness and materials parameters. Part 1: Test Method. ISO: Geneva, Switzerland, 2015.

52. Oliver, W.C.; Pharr, G.M. An improved technique for determining hardness and elastic modulus using load and displacement sensing indentation experiments. *J. Mater. Res.* **1992**, *7*, 1564–1583. [CrossRef]

53. Bargellini, A.; Graziano, V.; Cugliari, G.; Deregibus, A.; Castroflorio, T. Effects on Sleep Bruxism Activity of Three Different Oral Appliances: One Year Longitudinal Cohort Study. *Curr. Drug Deliv.* **2022**, *epub ahead of printing*. [CrossRef]

54. Matsumoto, H.; Tsukiyama, Y.; Kuwatsuru, R.; Koyano, K. The effect of intermittent use of occlusal splint devices on sleep bruxism: A 4-week observation with a portable electromyographic recording device. *J. Oral. Rehabil.* **2015**, *42*, 251–258. [CrossRef]

55. Lobbezoo, F.; Ahlberg, J.; Glaros, A.G.; Kato, T.; Koyano, K.; Lavigne, G.J.; de Leeuw, R.; Manfredini, D.; Svensson, P.; Winocur, E. Bruxism defined and graded: An international consensus. *J. Oral. Rehabil.* **2013**, *40*, 2–4. [CrossRef]

56. Taira, M.; Maeda, Y.; Sawada, T.; Komatsu, M.; Kondo, H. Mechanical properties of 3D-printed thermoplastic materials for orthodontic retainers. *Dent. Mat. J.* **2020**, *39*, 607–613.

57. Monzavi, A.; Li, W.; Li, Q.; Swain, M.V. Mechanical properties of 3D printed polymeric and ceramic orthodontic brackets. *J. Mech. Behav. Biomed. Mat.* **2018**, *87*, 277–285.

58. Matinlinna, J.P.; Alomari, S.; Sadasivan, M.; Salehi, H.; Alagl, A.S.; Hamedani, S. Effect of material properties and manufacturing process of occlusal splints on wear resistance: A literature review. *J. Prosth. Dent.* **2020**, *123*, 46–52.

59. Cheah, C.M.; Chua, C.K.; Tan, K.H.; Abu Bakar, M.S. Evaluation of mechanical properties and dimensional accuracy of 3D-printed occlusal splints. *Int. J. Prosth.* **2019**, *32*, 288–290.

60. Hwang, Y.H.; Song, J.H.; Hong STKim, H.Y. Evaluation of the mechanical properties of 3D-printed occlusal splints made of biocompatible resin. *Materials* **2021**, *14*, 3867.

61. Li, W.; Bai, S.; Zhou, J.; Xu, X.; Wang, Y. Comparison of mechanical properties between 3D-printed and traditionally polymerized occlusal splint materials. *J. Prosth. Dent.* **2020**, *124*, 98–105.

62. Nayyer, M.; Savabi, O. Wear comparison of CAD/CAM milled and 3D printed occlusal splints. *J. Prosth. Dent.* **2020**, *124*, 639–645.

63. Paradowska-Stolarz, A.; Wezgowiec, J.; Malysa, A.; Wieckiewicz, M. Effects of Polishing and Artificial Aging on Mechanical Properties of Dental LT Clear® Resin. *J. Funct. Biomater.* **2023**, *14*, 295. [CrossRef] [PubMed]

64. Raphael, K.G.; Janal, M.N.; Sirois, D.A.; Dubrovsky, B.; Wigren, P.E.; Klausner, J.J.; Krieger, A.C.; Lavigne, G.J. Masticatory muscle sleep background electromyographic activity is elevated in myofascial temporomandibular disorder patients. *J. Oral. Rehabil.* **2013**, *40*, 883–891. [CrossRef] [PubMed]

65. Greene, C.; Manfredini, D.; Ohrbach, R. Creating patients: How technology and measurement approaches are misused in diagnosis and convert healthy individuals into TMD patients. *Front. Dent. Med.* **2023**, *4*, 1183327. [CrossRef]

Journal of
Clinical Medicine

Article

Influence of Soft Stabilization Splint on Electromyographic Patterns in Masticatory and Neck Muscles in Healthy Women

Grzegorz Zieliński [1], Marcin Wójcicki [2], Michał Baszczowski [3], Agata Żyśko [3], Monika Litko-Rola [2], Jacek Szkutnik [2], Ingrid Różyło-Kalinowska [4] and Michał Ginszt [5,*]

[1] Department of Sports Medicine, Medical University of Lublin, 20-093 Lublin, Poland
[2] Independent Unit of Functional Masticatory Disorders, Medical University of Lublin, 20-093 Lublin, Poland
[3] Interdisciplinary Scientific Group of Sports Medicine, Department of Sports Medicine,
    Medical University of Lublin, 20-093 Lublin, Poland
[4] Department of Dental and Maxillofacial Radiodiagnostics, Medical University of Lublin,
    20-093 Lublin, Poland
[5] Department of Rehabilitation and Physiotherapy, Medical University of Lublin, 20-093 Lublin, Poland
* Correspondence: michal.ginszt@umlub.pl

Abstract: This study investigates the influence of soft stabilization splints on electromyographic patterns in masticatory and neck muscles in healthy women. A total of 70 healthy women were qualified for the research. The resting and clenching electromyographic patterns of the temporalis (TA), masseter (MM), digastric (DA), and sternocleidomastoid (SCM) muscles were measured using the BioEMG III™ apparatus. The interaction between splint application and resting muscle activity affected the results in all examined muscles except the temporalis muscle. A large effect size was observed in masseter (2.19 µV vs. 5.18 µV; $p = 0.00$; ES = 1.00) and digastric (1.89 µV vs. 3.17 µV; $p = 0.00$; ES = 1.00) both-sided RMS activity. Significant differences between the two conditions were observed in all Functional Clenching Indices (FCI) for MM, SDM, and DA muscles. All FCI values for the MM and DA muscles were significantly lower with than without the splint. We observed an increase in all activity indices due to splint application, which suggests a masseter muscle advantage during measurement. The soft stabilization splint influenced resting and functional activity in the MM, SDM, and DA muscles. During tooth clenching, a soft stabilization splint changed the involvement proportions of the temporalis and masseter muscles, transferring the main activity to the masseter muscles. Using a soft stabilization splint did not affect the symmetry of the electromyographic activity of the masticatory and neck muscles.

Keywords: stabilization splint; surface electromyography; temporalis; masseter; digastric; sternocleidomastoid; functional indices

Citation: Zieliński, G.; Wójcicki, M.;
Baszczowski, M.; Żyśko, A.;
Litko-Rola, M.; Szkutnik, J.;
Różyło-Kalinowska, I.; Ginszt, M.
Influence of Soft Stabilization Splint
on Electromyographic Patterns in
Masticatory and Neck Muscles in
Healthy Women. *J. Clin. Med.* **2023**,
12, 2318. https://doi.org/
10.3390/jcm12062318

Academic Editor: Luís Eduardo
Almeida

Received: 16 February 2023
Revised: 10 March 2023
Accepted: 15 March 2023
Published: 16 March 2023

1. Introduction

Managing dysfunctions of the stomatognathic system requires a comprehensive multi-directional approach, and is still challenging for clinicians worldwide. Temporomandibular disorders (TMDs) are the most common form of non-odontogenic orofacial pain, and drastically reduce life quality [1]. It has been estimated that 4% of adults develop clinically confirmed and painful TMDs each year. Moreover, the occurrence of TMDs increases with age, with peak incidence being reported as 4.5% in the 35–44 age group [2]. Treatment options for patients with TMDs, bruxism, and frequent headaches associated with stomatognathic system disorders include pharmacotherapy [3], physiotherapy [4], patient education [5], behavior therapy [6], and removable appliances called occlusal or stabilization splints [7]. There are many types of splints varying in design, e.g., the material from which they are made (hard and soft splints), the position of the splint (maxillar and mandibular), and the extent of coverage (full-arch-covering type and partial type covering only the central incisors) [8,9]. The effectiveness of the use of stabilization splints, both soft and hard, has been demonstrated in various scientific reports. Stabilization splint therapy has been described as a well-established treatment for TMDs [10,11], bruxism [12,13], and headache [14]. Soft stabilization splints are effective in the symptomatic management of TMDs, especially for symptoms such as temporomandibular joint (TMJ) clicking, TMJ pain, and masticatory muscle pain [15]. Moreover, stabilization splint therapy may reduce pain severity at rest and on palpation in patients with temporomandibular myofascial pain [16]. Therefore, occlusal splint therapy seems to be an efficient treatment for TMD patients, as proved by several studies with a success rate of 70–90% [17]. In addition, stabilization splints are used to control bruxism, prevent tooth abrasion, and stimulate muscular relaxation [12].

On the other hand, the effectiveness of stabilization splints is controversial. A systematic review found that a hard stabilization splint does not appear more effective than a soft splint, a non-occluding palatal splint, or physical therapy for managing masticatory muscle pain [18]. Based on a randomized controlled trial, stabilization splint treatment in combination with counseling and masticatory muscle exercises does not offer any additional benefit in relieving masticatory muscle pain and increasing mandible mobility compared to counseling and masticatory muscle exercises alone over a short time interval [19]. In addition, the positive effect of a stabilization splint on signs and symptoms of TMDs could not be confirmed or refuted based on a systematic review [20]. Moreover, soft splints show some disadvantages. A soft occlusal splint can encourage muscle hyperactivity, and deteriorates more quickly than a hard splint [21,22]. It has been reported that occlusal stabilization splints increased surface electromyographic (sEMG) values during clenching activity [23,24]. Despite many studies in this area, the clinical effectiveness of soft stabilization splints remains uncertain [7,11,20]. Moreover, the way a splint affects the proportions of the activity of the temporalis muscle and the masseter muscle, and whether it affects the activity of the cervical spine muscles, have yet to be clarified. The way a splint affects the activity of antagonist muscles, both at rest and during tooth clenching, also requires explanation. Despite many questions and a lack of clear scientific evidence for the healing effect of soft stabilization splints in bruxism and TMDs management, they are widely used in clinical practice.

Therefore, this study investigates the influence of soft stabilization splints on electromyographic patterns in the masticatory and cervical spine muscles of healthy women. We decided to apply sEMG measurement because it is commonly used in dentistry to analyze the myoelectric signals of masticatory muscles [25]. Physiological variations in the state of muscle fiber membranes form myoelectric signals. Surface electrodes permit noninvasive measurement of bioelectrical phenomena of muscular activity [26]. The interpretation of sEMG records involves using electromyographic indices to increase the validity of electromyographic examination [27]. Using standardized and novel functional indices for masticatory and neck muscle activity, and the assessment of four muscle groups, this study aimed at a complete analysis of bioelectrical activity when a soft stabilization splint is applied. We assumed that the splint would affect the resting and functional activity of the masticatory muscles in functional, activity, and asymmetry indices. In addition, we assumed that applying the soft stabilization splint would affect the activity of the masticatory antagonistic muscles and the cervical spine muscles.

2. Materials and Methods

2.1. Study Population

Ethical approval was obtained from the Bioethical Committee of the Medical University of Lublin (KE-0254/81/2021). The objectives and methods of the study were fully explained to the participants, who provided written informed consent. The Strengthening the Reporting of Observational Studies in Epidemiology (STROBE) inventory was used to evaluate research quality [28]. The participants were 70 healthy young women (mean age 23.4 ± 2.2 years). Recruitment and measurements for the presented research were conducted at the Medical University of Lublin (Independent Unit of Functional Masticatory

Disorders) between November 2021 and July 2022. The inclusion criteria were the following: (a) female gender; (b) age between 18 and 35 years; and (c) absence of temporomandibular disorders (TMDs), which was assessed using the Research Diagnostic Criteria for Temporomandibular Disorders (RDC/TMD) protocol. Women with any of the following were excluded from the project: any temporomandibular disorders (e.g., temporomandibular joint pain, masticatory muscle pain, disc displacement, temporomandibular joint diseases), any pain condition within the stomatognathic system, fibromyalgia, regular headaches, Angle's Class II or III malocclusion, open bite, lack of at least four support zones in dental arches, lack of more than four teeth within both dental arches, periodontal diseases, orthodontic treatment, possession of dental prostheses, neurological disorders, history of Botulinum toxin therapy, or current pregnancy. Moreover, an ultrasound examination was conducted using an M-Turbo ultrasound device (SonoSite, Inc., Bothell, WA, USA) to assess the condition of subjects' temporomandibular joint and masticatory muscles.

2.2. Oral Appliance Design

The study protocol consisted of two phases: (1) non-splint measurement; (2) splint measurement. A random choice was performed for the initial measurement. The women completed two masticatory tasks (resting activity and tooth clenching) with and without soft stabilization splints between the teeth. Soft stabilization splints were fabricated directly in the women's mouths using Variotime® Easy Putty material [23,29]. Each splint was 4 mm thick, measuring between the upper and lower premolars. The base and catalyst of the silicone material were mixed by hand according to the manufacturer's directions and formed into a cylinder. The material was then placed onto the patient's lower arch covering every tooth. To ensure proper thickness, two 2-mm thick Fleximeter® Strips were placed on both sides of the cylinder in the premolar region. Then the patient was asked to close her mouth until the teeth touched the Fleximeter® Strips [10,23]. The excess of each splint was cut off using a scalpel to provide maximum comfort to the patient while maintaining a stable position for the mandibular and maxillary arches (Figure 1).

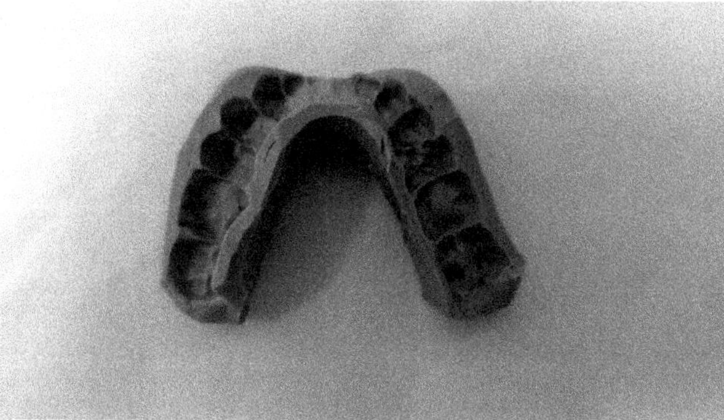

Figure 1. Soft stabilization splint example.

2.3. Electromyographic Examination

Resting (10 s) and maximum-clenching (3 times for 3 s, with 2 s break) bioelectric activity of the temporalis anterior (TA), superficial masseter (MM), anterior belly of the digastric muscle (DA), and middle part of the sternocleidomastoid muscle (SCM) was recorded using an 8-channel sEMG device (BioEMG III™, BioResearch Associates, Inc., Milwaukee, WI, USA). During electromyographic measurement, the participant sat in a dental chair with her head on the headrest and her torso perpendicular to the ground. Surface electrodes (Ag/AgCl, 30 mm diameter, 16 mm conductive surface, SORIMEX, Toruń, Poland) were placed bilaterally following the course of the fibers of the muscles, as previously described [30].

Microvolt values were amplified with minimal noise to 5000 times their original levels. Moreover, the sEMG values were reduced by 40 dB with a Noise Buster filter, eliminating 99% of 50/60 Hz sEMG noise. The electromyographic potentials based on root mean square (RMS) calculations were used to obtain the mean sEMG outcomes.

$$\text{Mean RMS TA (TA tot)} = (\text{TA-R} + \text{TA-L})/2 \tag{1}$$

$$\text{Mean RMS MM (MM tot)} = (\text{MM-R} + \text{MM-L})/2 \tag{2}$$

$$\text{Mean RMS SCM (SCM tot)} = (\text{SCM-R} + \text{SCM-L})/2 \tag{3}$$

$$\text{Mean RMS DA (DA tot)} = (\text{DA-R} + \text{DA-L})/2 \tag{4}$$

Functional Clenching Indices (FCI), Functional Clenching Activity Indices (FCAI), and Functional Clenching Symmetry Indices (FCSI) were used to normalize the mean bioelectric potentials. Indices were calculated based on mean RMS clenching (CL) and resting (REST) activity, according to our previous sEMG protocol [27]:

$$\text{FCI for TA right-sided (FCI TA-R)} = \text{CL TA-R}/\text{REST TA-R} \tag{5}$$

$$\text{FCI for TA left-sided (FCI TA-L)} = \text{CL TA-L}/\text{REST TA-L} \tag{6}$$

$$\text{FCI for TA both-sided (FCI TA tot)} = (\text{CL TA-R} + \text{CL TA-L})/(\text{REST TA-R} + \text{REST TA-L}) \tag{7}$$

$$\text{FCI for MM right-sided (FCI MM-R)} = \text{CL MM-R}/\text{REST MM-R} \tag{8}$$

$$\text{FCI for MM left-sided (FCI MM-L)} = \text{CL MM-L}/\text{REST MM-L} \tag{9}$$

$$\text{FCI for MM both-sided (FCI MM tot)} = (\text{CL MM-R} + \text{CL MM-L})/(\text{REST MM-R} + \text{REST MM-L}) \tag{10}$$

$$\text{FCI for SCM right-sided (FCI SCM-R)} = \text{CL SCM-R}/\text{REST SCM-R} \tag{11}$$

$$\text{FCI for SCM left-sided (FCI SCM-L)} = \text{CL SCM-L}/\text{REST SCM-L} \tag{12}$$

$$\text{FCI for SCM both-sided (FCI SCM tot)} = (\text{CL SCM-R} + \text{CL SCM-L})/(\text{REST SCM-R} + \text{REST SCM-L}) \tag{13}$$

$$\text{FCI for DA right-sided (FCI DA-R)} = \text{CL DA-R}/\text{REST DA-R} \tag{14}$$

$$\text{FCI for DA left-sided (FCI DA-L)} = \text{CL DA-L}/\text{REST DA-L} \tag{15}$$

$$\text{FCI for DA both-sided (FCI DA tot)} = (\text{CL DA-R} + \text{CL DA-L})/(\text{REST DA-R} + \text{REST DA-L}) \tag{16}$$

$$\text{FCAI right-sided (FCAI-R)} = (\text{FCI MM-R} - \text{FCI TA-R})/(\text{FCI MM-R} + \text{FCI TA-R}) \times 100 \tag{17}$$

$$\text{FCAI left-sided (FCAI-L)} = (\text{FCI MM-L} - \text{FCI TA-L})/(\text{FCI MM-L} + \text{FCI TA-L}) \times 100 \tag{18}$$

$$\text{FCAI both-sided (FCAI tot)} = (\text{FCI MM tot} - \text{FCI TA tot})/(\text{FCI MM tot} + \text{FCI TA tot}) \times 100 \tag{19}$$

$$\text{FCSI TA} = (\text{FCI TA-R} - \text{FCI TA-L})/(\text{FCI TA-R} + \text{FCI TA-L}) \times 100 \tag{20}$$

$$\text{FCSI MM} = (\text{FCI MM-R} - \text{FCI MM-L})/(\text{FCI MM-R} + \text{FCI MM-L}) \times 100 \tag{21}$$

$$\text{FCSI SCM} = (\text{FCI SCM-R} - \text{FCI SCM-L})/(\text{FCI SCM-R} + \text{FCI SCM-L}) \times 100 \tag{22}$$

$$\text{FCSI DA} = (\text{FCI DA-R} - \text{FCI DA-L})/(\text{FCI DA-R} + \text{FCI DA-L}) \times 100 \tag{23}$$

The following formulas were used to calculate activity (ACI) and asymmetry (ASI) indices based on the mean RMS potentials recorded during resting and functional activity, as specified by Naeije et al. and Ferrario et al. [31,32]:

$$\text{ACI right-sided (ACI-R)} = (\text{MM-R} - \text{TA-R})/(\text{MM-R} + \text{TA-R}) \times 100 \tag{24}$$

$$\text{ACI left-sided (ACI-L)} = (\text{MM-L} - \text{TA-L})/(\text{MM-L} + \text{TA-L}) \times 100 \tag{25}$$

$$\text{ACI both-sided (ACI tot)} = (\text{MM-R} + \text{MM-L} - \text{TA-R} - \text{TA-L})/(\text{MM-R} + \text{MM-L} + \text{TA-R} + \text{TA-L}) \times 100 \quad (26)$$

$$\text{ASI TA} = (\text{TA-R} - \text{TA-L})/(\text{TA-R} + \text{TA-L}) \times 100 \quad (27)$$

$$\text{ASI MM} = (\text{MM-R} - \text{MM-L})/(\text{MM-R} + \text{MM-L}) \times 100 \quad (28)$$

$$\text{ASI SCM} = (\text{SCM-R} - \text{SCM-L})/(\text{SCM-R} + \text{SCM-L}) \times 100 \quad (29)$$

$$\text{ASI DA} = (\text{DA-R} - \text{DA-L})/(\text{DA-R} + \text{DA-L}) \times 100 \quad (30)$$

2.4. Statistical Calculations

The repeatability of the sEMG procedure was verified with duplicate sEMG examinations on 10 participants, as previously reported [33]. An analysis of power was conducted using the G*Power 3.1.9.7 program (Heinrich Heine University Düsseldorf, Germany) [34]. The sample size was calculated based on previous studies [27]. The calculations indicated that a sample size of 68 participants would be sufficient to notice a significant difference between two independent means (t-test) with an α value of 0.05, a power value of 0.90, and an estimated medium effect size of 0.56.

The data comparison was performed using the GraphPad Prism 9.4.1 program (GraphPad Software, Inc., San Diego, CA, USA). The normality of the distribution of variables was verified using the Shapiro–Wilk test and the Kolmogorov–Smirnov test (with the Lilliefors correction). The Student's t-test (T) or Mann–Whitney U test (Z) was used to compare the differences between groups. The results were presented in the form of minimum (Min), maximum (Max), mean, and standard deviation values (SD). Effect sizes were determined for the t-test using the Cohen d method and interpreted as small (0.2), medium (0.5), or large (0.8) effect sizes [35,36]. A confidence interval (CI 95%) was calculated for results at a level of 95% [37]. Statistical significance was set at $p \leq 0.05$.

3. Results

Based on the study criteria, 70 women were qualified for the research. The participants' general characteristics, including age, height, weight, and mandibular range of motion, are presented in Table 1.

Table 1. Participants' general characteristics, including age, height, weight, and mandibular range of motion.

Variable	Min.	Max.	Mean	SD
Age (years)	19.00	34.00	23.43	2.27
Height (cm)	155.00	183.00	167.93	6.90
Weight (kg)	44.00	84.50	59.94	8.32
Active maximum mouth opening (mm)	32.00	62.00	49.77	5.90
Passive maximum mouth opening (mm)	35.00	65.00	52.49	5.83
Active laterotrusion right (mm)	0.00	13.00	9.51	2.17
Active laterotrusion left (mm)	0.00	15.00	9.99	2.35
Active protrusion (mm)	3.00	14.00	9.11	2.38

Min.—minimum; Max.—maximum; SD—standard deviation.

The interaction between splint application and resting muscle activity affected the results within all examined muscles except the temporalis muscle (Table 2). A large effect size was observed in masseter (2.19 µV vs. 5.18 µV; $p = 0.00$; ES = 1.00) and digastric (1.89 µV vs. 3.17 µV; $p = 0.00$; ES = 1.00) both-sided RMS activity. In all cases, the splint application caused a significant increase in the resting RMS activity of the examined muscle groups. However, significant differences were observed only for the masseter and sternocleidomastoid muscle during clenching activity, with a small effect size (Table 3). In these measurements, an increase in the functional activity of the examined muscles was also observed after placing the stabilization splint.

Table 2. Comparison of root mean square (RMS) bioelectric resting potentials with and without stabilization splint.

Muscle	Resting RMS Values without Stabilization Splint (µV)				Resting RMS Values with Stabilization Splint (µV)				Test	*p*	ES	CI 95%	
	Min.	Max.	Mean	SD	Min.	Max.	Mean	SD					
TA-R	0.85	10.54	2.46	1.59	0.59	19.42	3.26	2.97	Z	−1.63	0.10	0.16	−0.08	0.75
TA-L	0.80	6.82	2.56	1.46	0.64	13.54	3.12	2.44	Z	−1.02	0.31	0.10	−0.20	0.71
TA tot	0.83	7.39	2.51	1.26	0.66	15.34	3.19	2.58	Z	−1.27	0.20	0.12	−0.14	0.71
MM-R	0.71	5.73	2.11	1.09	0.77	30.30	5.41	5.65	Z	−5.38	0.00 *	0.53	0.90	2.14
MM-L	0.71	7.72	2.27	1.25	0.83	22.28	4.95	4.45	Z	−4.24	0.00 *	0.42	0.61	1.98
MM tot	0.78	5.19	2.19	1.05	0.80	26.25	5.18	4.83	Z	−5.06	0.00 *	1.00	0.83	2.13
SCM-R	0.68	3.34	1.20	0.41	0.90	6.89	1.70	1.00	Z	−4.34	0.00 *	0.43	0.16	0.43
SCM-L	0.59	2.65	1.34	0.42	0.78	7.42	1.75	0.92	Z	−3.64	0.00 *	0.36	0.13	0.44
SCM tot	0.76	2.71	1.27	0.36	0.91	5.48	1.73	0.84	Z	−4.17	0.00 *	0.76	0.18	0.44
DA-R	0.86	6.46	1.92	0.99	1.14	9.08	3.12	1.56	Z	−5.83	0.00 *	0.57	0.67	1.32
DA-L	0.92	7.12	1.86	1.00	1.16	10.80	3.22	1.73	Z	−6.24	0.00 *	0.61	0.74	1.46
DA tot	0.90	6.79	1.89	0.98	1.23	9.94	3.17	1.57	Z	−5.79	0.00 *	1.00	0.76	1.44

RMS—root mean square; Z—Mann–Whitney U test; ES—effect size; CI—confidence interval; Min.—minimum; Max.—maximum; SD—standard deviation; TA—temporalis muscle; MM—masseter muscle; SCM—sternocleidomastoid muscle; DA—digastric muscle; R—right side; L—left side; tot—both-sided; µV—microvolt; *—significant difference.

Table 3. Comparison of root mean square (RMS) functional bioelectric potentials during maximal tooth clenching with and without stabilization splint.

Muscle	Clenching RMS Values without Stabilization Splint (µV)				Clenching RMS Values with Stabilization Splint (µV)				Test	*p*	ES	CI 95%	
	Min.	Max.	Mean	SD	Min.	Max.	Mean	SD					
TA-R	14.70	403.80	145.32	84.67	12.20	745.70	149.41	99.99	Z	−0.28	0.78	0.03	−21.40	25.70
TA-L	12.40	319.00	138.70	68.30	5.90	287.50	137.14	59.82	Z	0.14	0.89	0.02	−20.30	23.20
TA tot	22.60	361.40	142.01	73.42	9.05	411.85	143.27	69.99	T	−0.33	0.74	0.03	−18.85	25.50
MM-R	13.70	419.00	149.56	90.64	29.60	387.90	182.56	86.87	Z	−2.58	0.01 *	0.25	9.30	64.20
MM-L	5.10	526.10	147.99	99.49	26.30	463.20	178.23	94.73	Z	−2.15	0.03 *	0.21	3.70	63.10
MM tot	9.40	441.25	148.78	91.75	27.95	391.85	180.40	86.95	Z	−2.09	0.04 *	0.35	7.70	63.95
SCM-R	1.60	42.50	10.43	7.78	2.30	76.40	12.81	9.39	Z	−2.53	0.01 *	0.24	0.60	4.20
SCM-L	1.40	40.70	10.15	7.91	2.50	193.00	15.08	23.97	Z	−2.47	0.01 *	0.27	0.50	4.20
SCM tot	1.50	41.60	10.29	7.49	2.40	106.55	13.94	14.78	Z	−2.72	0.01 *	0.03	0.70	4.20
DA-R	4.10	66.50	21.74	14.25	4.20	58.00	22.52	11.43	Z	−1.12	0.26	0.09	−1.50	5.50
DA-L	4.60	106.00	22.88	18.63	5.80	91.80	23.59	15.97	Z	−0.90	0.37	0.10	−1.80	5.00
DA tot	4.70	77.05	22.31	14.64	5.00	74.90	23.05	12.76	Z	−1.01	0.31	0.03	−1.65	5.00

RMS—root mean square; Z—Mann–Whitney U test; ES—effect size; CI—confidence interval; Min.—minimum; Max.—maximum; SD—standard deviation; TA—temporalis muscle; MM—masseter muscle; SCM—sternocleidomastoid muscle; DA—digastric muscle; R—right side; L—left side; tot—both-sided; µV—microvolt; *—significant difference.

Table 4 shows significant differences between the two conditions in all functional clenching indices for masseter, sternocleidomastoid, and digastric muscles. All FCI values for the masseter (small effect size) and digastric muscles (large effect size) were significantly lower with than without the splint. The opposite tendency was determined in the sternocleidomastoid muscle, where splint application caused an increase in functional clenching indices with large effect sizes in the right and left muscle groups. We also observed a significant decrease in functional clenching activity indices for the right and both-sided values with small effect sizes. Significant differences between the two measurements were observed in all activity indices in resting and clenching conditions with small to medium effect sizes (Table 5). In all cases we observed increased activity indices due to splint application, which suggests a masseter muscle advantage during measurement.

Table 4. Comparison of functional indices with and without stabilization splint.

Indices	Without Stabilization Splint				With Stabilization Splint				Test	*p*	ES	CI 95%	
	Min.	Max.	Mean	SD	Min.	Max.	Mean	SD					
FCI TA-R	4.44	281.80	73.00	51.78	6.74	388.47	67.03	57.52	Z	0.99	0.32	0.10	−19.37	6.29
FCI TA-L	7.87	230.45	70.30	47.74	3.92	280.00	70.41	58.15	Z	0.62	0.53	0.06	−18.68	10.77
FCI tot	8.68	204.60	66.97	42.44	5.41	294.00	65.59	51.18	Z	0.17	0.86	0.03	−16.64	7.89
FCI MM-R	8.09	524.46	90.97	84.91	3.39	239.20	63.92	58.41	Z	2.59	0.01 *	0.25	−33.11	−4.48
FCI MM-L	3.03	395.56	84.64	75.72	3.66	449.71	67.47	73.46	Z	2.08	0.04 *	0.20	−31.31	−1.00
FCI MM tot	5.01	407.87	85.47	75.38	3.77	303.76	61.94	57.68	Z	2.44	0.01 *	0.24	−31.23	−3.05
FCI SCM-R	−24.72	74.01	0.54	10.86	1.85	41.98	8.70	6.35	Z	−9.15	0.00 *	0.94	6.83	8.91
FCI SCM-L	−80.91	24.83	−1.10	10.65	1.45	94.61	8.98	11.54	Z	−9.58	0.00 *	0.96	5.74	8.33
FCI SCM tot	−54.62	10.83	−0.96	7.60	1.62	57.13	8.77	7.89	Z	−9.76	0.00 *	0.25	6.52	8.26
FCI DA-R	1.83	45.50	13.09	9.77	0.01	0.56	0.18	0.10	Z	10.21	0.00 *	1.00	−12.74	−8.51
FCI DA-L	1.97	82.81	13.93	13.50	0.05	1.51	0.19	0.18	Z	10.21	0.00 *	1.00	−11.14	−8.64
FCI DA tot	1.91	59.96	13.48	10.42	0.03	0.86	0.18	0.11	Z	10.21	0.00 *	1.00	−12.17	−9.55
FCSI TA	−86.57	66.08	2.63	30.73	−47.26	82.69	−0.17	24.30	T	0.68	0.50	0.07	−12.06	6.47
FCSI MM	−46.28	63.44	5.59	25.66	−58.70	68.68	−0.76	24.42	Z	1.50	0.14	0.25	−13.55	4.35
FCSI SCM	−1048.60	11,407.25	240.19	1439.96	−77.67	44.46	3.82	21.27	Z	2.92	0.00 *	0.11	−60.62	−12.75
FCSI DA	−87.75	35.52	−1.00	21.27	−85.42	39.99	−1.29	24.43	T	0.08	0.94	0.01	−7.95	7.36
FCAI-R	−67.24	81.20	5.22	34.23	−81.87	71.72	−6.92	34.47	T	2.09	0.04 *	0.35	−25.61	−0.90
FCAI-L	−72.04	68.19	2.36	36.68	−68.18	63.36	−5.87	33.47	T	1.39	0.17	0.23	−21.03	3.60
FCAI tot	−63.98	73.90	4.57	31.66	−70.17	63.09	−7.08	30.90	T	2.22	0.03 *	0.22	−22.10	−1.19

Z—Mann–Whitney U test; T—Student's *t*-test; ES—effect size; CI—confidence interval; Min.—minimum; Max.—maximum; SD—standard deviation; FCI—Functional Clenching Index; FCSI—Functional Clenching Symmetry Index; FCAI—Functional Clenching Activity Index; TA—temporalis muscle; MM—masseter muscle; SCM—sternocleidomastoid muscle; DA—digastric muscle; R—right side; L—left side; tot—both-sided; *—significant difference.

Table 5. Comparison of asymmetry and activity indices at rest and during maximal tooth clenching with and without stabilization splint.

Activity	Indices	Without Stabilization Splint				With Stabilization Splint				Test	*p*	ES	CI 95%	
		Min.	Max.	Mean	SD	Min.	Max.	Mean	SD					
Rest	ASI TA	−64.44	54.61	−1.86	25.92	−45.30	53.56	1.56	21.83	Z	−0.71	0.48	0.07	−5.31	11.16
	ASI MM	−60.83	35.71	−2.76	17.75	−69.71	60.03	3.04	24.02	T	−1.62	0.11	0.28	−1.26	12.86
	ASI SCM	−37.69	33.33	−5.48	13.82	−35.40	46.97	−2.55	17.04	T	−0.79	0.43	0.08	−2.26	8.11
	ASI DA	−18.22	28.70	1.13	9.66	−37.86	37.13	−0.89	13.99	T	0.75	0.45	0.07	−6.04	2.00
	ACI-R	−77.89	68.06	−5.14	31.65	−49.32	80.16	16.81	33.47	T	−3.99	0.00 *	0.67	11.07	32.85
	ACI-L	−76.70	53.92	−4.48	32.51	−54.25	80.43	15.48	33.49	T	−3.58	0.00 *	0.60	8.93	30.99
	ACI tot	−71.79	58.97	−5.82	29.78	−38.96	78.43	16.86	29.80	T	−3.95	0.00 *	0.39	12.73	32.64
Clenching	ASI TA	−74.21	69.31	0.50	18.49	−29.00	81.06	1.47	16.11	Z	−0.10	0.92	0.01	−4.52	5.15
	ASI MM	−32.62	45.74	3.15	16.78	−25.95	32.04	2.12	14.45	T	0.39	0.70	0.07	−6.26	4.20
	ASI SCM	−47.22	47.95	2.28	18.35	−81.14	51.49	1.44	21.23	Z	0.26	0.80	0.91	−6.41	5.54
	ASI DA	−88.11	35.09	0.07	21.73	−33.75	36.90	−0.01	19.41	Z	0.31	0.76	1.00	−7.87	5.85
	ACI-R	−47.73	79.27	−0.11	24.53	−84.60	55.88	11.79	24.14	T	−2.89	0.00 *	0.49	3.77	20.04
	ACI-L	−68.62	77.06	−2.64	24.34	−33.77	79.76	10.97	19.59	T	−3.65	0.00 *	0.62	6.23	21.00
	ACI tot	−44.63	51.76	−1.45	20.87	−72.57	67.71	11.27	20.63	T	−3.75	0.00 *	0.37	5.78	19.65

Z—Mann–Whitney U test; T—Student's *t*-test; ES—effect size; CI—confidence interval; Min.—minimum; Max.—maximum; SD—standard deviation; ASI—Asymmetry Index; ACI—Activity Index; TA—temporalis muscle; MM— masseter muscle; SCM—sternocleidomastoid muscle; DA—digastric muscle; R—right side; L—left side; tot—both-sided; *—significant difference.

4. Discussion

This investigation aimed to examine the influence of soft stabilization splints on electromyographic patterns in masticatory and neck muscles in healthy women. We assumed that the splint would affect resting and functional activity in the masticatory muscles. In addition, we thought that applying the soft stabilization splint would affect the activity of the masticatory antagonistic muscles and the cervical spine muscles. Using hard stabilization splints seems more reasonable than soft splints for TMDs and bruxism. In several reports, hard and soft stabilization splints effectively treat TMDs. However, hard stabilization splints provide a quicker decrease in TMD symptoms [15]. Moreover, compared to the absence of a splint, hard splints produced less strain on molar teeth but more strain on premolar teeth during maximum voluntary tooth clenching. In contrast, soft occlusal splints did not reduce the strain on all target teeth significantly during clenching tasks [38]. Hard stabilization splints significantly reduced the number of sEMG high-activity events per hour of sleep. In contrast, soft occlusal splints do not inhibit jaw muscle activity compared to baseline values [39]. Despite the advantage of a hard splint, soft splints are used to ease implementation and reduce financial costs to the patient [7]. However, the effect of using soft splints on masticatory muscle activity has yet to be unequivocally evaluated in current studies.

This study showed increased functional activity during tooth clenching in splint condition and increased resting activity of masticatory muscles during splint application within RMS values. The increased resting electromyographic RMS activity was significant in the masseter, digastric, and sternocleidomastoid muscle groups. The increase in the RMS values was also detectable in the temporalis muscle at rest, but the differences were not statistically significant. This may be related to lower reaction of the temporalis muscle to the mandibular position change compared to the other muscle groups. Applying the splint did not change the asymmetry of the masticatory muscles at rest in ASI indices. The Gholampour et al. study showed a considerable symmetry effect in splint condition [13]. In the abovementioned investigation, the splint allowed for asymmetric and non-uniform loading. However, the splints were produced using a hard polymerized colorless acrylic resin, compared to the silicone material used in our investigation. Splint application significantly affected the ratio of the temporalis muscle's involvement with the masseter muscle in ACI indices. In resting activity and during tooth clenching, a soft stabilization splint changed the involvement proportions of the temporalis and masseter muscles, transferring the main activity to the masseter muscles. The functional RMS activity during tooth clenching with the soft splint was significantly higher in the masseter and the sternocleidomastoid muscle than without the splint. This may indicate the need for greater stabilization of the cervical spine during clenching due to the phenomenon of co-contraction [40]. The results of the Akat et al. study identified significant differences in sEMG parameters with hard and soft occlusal splints after three months of treatment [12]. In the study, sEMG activity decreased with all splint types, most prominently in the hard occlusal splint group. The differences in our observations may result from a different population (subjects with diagnosed bruxism, mixed population vs. healthy young women), a more extended observation period (3 months vs. immediate effect), and methodological assumptions of the sEMG signal analysis (raw sEMG signal vs. sEMG normalization). Our experiment used RMS data, validated ASI and ACI indices, and Functional Indices for the electromyographic assessment. We observed a significant decrease for all Functional Clenching Indices and Functional Clenching Activity Indices in the masseter, sternocleidomastoid, and digastric muscle groups. This decrease may indicate a disturbed proportion between the resting and functional activity of the examined muscles [27]. However, the soft stabilization splint did not affect the results in terms of asymmetry indices in Functional Clenching Symmetry Indices. Therefore, the proportions of masticatory muscle involvement at rest and during activity on the left and right sides do not seem to change under stabilization splint application.

The presented research has several limitations that should be addressed in future investigations. Firstly, our results are limited by the immediate follow-up. Secondly, in our study, each splint was 4 mm thick, measuring between the upper and lower premolars. According to the current report, 2 mm and 4 mm splints effectively treat muscle disorders and disc displacements, especially for masticatory muscle pain and TMJ acoustic symptoms [10]. Therefore, we decided on a standard size for all patients to standardize the measurement results. In temporomandibular disorder therapy, a splint that is individually adjusted to the current resting position of the mandible and minimal resting activity of the masticatory muscles should be used. Thirdly, we tested our splint on healthy adult women. It was estimated that TMDs affect women more often than men [41,42]. Moreover, we standardized our study group to eliminate the influence on the results of the experiment of gender, age, and changes in muscle activity patterns in response to pain. Therefore, future research should be performed in patients with stomatognathic system disorders, e.g., TMDs, or bruxism populations.

Finally, we should stress that we are not claiming that a soft stabilization splint is ineffective for treating TMDs and bruxism. In this paper, we have reported the results of the immediate effect of soft splint use on sEMG muscle activity. However, our study indicates the consideration of the adverse effects of long-term use of soft stabilization splints in the treatment of masticatory system disorders, as muscle hyperactivity and changes in electromyographic patterns over a more extended period may cause adverse effects within the stomatognathic system. Therefore, the long-term impact of soft stabilization splint use on masticatory and neck muscle activity requires further research. Moreover, additional objective measurement methods, e.g., computer simulations, can be used to evaluate splint efficiency in further studies [43].

5. Conclusions

Soft stabilization splint use influences resting and functional activity within the MM, SCM, and DA muscles. During tooth clenching activity, a soft stabilization splint changes the involvement proportions of the temporalis and masseter muscles, transferring the main activity to the masseter muscles. Using a soft stabilization splint does not affect the symmetry of the electromyographic activity of the masticatory and neck muscles.

Author Contributions: Conceptualization, M.G. and G.Z.; methodology, M.G., G.Z., J.S. and I.R.-K.; formal analysis, M.G. and G.Z.; investigation, M.G., G.Z., J.S., M.W., M.B., A.Ż. and M.L.-R.; resources, M.G., G.Z. and I.R.-K.; data curation, M.G. and G.Z.; writing—original draft preparation, M.G. and G.Z.; supervision, M.G.; project administration, M.G. funding acquisition, M.G. All authors have read and agreed to the published version of the manuscript.

Funding: This research received no external funding.

Institutional Review Board Statement: The study was conducted according to the Declaration of Helsinki guidelines and approved by the Bioethics Committee of the Medical University of Lublin (KE-0254/81/2021).

Informed Consent Statement: Informed consent was obtained from all subjects involved in the study. Written informed consent has been obtained from the patients to publish this paper.

Data Availability Statement: The data presented in this study are available upon request from the corresponding author.

Conflicts of Interest: The authors declare no conflict of interest.

References

1. Palmer, J.; Durham, J. Temporomandibular disorders. *BJA Educ.* **2021**, *21*, 44–50. [CrossRef]
2. Slade, G.D.; Fillingim, R.B.; Sanders, A.E.; Bair, E.; Greenspan, J.D.; Ohrbach, R.; Dubner, R.; Diatchenko, L.; Smith, S.B.; Knott, C.; et al. Summary of Findings From the OPPERA Prospective Cohort Study of Incidence of First-Onset Temporomandibular Disorder: Implications and Future Directions. *J. Pain* **2013**, *14*, T116–T124. [CrossRef]
3. Gauer, R.L.; Semidey, M.J. Diagnosis and treatment of temporomandibular disorders. *Am. Fam. Physician* **2015**, *91*, 378–386.
4. Ginszt, M.; Zieliński, G.; Berger, M.; Szkutnik, J.; Bakalczuk, M.; Majcher, P. Acute Effect of the Compression Technique on the Electromyographic Activity of the Masticatory Muscles and Mouth Opening in Subjects with Active Myofascial Trigger Points. *Appl. Sci.* **2020**, *10*, 7750. [CrossRef]
5. Xu, L.; Cai, B.; Lu, S.; Fan, S.; Dai, K. The Impact of Education and Physical Therapy on Oral Behaviour in Patients with Temporomandibular Disorder: A Preliminary Study. *BioMed Res. Int.* **2021**, *2021*, 6666680. [CrossRef]
6. Litt, M.D.; Shafer, D.M.; Kreutzer, D.L. Brief cognitive-behavioral treatment for TMD pain: Long-term outcomes and moderators of treatment. *Pain* **2010**, *151*, 110–116. [CrossRef] [PubMed]
7. Riley, P.; Glenny, A.-M.; Worthington, H.V.; Jacobsen, E.; Robertson, C.; Durham, J.; Davies, S.; Petersen, H.; Boyers, D. Oral splints for patients with temporomandibular disorders or bruxism: A systematic review and economic evaluation. *Health Technol. Assess.* **2020**, *24*, 1–224. [CrossRef] [PubMed]
8. Lukic, N.; Saxer, T.; Hou, M.-Y.; Zumbrunn Wojczyńska, A.; Gallo, L.M.; Colombo, V. Short-term effects of NTI-tss and Michigan splint on nocturnal jaw muscle activity: A pilot study. *Clin. Exp. Dent. Res.* **2021**, *7*, 323–330. [CrossRef] [PubMed]
9. Albagieh, H.; Alomran, I.; Binakresh, A.; Alhatarisha, N.; Almeteb, M.; Khalaf, Y.; Alqublan, A.; Alqahatany, M. Occlusal splints-types and effectiveness in temporomandibular disorder management. *Saudi Dent. J.* **2023**, *35*, 70–79. [CrossRef] [PubMed]
10. Bilir, H.; Kurt, H. Influence of Stabilization Splint Thickness on Temporomandibular Disorders. *Int. J. Prosthodont.* **2022**, *35*, 163–173. [CrossRef]
11. Zhang, C.; Wu, J.-Y.; Deng, D.-L.; He, B.-Y.; Tao, Y.; Niu, Y.-M.; Deng, M.-H. Efficacy of splint therapy for the management of temporomandibular disorders: A meta-analysis. *Oncotarget* **2016**, *7*, 84043–84053. [CrossRef]

12. Akat, B.; Görür, S.A.; Bayrak, A.; Eren, H.; Eres, N.; Erkcan, Y.; Kılıçarslan, M.A.; Orhan, K. Ultrasonographic and electromyographic evaluation of three types of occlusal splints on masticatory muscle activity, thickness, and length in patients with bruxism. *Cranio* **2023**, *41*, 59–68. [CrossRef] [PubMed]

13. Gholampour, S.; Gholampour, H.; Khanmohammadi, H. Finite element analysis of occlusal splint therapy in patients with bruxism. *BMC Oral Health* **2019**, *19*, 205. [CrossRef]

14. Manrriquez, S.L.; Robles, K.; Pareek, K.; Besharati, A.; Enciso, R. Reduction of headache intensity and frequency with maxillary stabilization splint therapy in patients with temporomandibular disorders-headache comorbidity: A systematic review and meta-analysis. *J. Dent. Anesth. Pain Med.* **2021**, *21*, 183–205. [CrossRef]

15. Anish Poorna, T.; John, B.; Joshna, E.K.; Rao, A. Comparison of the effectiveness of soft and hard splints in the symptomatic management of temporomandibular joint disorders: A randomized control study. *Int. J. Rheum. Dis.* **2022**, *25*, 1053–1059. [CrossRef]

16. Al-Ani, Z.; Gray, R.J.; Davies, S.J.; Sloan, P.; Glenny, A.-M. Stabilization Splint Therapy for the Treatment of Temporomandibular Myofascial Pain: A Systematic Review. *J. Dent. Educ.* **2005**, *69*, 1242–1250. [CrossRef] [PubMed]

17. Littner, D.; Perlman-Emodi, A.; Vinocuor, E. Efficacy of treatment with hard and soft occlusal appliance in TMD. *Refu'at Ha-peh Veha-shinayim* **2004**, *21*, 52–58.

18. Türp, J.C.; Komine, F.; Hugger, A. Efficacy of stabilization splints for the management of patients with masticatory muscle pain: A qualitative systematic review. *Clin. Oral Investig.* **2004**, *8*, 179–195. [CrossRef] [PubMed]

19. Niemelä, K.; Korpela, M.; Raustia, A.; Ylöstalo, P.; Sipilä, K. Efficacy of stabilisation splint treatment on temporomandibular disorders. *J. Oral Rehabil.* **2012**, *39*, 799–804. [CrossRef]

20. Honnef, L.R.; Pauletto, P.; Conti Réus, J.; Massignan, C.; de Souza, B.D.M.; Michelotti, A.; Flores-Mir, C.; De Luca Canto, G. Effects of stabilization splints on the signs and symptoms of temporomandibular disorders of muscular origin: A systematic review. *Cranio* **2022**, 1–12. [CrossRef]

21. Okeson, J.P. The effects of hard and soft occlusal splints on nocturnal bruxism. *J. Am. Dent. Assoc.* **1987**, *114*, 788–791. [CrossRef] [PubMed]

22. Ré, J.-P.; Perez, C.; Darmouni, L.; Carlier, J.F.; Orthlieb, J.-D. The occlusal splint therapy. *Int. J. Stomatol. Occlusion Med.* **2009**, *2*, 82–86. [CrossRef]

23. Berger, M.; Ginszt, M.; Suwała, M.; Szkutnik, J.; Gawda, P.; Ginszt, A.; Tarkowski, Z. The immediate effect of temporary silicone splint application on symmetry of masticatory muscle activity evaluated using surface electromyography. *Pol. Ann. Med.* **2017**, *24*, 19–23. [CrossRef]

24. Cruz-Reyes, R.A.; Martínez-Aragón, I.; Guerrero-Arias, R.E.; García-Zura, D.A.; González-Sánchez, L.E. Influence of occlusal stabilization splints and soft occlusal splints on the electromyographic pattern, in basal state and at the end of six weeks treatment in patients with bruxism. *Acta Odontol. Latinoam.* **2011**, *24*, 66–74.

25. Nishi, S.E.; Basri, R.; Alam, M.K. Uses of electromyography in dentistry: An overview with meta-analysis. *Eur. J. Dent.* **2016**, *10*, 419–425. [CrossRef] [PubMed]

26. Nazmi, N.; Rahman, M.A.A.; Yamamoto, S.-I.; Ahmad, S.A.; Zamzuri, H.; Mazlan, S.A. A Review of Classification Techniques of EMG Signals during Isotonic and Isometric Contractions. *Sensors* **2016**, *16*, 1304. [CrossRef]

27. Ginszt, M.; Zieliński, G. Novel Functional Indices of Masticatory Muscle Activity. *J. Clin. Med.* **2021**, *10*, 1440. [CrossRef]

28. Von Elm, E.; Altman, D.G.; Egger, M.; Pocock, S.J.; Gøtzsche, P.C.; Vandenbroucke, J.P. The Strengthening the Reporting of Observational Studies in Epidemiology (STROBE) statement: Guidelines for reporting observational studies. *J. Clin. Epidemiol.* **2008**, *61*, 344–349. [CrossRef] [PubMed]

29. Botelho, A.L.; Silva, B.C.; Gentil, F.H.; Sforza, C.; Da Silva, M.A.M. Immediate Effect of the Resilient Splint Evaluated Using Surface Electromyography in Patients with TMD. *Cranio* **2010**, *28*, 266–273. [CrossRef]

30. Ginszt, M.; Zieliński, G.; Szkutnik, J.; Wójcicki, M.; Baszczowski, M.; Litko-Rola, M.; Rózyło-Kalinowska, I.; Majcher, P. The Effects of Wearing a Medical Mask on the Masticatory and Neck Muscle Activity in Healthy Young Women. *J. Clin. Med.* **2022**, *11*, 303. [CrossRef]

31. Naeije, M.; McCarroll, R.S.; Weijs, W.A. Electromyographic activity of the human masticatory muscles during submaximal clenching in the inter-cuspal position. *J. Oral Rehabil.* **1989**, *16*, 63–70. [CrossRef]

32. Ferrario, V.F.; Sforza, C.; Miani, A., Jr.; D'Addona, A.; Barbini, E. Electromyographic activity of human masticatory muscles in normal young people. Statistical evaluation of reference values for clinical applications. *J. Oral Rehabil.* **1993**, *20*, 271–280. [CrossRef]

33. Ginszt, M.; Szkutnik, J.; Zieliński, G.; Bakalczuk, M.; Stodółkiewicz, M.; Litko-Rola, M.; Ginszt, A.; Rahnama, M.; Majcher, P. Cervical Myofascial Pain Is Associated with an Imbalance of Masticatory Muscle Activity. *Int. J. Environ. Res. Public Health* **2022**, *19*, 1577. [CrossRef]

34. Faul, F.; Erdfelder, E.; Lang, A.-G.; Buchner, A. G*Power 3: A flexible statistical power analysis program for the social, behavioral, and biomedical sciences. *Behav. Res. Methods* **2007**, *39*, 175–191. [CrossRef] [PubMed]

35. Lakens, D. Calculating and reporting effect sizes to facilitate cumulative science: A practical primer for t-tests and ANOVAs. *Front. Psychol.* **2013**, *4*, 863. [CrossRef] [PubMed]

36. Fritz, C.O.; Morris, P.E.; Richler, J.J. Effect size estimates: Current use, calculations, and interpretation. *J. Exp. Psychol. Gen.* **2012**, *141*, 2–18. [CrossRef] [PubMed]

37. Hazra, A. Using the confidence interval confidently. *J. Thorac. Dis.* **2017**, *9*, 4125–4130. [CrossRef]
38. Tanaka, Y.; Yoshida, T.; Ono, Y.; Maeda, Y. The effect of occlusal splints on the mechanical stress on teeth as measured by intraoral sensors. *J. Oral Sci.* **2021**, *63*, 41–45. [CrossRef]
39. Arima, T.; Takeuchi, T.; Tomonaga, A.; Yachida, W.; Ohata, N.; Svensson, P. Choice of biomaterials—Do soft occlusal splints influence jaw-muscle activity during sleep? A preliminary report. *Appl. Surf. Sci.* **2012**, *262*, 159–162. [CrossRef]
40. Giannakopoulos, N.N.; Schindler, H.J.; Rammelsberg, P.; Eberhard, L.; Schmitter, M.; Hellmann, D. Co-activation of jaw and neck muscles during submaximum clenching in the supine position. *Arch. Oral Biol.* **2013**, *58*, 1751–1760. [CrossRef]
41. Zieliński, G.; Byś, A.; Szkutnik, J.; Majcher, P.; Ginszt, M. Electromyographic Patterns of Masticatory Muscles in Relation to Active Myofascial Trigger Points of the Upper Trapezius and Temporomandibular Disorders. *Diagnostics* **2021**, *11*, 580. [CrossRef] [PubMed]
42. Slade, G.D.; Bair, E.; Greenspan, J.D.; Dubner, R.; Fillingim, R.B.; Diatchenko, L.; Maixner, W.; Knott, C.; Ohrbach, R. Signs and Symptoms of First-Onset TMD and Sociodemographic Predictors of Its Development: The OPPERA Prospective Cohort Study. *J. Pain* **2013**, *14*, T20–T32.e3. [CrossRef] [PubMed]
43. Gholampour, S.; Jalali, A. Thermal analysis of the dentine tubule under hot and cold stimuli using fluid–structure interaction simulation. *Biomech. Model. Mechanobiol.* **2018**, *17*, 1599–1610. [CrossRef] [PubMed]

Journal of
Clinical Medicine

Systematic Review

Analysis of Mandibular Muscle Variations Following Condylar Fractures: A Systematic Review

Francesco Inchingolo [1,*,†], Assunta Patano [1,†], Angelo Michele Inchingolo [1], Lilla Riccaldo [1,*], Roberta Morolla [1], Anna Netti [1], Daniela Azzollini [1], Alessio Danilo Inchingolo [1], Andrea Palermo [2], Alessandra Lucchese [3,4], Daniela Di Venere [1] and Gianna Dipalma [1]

1   Department of Interdisciplinary Medicine, University of Bari "Aldo Moro", 70121 Bari, Italy;
    assuntapatano@gmail.com (A.P.); angeloinchingolo@gmail.com (A.M.I.); robertamorolla@gmail.com (R.M.);
    annanetti@inwind.it (A.N.); daniela.azzollini93@gmail.com (D.A.); ad.inchingolo@libero.it (A.D.I.);
    daniela.divenere@uniba.it (D.D.V.); giannadipalma@tiscali.it (G.D.)
2   College of Medicine and Dentistry, Birmingham B4 6BN, UK; andrea.palermo2004@libero.it
3   Unit of Dentistry-Orthodontics, IRCCS San Raffaele Scientific Institute, 20132 Milan, Italy;
    lucchese.alessandra@unisr.it
4   Unit of Dentistry, Research Center for Oral Pathology and Implantology, IRCCS San Raffaele
    Scientific Institute, 20132 Milan, Italy
*   Correspondence: francesco.inchingolo@uniba.it (F.I.); l.riccaldo@studenti.uniba.it (L.R.);
    Tel.: +39-331-211-1104 (F.I.); +39-327-023-8282 (L.R.)
†   These authors contributed equally to this work.

Abstract: This review analyzes muscle activity following mandibular condylar fracture (CF), with a focus on understanding the changes in masticatory muscles and temporomandibular joint (TMJ) functioning. Materials and Methods: The review was conducted following the preferred reporting items for systematic reviews and meta-analyses (PRISMA) guidelines. A search was performed on online databases using the keywords "masticatory muscles" AND ("mandibular fracture" OR "condylar fracture"). The eligibility criteria included clinical trials involving human intervention and focusing on muscle activity following a condylar fracture. Results: A total of 13 relevant studies were reviewed. Various studies evaluated muscle activity using clinical evaluation, bite force measurement, electromyography (EMG), magnetic sensors and radiological examinations to assess the impact of mandibular fractures on masticatory muscles. Conclusions: Mandibular condylar fractures can lead to significant changes in muscle activity, affecting mastication and TMJ functioning. EMG and computed tomography (CT) imaging play crucial roles in assessing muscle changes and adaptations following fractures, providing valuable information for treatment planning and post-fracture management. Further research is required to explore long-term outcomes and functional performance after oral motor rehabilitation in patients with facial fractures. Standardized classifications and treatment approaches may help improve the comparability of future studies in this field.

Keywords: condylar fracture; mandibular fracture; masticatory muscles; electromyography; muscles activity; bite forces; temporomandibular joint

Citation: Inchingolo, F.; Patano, A.; Inchingolo, A.M.; Riccaldo, L.; Morolla, R.; Netti, A.; Azzollini, D.; Inchingolo, A.D.; Palermo, A.; Lucchese, A.; et al. Analysis of Mandibular Muscle Variations Following Condylar Fractures: A Systematic Review. *J. Clin. Med.* **2023**, *12*, 5925. https://doi.org/10.3390/jcm12185925

Academic Editors: Eiji Tanaka and Luís Eduardo Almeida

Received: 1 August 2023
Revised: 30 August 2023
Accepted: 11 September 2023
Published: 12 September 2023

1. Introduction

1.1. Incidence and Pathogenesis

The mandible is the second most commonly fractured part of the maxillofacial skeleton after the nasal bone due to its position and prominence. The angle (27.0%), symphysis (21.3%), body (16.8%), ramus (5.4%), and coronoid (1.0%) were the most frequently injured sites, independent of mechanism [1]. Condylar fractures (CFs), instead, account for 17.5% to 52% of all mandibular fractures [2–4] (Figure 1).

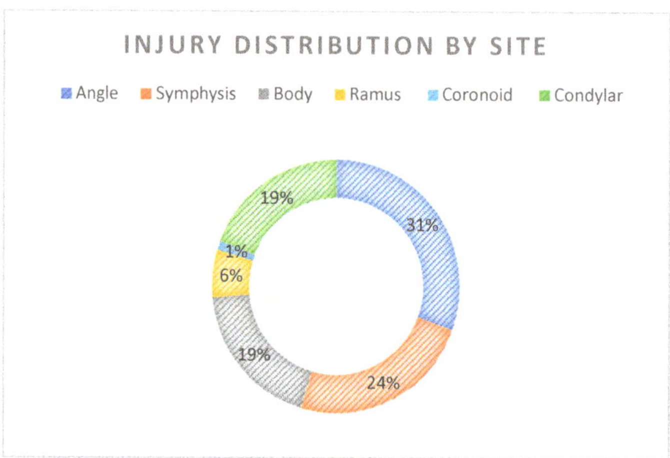

Figure 1. Graphic representation of injury distribution by site.

The relative frequency of such fractures is determined by the particular morphology of the condyle, which makes it the most fragile point of the mandibular bone [5,6]. On the one hand, its slenderness tends to make it more fragile and more prone to fracture, but on the other hand, fracturing and disrupting the propulsive forces allows it to protect the glenoid cavity and skull bones [7,8]. Most are induced by indirect pressures sent to the condyle from a blow elsewhere, rather than by direct trauma. Because the coronoid process (CP) is physically protected by the zygomatico-malar complex and its accompanying muscles, isolated coronoid fractures are extremely rare [9,10]. The majority of coronoid fractures are caused by indirect blunt or penetrating trauma. Iatrogenic fractures of the CP have been reported after extractions of maxillary and mandibular third molars, sagittal split osteotomies, and cystectomies [11,12]. CF are rather common injuries; however, fractures of the CP are extremely rare, accounting for only 1% of all mandible fractures [13]. The presence of dental elements is a protective factor against traumatic impacts to the jaw, especially if the teeth are in the position of maximal intercuspation at the time of the trauma [14–18]. On the contrary, if some teeth are absent at the time of the trauma, for example, if the posterior sectors or the mouth is disclosed, the force is transmitted directly to the condyle with subsequent risk of fracture of the condyle [19]. In addition, the position of the muscles at the time the injury occurs is crucial in determining the direction and extent of condylar dislocation [20]. Mandibular condylar process fractures with condyle displacement cause rapid disruption of the articulating surfaces, intra-articular disc, ligaments, and muscle attachments [21]. These disturbances are followed by changes in typical maximum excursion ranges, reductions in maximum biting forces, and changes in muscle activity patterns [22,23]. CF can be either unilateral or bilateral, and a correct diagnosis based also on the muscles involved allows their proper clinical management. Among the main signs and symptoms of mandibular fracture are pain above the preauricular zone and reduced anterior opening [23,24]. The signs change depending on whether the fracture is lateral or bilateral. The unilateral CF causes a homolateral prematurity occlusion with a resulting contralateral open bite and vertical dimension loss [25,26]. Movements of laterality may appear reduced just from the contralateral side to the fractured side [27]. In most cases, bilateral CF caused by an indirect head tilt cause an open bite in addition to a loss of vertical dimension with posterior dental precontact [28].

1.2. Classifications of Condylar Fractures

There are numerous classifications of CF in use internationally [29,30]. As a result, the conclusions of different authors are frequently incompatible [31]. Lindahl's classification is one of the best-known in which CF is defined according to its location [32] (Figure 2).

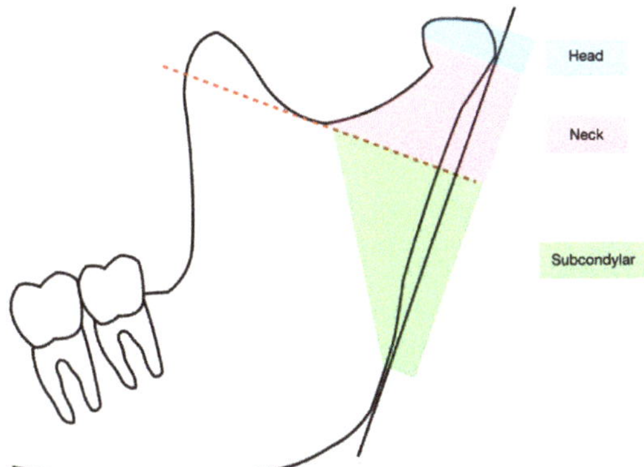

Figure 2. Graphic representation of the three localizations of condylar fracture. The red dotted line is the sigmoid notch line that divides the neck of the condyle from the subcondylar area.

Thus, based on the location of the fracture they are distinguished into:

1. Head-condyle fracture: located at the level above or at the level of the joint capsule and according to its course can be defined as vertical or horizontal;
2. Condylar-neck fracture: located in the area below the head of the condyle;
3. Subcondylar fracture: located below the neck of the condyle.

1.3. Clinical, Instrumental Diagnosis and Treatment

Early diagnosis of CF is critical in deciding the correct treatment plan with the aim of avoiding the occurrence of subsequent complications. In case of suspected mandibular fracture, very often as a result of sustained trauma, it is necessary to proceed with an objective clinical examination based on palpation of the area, both intra-oral and extra-oral inspection, and assessment of joint function [33,34]. Radiographic investigations are then required for diagnostic confirmation. For the study of the TMJ, the specific examinations are directly computed tomography (CT) in axial and coronal projection and orthopantomography (OPT) [35]. In the diagnostic phase, CT with tridimensional reconstructions can also be of great help [36]. To assess the function of the masticatory muscles, electromyography (EMG) can be performed, which uses skin electrodes to record the activity of the muscle fibers both in activity and at rest [37,38].

For many years, conservative therapy was considered the gold standard for the treatment of the mandibular CF [39]. In the latest period, however, many surgeons tend to prefer surgical treatment as the best solution, probably because of the new technologies introduced [40–42]. In fact, the surgical procedure achieves fracture reduction, which together with internal fixation, allows a good anatomical repositioning to be achieved [43]. Usually, surgical treatment is chosen in cases where conservative treatment fails to achieve proper "restitutio ad-integrum" of the fracture site [44,45].

Muscular therapy following temporomandibular disorders caused by condyle fracture is critical for proper masticatory function recovery. It tries to restore muscle balance and enhance joint mobility with focused workouts, manual treatment, and specialized equipment [46]. The objective is to alleviate discomfort, restore normal biomechanics, and avoid muscle compensation. Major treatments for the management of pain caused by temporomandibular disorders include physical therapy, drug therapy, laser therapy, occlusal therapy, oxygen–ozone therapy, extracorporeal shock wave therapy, and transcutaneous electrical stimulation [47].

The aim of this review is to analyze muscle activity following mandibular CF.

2. Materials and Methods

2.1. Protocol and Registration

This study was carried out in accordance with the Preferred Reporting Items for Systematic Reviews and Meta-Analyses (PRISMA) guidelines and submitted to PROSPERO (International Prospective Register of Reviews) with 448,110 [48].

2.2. Search Processing

The search was conducted on 13 July 2023 on the PubMed, Scopus, and Web of Science databases without the inclusion of any time interval. The search approach included the following Boolean keywords: "masticatory muscles" AND ("mandibular fracture" OR "condylar fracture"). These keywords were chosen because they most accurately reflected the aim of our investigation, which was to find out more about the activity and function of the masticatory muscles following a unilateral or bilateral mandibular CF.

2.3. Eligibility Criteria and Study Selection

The two steps of the selection process were the appraisal of the title and abstract and the complete text. Any article that fit the following requirements was taken into consideration: (a) clinical trials including human intervention; (b) muscle activity following a condylar fracture; (c) free full text. Publications (such as meta-analyses, research methods, conference papers, in vitro or animal experiments) that lacked original data were not included. Titles and abstracts from the preliminary search were retrieved and evaluated for relevance. Full articles from pertinent research were acquired for further analysis. The retrieved studies were assessed for inclusion using the aforementioned criteria by two different reviewers (R.M. and A.P.).

2.4. Data Processing

R.M. and A.P., the two reviewers, independently evaluated the quality of the studies, based on selection criteria after performing a database search to extrapolate the findings. In order to use with Zotero, the chosen articles were downloaded in the 6.0.15 version. A senior reviewer (F.I.) was consulted in order to address any disagreements between the two reviewers.

2.5. PICOS Requirements

The PICOS (population, intervention, comparison, outcome, study design) criteria, which are used in this evaluation, encompass population, intervention, comparison, outcomes, and study design (Table 1).

Table 1. PICOS criteria.

Criteria	Application in the present study
Population	Subjects suffered CF.
Intervention	Surgical or conservative treatment of condylar mandibular fracture (unilateral or bilateral).
Comparisons	Comparison of different methods of recording muscle activity (EMG, CT scans, clinical palpation).
Outcomes	Changes in masticatory muscle activity following condylar fracture detected by: clinical palpation, bite force, electromyography, CT scan, and magnetic jaw track device.
Study design	Case control studies, observational studies, prospective cohort studies, retrospective studies.

2.6. Quality Assessment

The quality of the included papers was assessed by two reviewers, RF and EI, using the reputable Cochrane risk-of-bias assessment for randomized trials (RoB 2). The following six areas of possible bias are evaluated by this tool: random sequence generation, allocation concealment, participant and staff blinding, outcome assessment blinding, inadequate outcome data, and selective reporting. A third reviewer (FI) was consulted in the event of a disagreement until an agreement was reached.

3. Results

3.1. Selection and Characteristics of the Study

A total of 485 publications were found in the online database (PubMed n = 211, Scopus n = 199, and Web of Science n = 75); no papers were found using a manual search. After 259 duplicate studies were removed, 226 studies were evaluated by looking at the title and abstract. From here, 37 records were chosen out of 189 items that failed to fulfill the requirements for inclusion. Subsequently, 21 non-retrieved records were excluded. There were 16 reports evaluated for eligibility, and three reports were removed. In the end, 13 studies were reviewed for the qualitative analysis. The selection process and the summary of selected records are shown in Figure 3. The study characteristics are summarized in Table 2.

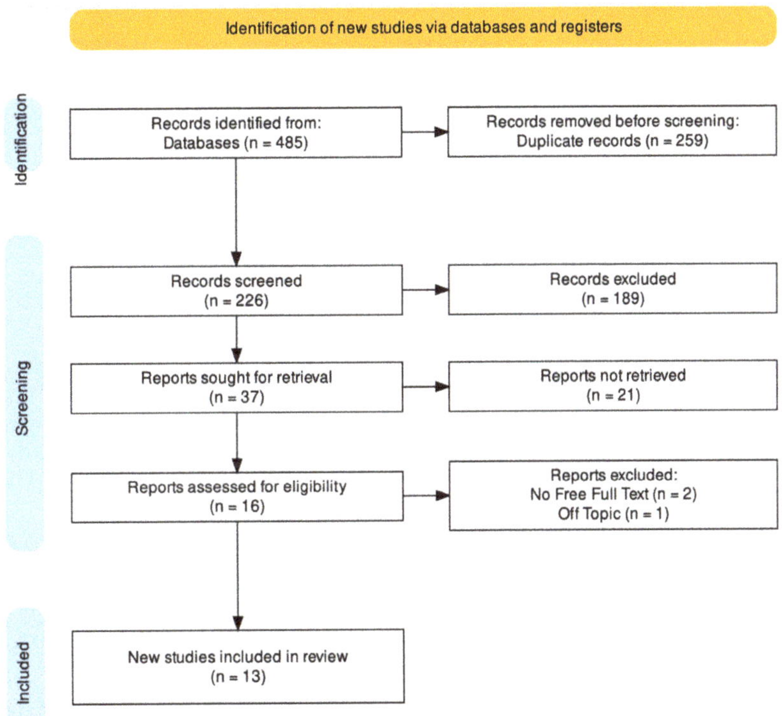

Figure 3. PRISMA ScR flowchart diagram of the inclusion process.

The following review analyzes a total of 13 papers including four case control (30%), four observational studies (30%), three prospective cohort studies (24%), and two retrospective studies (16%). Among the analyzed studies, three deal exclusively with bilateral fractures, five deal exclusively with unilateral fractures, and the last five consider both unilateral and bilateral fractures.

Table 2. Characteristics of the studies included in the qualitative analysis.

Author	Study Design	Number of Patients/Gender	Country	Type of Fracture (Unilateral/Bilateral)	Matherial and Methods	Outcomes
1-Talwar et al. (1998) [49]	Case-control	22 (15 M–7 F)	USA	Bilateral	Compared open reduction (n = 6), closed therapy (n = 14), or a combination (n = 2) with 22 controls	Reduced posterior face height and moment arm length in masseter and pterygoid muscles. Different temporalis muscle orientations. Restricted movement during the procedure. Reduced bite forces in patients. Temporalis muscle used more during maximal biting
2-Ellis (2001) [21]	Observational study	155 (127 M–28 F)	USA	Unilateral	Bite force of masseter muscles at intervals	No significant differences in bite forces between treatment groups. Improved bite force over time
3-Throckmorton et al. (1999) [50]	Case-control	22 (15 M–7 F)	USA	Bilateral	Examined patients and controls over time. Recorded incisor movements and muscle activity.	Reduced anterior translation and lateral excursion due to anatomical disruption and poor lateral pterygoid function. Reduced muscle activity during closure stages. Most individuals resumed regular eating within a year
4-Choi et al. (1996) [51]	Observational study	10 M	Korea	Bilateral	Patients with various symptoms and CT scans showing bilateral CF	Jaw exercises and mandibular manipulation eliminated open bite after IME release. Restoration of occlusion and function through masticatory muscle adaptation
5-Lindahl et al. (1977) [32]	Observational study	67 patients	Sweden	Bilateral and unilateral	Radiographic and clinical examination at intervals	Maximal opening returned to normal in most individuals after two years Children showed fewer persistent joint and muscle complaints than adults
6-Kahl-Nieke et al. (1999) [52]	Observational study	19 patients (9 F and 10 M)	Germany	Unilateral	Analyzed post-traumatic and post-therapeutic state of soft tissue in children with CF	Functional restoration was good or very good after an average of 5 years. Lateral pterygoid muscle diminished in nearly two-thirds of patients. Changes in volume between healthy and damaged sides
7-Kuntamukkula et al. (2018) [53]	Prospective cohort study	30 patients	India	Unilateral	Evaluated TMJ dynamics and muscle EMG	TMJ abnormalities persisted six months after the treatment of CF with open reduction and internal fixation. Long-term studies needed for accurate timeline

Table 2. *Cont.*

Author	Study Design	Number of Patients/Gender	Country	Type of Fracture (Unilateral/Bilateral)	Matherial and Methods	Outcomes
8-Raustia et al. (1990) [54]	Restrospective study	17 patients	Finland	Unilateral/bilateral	Evaluated muscle density with CT after mean period of 33 months	No significant differences in muscle density, but lateral pterygoid muscle is smaller on the fractured side. Difference increased with time
9-Salunkhe et al. (2022) [55]	Case-control study	30 patients	India	Unilateral/bilateral	Divided into groups based on fracture type and compared masticatory forces with a control group	Temporary reduction in chewing force observed in unilateral fracture cases, gradually restored over time. Bilateral fractures took longer to restore the chewing force
10-Sforza et al. (2009) [56]	Case-control study	9 patients (8 M, 1 F)	Italy	3 unilateral and 6 bilateral	Evaluated mandibular movements and EMG activity during clenching	Patients had a significantly higher percentage of rotational movement. Changes in mandibular movements and EMG indices compared to healthy subjects
11-Throckmorton et al. (2004) [57]	Prospective study	81 male patients	USA	Unilateral	Recorded chewing cycles with sensor array during mastication on both sides. Surgical correction normalized incisor pathways on the opposite side	Surgical correction better normalizes opening incisor pathways during mastication on the side opposite the fracture.
12-Meller et al. (1997) [58]	Retrospective study	9 adult men	Denmark	Unilateral	Evaluated muscle contractions during chewing before and after treatment.	Significant increase in muscle contractions on impaired joint side during chewing after treatment.
13-Pagliotto da Silva et al. (2016) [59]	Prospective cohort study	26 adults, both gender.	Brasil	Unilateral and bilateral	Divided into groups based on treatment. Assessed orofacial myofunctional system and mandibular range of motion	Surgical open reduction showed better symmetry in masseter muscle activation compared to closed reduction treatment

3.2. Quality Assessment and Risk of Bias

The risk of bias in the included studies is reported in Figure 4. Regarding the randomization process, 50% of studies present a high risk of bias and allocation concealment. All other studies ensure a low risk of bias: 75% of studies exclude a performance; 50% of studies confirm an increased risk of detection bias (self-reported outcome); and 50% of the included studies present a low detection bias (objective measures) (Figure 4). Two studies ensure a low risk regarding attrition and reporting bias.

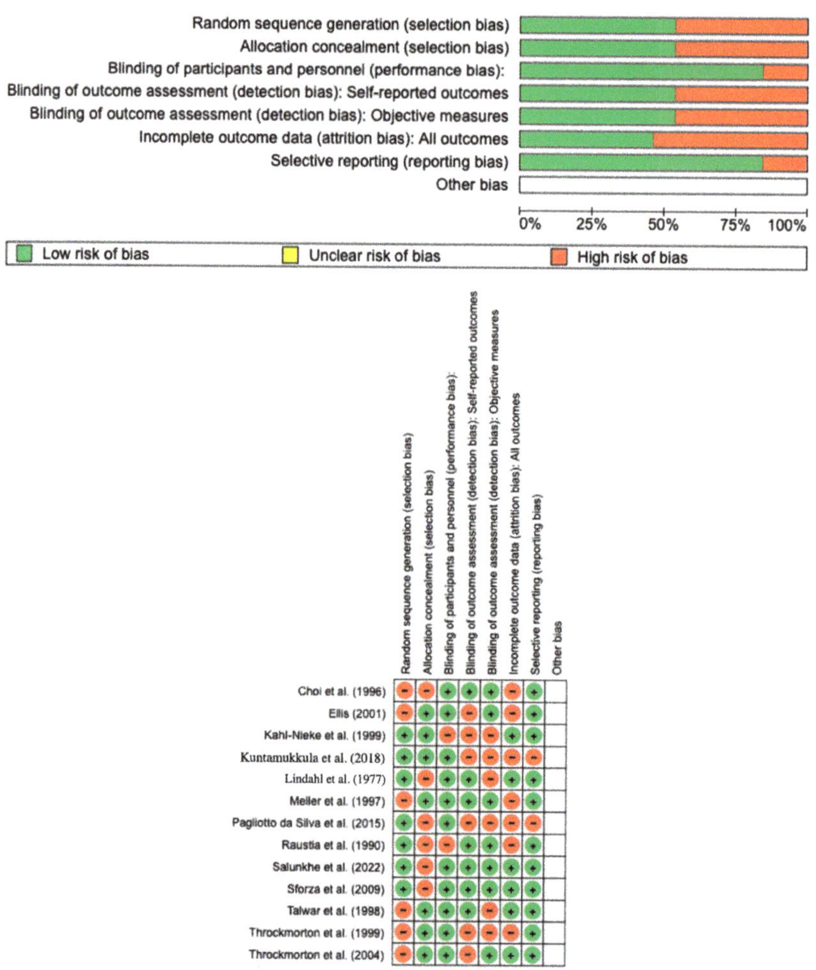

Figure 4. Risk of bias; red indicates high risk, and green indicates low risk of bias. Choi et al. [51]; Ellis et al. [21]; Kahl-Nieke et al. [52]; Kuntamukkula et al. [53]; Lindahl et al. [32]; Meller et al. [58]; Pagliotto da Silva et al. [59]; Raustia et al. [54]; Salunkhe et al. [55]; Sforza et al. [56]; Talwar et al. [49]; Throckmorton et al. [50]; Throckmorton et al. [57].

4. Discussion

Following mandibular condylar fractures, alterations in masticatory muscles occur, impacting temporomandibular joint function. Assessing these changes requires comprehensive methods. Clinical palpation, bite force analysis, electromyography, CT scans, and magnetic jaw tracking are employed to scrutinize muscular alterations. These techniques offer insights into muscle functionality, symmetry, and coordination post-fracture, aiding a comprehensive understanding of post-traumatic masticatory muscle adaptations and their implications on mandibular function.

4.1. Clinical Palpation

Lindahl's study proposed to understand the recovery process and the potential development of dysfunction in the masticatory system following CF. Examiners performed a comprehensive clinical examination, combined with radiographic evaluations, on a group of 67 individuals of different age groups [32]. During clinical examinations, several parameters related to muscle function were recorded, such as palpability of the condylar head to determine the presence of any dislocations, measurement of mandibular movements (maximum opening, laterotrusion and protrusion), and detection of functional symmetry [32]. The results provide valuable information about the long-term consequences of CF and that there are significant variations among different age groups. Indeed, in children, most fractures heal without causing significant masticatory dysfunction. In adolescents, dysfunction is more frequent but usually less severe than in adults. The latter, however, show a higher incidence of symptoms of masticatory dysfunction, potentially related to persistent dislocation of the condylar fragment [32].

4.2. Bite Force

Mastication is a crucial function of the stomatognathic system, and chewing force is an important parameter for assessing the restoration of function after surgery.

Talwar's study analyzes the effects of bilateral fractures of condylar processes on chewing force. Through the analysis of data from cephalograms and biomechanical studies, significant morphological and biomechanical changes were found in patients with CF [49]. A decrease in bite force was observed in the first few months after the fracture, with a gradual recovery over time. This decrease in force appears to be related to complex adaptations of the masticatory system, including changes in craniofacial morphology and muscle activity [49]. Indeed, the results showed a hyperdivergent craniofacial morphology in bilateral fractures, with lower posterior facial height, higher mandibular and genial angles. In addition, temporalis muscle activity was found to be more involved in chewing in patients with CF, suggesting an adaptation strategy to reduce the load on the fractured joints [49].

In an American study, the authors investigated whether chewing force in patients with unilateral fractures of the condylar process of the mandible varied according to the type of restorative treatment whether with open or closed techniques. Data from 155 patients with unilateral fractures of the condylar process of the mandible treated with open or closed techniques were analyzed [21]. Maximum clamping force measurements were made using a force transducer. Despite the different treatment approaches, both groups of patients show a similar ability to generate occlusal forces. However, neuromuscular adaptations in muscle recruitment during chewing on the side opposite the fracture are observed, indicating a strategy to protect the injury site [21]. In summary, patients adjust to reduce the load on the fracture, but chewing ability does not vary significantly between the two treatment approaches [21].

In contrast, Kuntamukkula's study evaluates the dynamic stability of the TMJ in patients with unilateral condyle fractures undergoing reduction surgery and open internal fixation. During the six-month recovery period, bite force was evaluated both statically and dynamically on both sides of the mandible [53]. The results indicate that although the maximum bite force is similar on both sides, the mean functional bite force is significantly higher on the unoperated side, suggesting lower masticatory efficiency on the injured side [53]. Neuromuscular adaptations and early movements of the condyle-disc complex on the operated side were also observed, indicating that despite restoration of TMJ anatomy with open treatment, dynamic stability may take more than six months to be achieved [53].

Salunkhe et al., also investigated significant changes in masticatory loads in patients with mandibular condyle fractures undergoing reduction surgery and open internal fixation. The results showed that, initially, the bite force in the molar region was significantly reduced in the first postoperative week, but gradually approached the levels of the control group by the ninth postoperative week [55]. This suggests that the healing and restoration of normal architecture led to a gradual and modest increase in biting forces. It was also observed that unilateral fractures of the mandibular condyle had a greater impact on biting force than bilateral fractures. The evolution of stability and masticatory function took a longer period in the case of bilateral fractures [55]. These results underscore the importance of careful evaluation and an adequate recovery period after surgery to ensure complete restoration of masticatory function in patients with mandibular condyle fractures [55].

Fractures of the mandibular condyle may temporarily affect chewing strength, but with proper surgical treatment and recovery period, chewing ability tends to improve over time.

4.3. Electromiography

Sforza's study aims to quantitatively assess the percentage contribution of rotation and translation movements of the mandible during maximum mouth opening in patients treated for CF [56]. The researchers used EMG to measure the activity of masticatory and neck muscles during maximum voluntary teeth clenching. They also calculated different EMG indices to develop simpler estimations of TMJ functioning [56].

The results indicated that patients showed altered patterns of mandibular motion during mouth opening, with a larger percentage of rotation and reduced vertical displacement compared to healthy individuals. EMG indices in patients also differed from those in the healthy subjects, indicating potential functional adaptations following the fractures [56]. Overall, the study provides valuable insights into the changes in masticatory muscles after mandibular fractures, especially in relation to TMJ functioning, which can aid in the treatment and management of such injuries [56].

Hjorth's study also evaluates parameters recorded through EMG: it reports the results of a longitudinal study conducted on patients with unilateral fractures of the mandibular condyle [58]. The aim of the study was to examine the effects of these fractures on the activity of the temporal and masseter muscles up to 6–12 months after the trauma, using EMG to record muscle changes [58].

During observation, the level of muscle activity at rest (postural) did not change significantly. However, there was a significant increase in muscle activity during maximal contraction of the contralateral temporalis and masseter muscles (on the side opposite the fracture) from the time immediately after fracture treatment (T0) until 6–12 months later (T2). Furthermore, during natural mastication, the maximal activity of the anterior temporalis muscle increased significantly over time [58].

During unilateral mastication (UM, performed on one side only), a significant increase in activity was noted in the anterior temporalis and contralateral masseter muscles from T0 to T2. Furthermore, muscle contractions became shorter and stronger in all muscles during UM, except for the ipsilateral masseter (on the same side as the fracture), which had a significant increase in activity only between the first and second examinations [58].

In summary, EMG revealed significant changes in muscle activity in patients with unilateral mandibular condyle fractures. Increases in the maximal muscle activity of the contralateral temporalis and masseter muscle were observed during contraction and natural chewing [58]. Furthermore, during UM, stronger and shorter contractions occurred in all muscles involved, except in the ipsilateral masseter. These results indicate a muscular adaptation in the course of time after the fracture and suggest a possible role of the suppression reflex in the muscle of the side of the fracture as a consequence of the trauma [58].

Pagliotto da Silva aimed to investigate the oral motor function in patients with facial fractures [59].

The study included 38 adult patients with facial fractures, divided into two groups based on the treatment received: Group 1 (G1) received open reduction of facial fractures, and Group 2 (G2) received closed reduction with maxillomandibular fixation. A control group of 19 healthy volunteers was also recruited [59].

The participants underwent oral motor clinical assessment, mandibular range of movement measurements, and surface electromyography (sEMG) evaluation of the anterior temporal and masseter muscles. The data were analyzed using non-parametric tests.

The results showed that patients with facial fractures had poorer performance in oral motor functions, including swallowing and mastication, compared to the control group. Both G1 and G2 exhibited a limited mandibular range of motion, particularly in maximal incisor distance. The electromyographic assessment revealed that both groups of patients with facial fractures had lower overall muscle activity compared to the control group [59].

However, there were no significant differences in oral motor function or muscle activity between G1 and G2, suggesting that the severity of facial fractures did not influence muscle function and performance four months after fracture correction.

Overall, the study indicates that patients with facial fractures experience deficits in oral motor function, but the type of fracture treatment (open or closed reduction) did not significantly affect the outcomes of muscle function and performance after four months of treatment. The electromyographic evaluation of the anterior and posterior heads of the temporalis, masseter and sternocleidomastoid muscles in bicondylar fractures showed a lower activity of the masseters than that of the temporals during the closing phase, especially on the working side, the opposite of what normally happens [57]. During the occlusal phase, on the other hand, the muscle force is reduced compared to healthy patients, partly because the reduced activity of the lateral pterygoid muscle tends to overload the balancing condyle [57].

4.4. CT Scan

A practical tool for evaluating soft tissue changes in the nervous system after a CF that is not visible on standard radiographs is CT. It has been used to determine the disc's position and assess how well it is functioning. It has also been shown to be superior to traditional radiography in terms of detecting subtle changes in the condylar head and the mandibular fossa [60–62]. Additionally, CT has been useful in examining the skeletal muscles to detect muscular atrophy, pseudohypertrophy, denervation atrophy and hypertrophy [62–64].

A CT scan analysis of the masseter, temporalis and lateral pterygoid muscle showed that the first two muscles on the fractured side have a similar density to those on the non-fractured side, although slightly lower. The only statistically significant difference is related to the lower density of the lateral pterygoid muscle on the fractured side compared to the non-fractured side, considering that the lateral pterygoid has absolutely less density than the rest of the other muscles. In some cases, the muscle shows a reduction in density in direct proportion to the passage of time [54].

Kahl-Nieke et al. analysed the changes undergone by the lateral pterygoid muscle in 19 child patients with a unilateral fracture during and after treatment with functional devices that promote condylar remodeling. The device was worn for approximately nine months and was intended to promote condyle formation with resorption of the fractured fragment. In the follow-up in more than 70% of the patients, there was a reduction in the muscle mass of the lateral pterygoid in the range of 13 to 69%. The extent of the mass reduction is directly proportional to the depth and severity of the fracture: deep fractures and fractures with complete dislocation of the disc suffered the greatest mass reduction [52].

In patients with bilateral CF due to muscle contraction, there is an open-bite and retroposed position of the mandible. After 10 days of intermaxillary fixation, months of jaw exercises, muscle training, and mandibular manipulation, the open bite was removed. Masticatory muscle adaptation allows for the restoration of occlusion and function [51].

4.5. Magnetic Jaw Track Device

Following a mandibular fracture, in addition to the alteration of the anatomy of the TMJ the function also changes: the masticatory cycles are altered, especially the changes that the two ends of the lateral pterygoid muscle undergo [65,66].

To assess the masticatory cycles, measurements were taken during the chewing of a gummy bear after the insertion of a small magnet at the level of the lower incisors, the patient's face with the frankfurter plane parallel to the floor. A magnetic sensor array (Sirognathograph; Siemens Corp, Bensheim, Germany) was attached to the patient's head to monitor the activity of the magnet. All these devices are connected to a computer that reports what occurs in the three planes of space [54,57,67–73].

In the case of bicondylar fractures, the muscles affected are the two ends of the lateral pterygoid, which makes the protrusion movement difficult since it fits over the head and neck of the condyle. The other muscles are not particularly affected [50].

The study proposed by Throckmorton hypothesizes that in unilateral CF treated with the closed technique, the alterations of the masticatory cycles persist even after bone healing, unlike what happens with the open technique [57].

The patients were divided into two groups: the first group was treated surgically, using a retromandibular approach for surgery in the condylar area. In these patients, plates were fixed without compression using at least two screws of 2 mm length. The second group included patients who did not receive any surgical treatment of the condyle but used Class II elastics to guide the mandible into correct occlusion and favor pseudocondylar formation. Patients in both groups followed the same physiotherapy exercises and the use of rubber bands to guide correct occlusion [57].

These patients were treated either surgically or in a combined manner (one side was treated with surgery and the other with a conservative approach [50].

At a follow-up of two years, the inferior excursion remains reduced compared to the control of healthy patients [57]. The excursion is lower if the fracture is bicondylar [50].

The posterior opening range in patients treated with both techniques is between 4.7 and 5.5 mm after six weeks to one year; in patients with a bicondylar fracture it is even wider. In the control group, on the other hand, the maximum posterior excursion is 4 mm. In the control two years after the end of treatment, however, the values tend to decrease and return to a normal range. Additionally, in the open group, condylar mobility on the fractured side was lower than that on the non-fractured side six months after the fracture, whereas in the closed group, condylar mobility on both sides was essentially comparable. Even while these variations were not statistically significant, they do at least imply that the open group's condylar mobility may have been slightly higher after six weeks [50,57].

A significant difference was evident one year post-treatment with better stabilization in the group treated with the conservative technique than in the group treated with the open technique: lateral excursion on the fractured side returns to the normal range as early as six months, while in surgically treated patients it takes at least one year [57].

The more used measures of masticatory function (duration and excursive ranges) are not significantly affected by surgical treatment of unilateral condylar process fractures. On the side opposite the fracture, however, surgical treatment better normalizes opening incisor channels during mastication. Similar to the opening phases, the patients' reduced excursion toward the working side during the fast-close phase is consistent with the inhibition of normal translation of the balancing side condyle after bilateral condylar process fractures [50].

The review has several limitations:

1. Heterogeneous study designs: The included studies vary in design, such as case-control, observational, retrospective, and prospective studies. Combining data from different study designs may introduce heterogeneity and affect the validity of the review's conclusions.
2. Limited sample sizes: Some of the included studies have small sample sizes, which could limit the statistical power and generalizability of the findings.
3. Lack of quality assessment: The review does not mention whether a quality assessment of the included studies was conducted. Evaluating the methodological quality of the studies is essential to assess the overall reliability of the evidence.
4. Limited scope of analysis: The review mainly focuses on muscle activity following mandibular CF, but it does not discuss other potential outcomes or complications related to these fractures, such as pain, joint function, or psychosocial impact. A more comprehensive analysis of the implications of these fractures would provide a more robust understanding of the topic.

5. Conclusions

In conclusion, the systematic review of literature has revealed a multifaceted landscape of evidence regarding the impact of condylar fractures on the masticatory system and the subsequent recovery processes. Clinical palpation studies underscore the age-dependent variations in the consequences of condylar fractures, with children generally experiencing milder dysfunction compared to adolescents and adults, potentially related to persistent condylar fragment dislocation. Bite-force analyses elucidate the dynamic adaptations of the masticatory system post-fracture, with reductions in force initially observed and subsequent adjustments to protect the injured site. Electromyography studies offer insights into muscle activity changes, revealing muscular adaptations and functional alterations, particularly in patients with unilateral fractures. CT scans provide valuable insights into soft tissue changes and muscular atrophy, emphasizing the importance of long-term evaluation. Magnetic jaw track device studies shed light on altered masticatory cycles following fractures, with differences observed between surgical and non-surgical treatments.

Overall, these findings underscore the complexity of the masticatory system's response to condylar fractures and the importance of tailored treatment approaches and long-term monitoring for optimal recovery. Further research is needed to enhance our understanding of these processes and to guide more effective clinical management strategies for patients with mandibular condylar fractures.

Author Contributions: Conceptualization, A.M.I., L.R., R.M., A.N. and D.A.; methodology, A.P. (Assunta Patano), F.I. and A.D.I.; software, D.A.; validation, L.R., R.M. and A.L.; formal analysis, D.D.V., F.I. and G.D.; investigation, R.M., L.R. and A.N.; resources, D.A., A.P. (Andrea Palermo) and A.M.I.; data curation, A.D.I., A.L.; G.D., F.I. and D.D.V.; writing—original draft preparation, G.D.; writing—review and editing, L.R., A.P. (Andrea Palermo) and F.I.; visualisation, A.P. (Assunta Patano) and D.D.V.; supervision, A.M.I., A.D.I. and D.A.; project administration, A.N., R.M. and D.A. All authors have read and agreed to the published version of the manuscript.

Funding: This research received no external funding.

Institutional Review Board Statement: Not applicable.

Informed Consent Statement: Not applicable.

Data Availability Statement: Not applicable.

Conflicts of Interest: The authors declare no conflict of interest.

Abbreviations

CF	Condylar fracture
CP	Coronoid process
CT	Computed tomography
EMG	Electromyography
F	Female
M	Male
OPT	Ortopantomography
sEMG	Surface electromyography
TMJ	Temporomandibular joint
UM	Unilateral mastication

References

1. Inchingolo, A.M.; Patano, A.; De Santis, M.; Del Vecchio, G.; Ferrante, L.; Morolla, R.; Pezzolla, C.; Sardano, R.; Dongiovanni, L.; Inchingolo, F.; et al. Comparison of Different Types of Palatal Expanders: Scoping Review. *Children* **2023**, *10*, 1258. [CrossRef] [PubMed]
2. Villarreal, P.M.; Monje, F.; Junquera, L.M.; Mateo, J.; Morillo, A.J.; González, C. Mandibular Condyle Fractures: Determinants of Treatment and Outcome. *J. Oral Maxillofac. Surg.* **2004**, *62*, 155–163. [CrossRef] [PubMed]
3. Ellis, E.; Throckmorton, G.S. Treatment of Mandibular Condylar Process Fractures: Biological Considerations. *J. Oral Maxillofac. Surg.* **2005**, *63*, 115–134. [CrossRef] [PubMed]

4.   Morris, C.; Bebeau, N.P.; Brockhoff, H.; Tandon, R.; Tiwana, P. Mandibular Fractures: An Analysis of the Epidemiology and Patterns of Injury in 4,143 Fractures. *J. Oral Maxillofac. Surg.* **2015**, *73*, 951.e1–951.e12. [CrossRef]
5.   Vincent, A.G.; Ducic, Y.; Kellman, R. Fractures of the Mandibular Condyle. *Facial Plast. Surg.* **2019**, *35*, 623–626. [CrossRef]
6.   Mooney, S.; Gulati, R.D.; Yusupov, S.; Butts, S.C. Mandibular Condylar Fractures. *Facial Plast. Surg. Clin. N. Am.* **2022**, *30*, 85–98. [CrossRef]
7.   Rozeboom, A.V.J.; Dubois, L.; Bos, R.R.M.; Spijker, R.; de Lange, J. Closed Treatment of Unilateral Mandibular Condyle Fractures in Adults: A Systematic Review. *Int. J. Oral Maxillofac. Surg.* **2017**, *46*, 456–464. [CrossRef]
8.   Inchingolo, A.M.; Malcangi, G.; Piras, F.; Palmieri, G.; Settanni, V.; Riccaldo, L.; Morolla, R.; Buongiorno, S.; de Ruvo, E.; Inchingolo, A.D.; et al. Precision Medicine on the Effects of Microbiota on Head-Neck Diseases and Biomarkers Diagnosis. *J. Pers. Med.* **2023**, *13*, 933. [CrossRef]
9.   Takenoshita, Y.; Enomoto, T.; Oka, M. Healing of Fractures of the Coronoid Process: Report of Cases. *J. Oral Maxillofac. Surg.* **1993**, *51*, 200–204. [CrossRef]
10. Park, S.-S.; Lee, K.-C.; Kim, S.-K. Overview of Mandibular Condyle Fracture. *Arch. Plast. Surg.* **2012**, *39*, 281–283. [CrossRef]
11. Sollazzo, V.; Pezzetti, F.; Massari, L.; Palmieri, A.; Brunelli, G.; Zollino, I.; Lucchese, A.; Caruso, G.; Carinci, F. Evaluation of Gene Expression in MG63 Human Osteoblastlike Cells Exposed to Tantalum Powder by Microarray Technology. *Int. J. Periodontics Restor. Dent.* **2011**, *31*, e17–e28.
12. Inchingolo, F.; Ballini, A.; Cagiano, R.; Inchingolo, A.D.; Serafini, M.; De Benedittis, M.; Cortelazzi, R.; Tatullo, M.; Marrelli, M.; Inchingolo, A.M.; et al. Immediately Loaded Dental Implants Bioactivated with Platelet-Rich Plasma (PRP) Placed in Maxillary and Mandibular Region. *Clin. Ter.* **2015**, *166*, e146–e152. [CrossRef] [PubMed]
13. Shen, L.; Li, J.; Li, P.; Long, J.; Tian, W.; Tang, W. Mandibular Coronoid Fractures: Treatment Options. *Int. J. Oral Maxillofac. Surg.* **2013**, *42*, 721–726. [CrossRef]
14. Minervini, G.; Franco, R.; Marrapodi, M.M.; Crimi, S.; Badnjević, A.; Cervino, G.; Bianchi, A.; Cicciù, M. Correlation between Temporomandibular Disorders (TMD) and Posture Evaluated trough the Diagnostic Criteria for Temporomandibular Disorders (DC/TMD): A Systematic Review with Meta-Analysis. *J. Clin. Med.* **2023**, *12*, 2652. [CrossRef]
15. Minervini, G.; Franco, R.; Marrapodi, M.M.; Fiorillo, L.; Cervino, G.; Cicciù, M. Economic Inequalities and Temporomandibular Disorders: A Systematic Review with Meta-Analysis. *J. Oral Rehabil.* **2023**, *50*, 715–723. [CrossRef]
16. Qamar, Z.; Alghamdi, A.M.S.; Haydarah, N.K.B.; Balateef, A.A.; Alamoudi, A.A.; Abumismar, M.A.; Shivakumar, S.; Cicciù, M.; Minervini, G. Impact of Temporomandibular Disorders on Oral Health-related Quality of Life: A Systematic Review and Meta-Analysis. *J. Oral Rehabil.* **2023**, *50*, 706–714. [CrossRef]
17. Minervini, G.; Franco, R.; Marrapodi, M.M.; Fiorillo, L.; Cervino, G.; Cicciù, M. Prevalence of Temporomandibular Disorders (TMD) in Pregnancy: A Systematic Review with Meta-Analysis. *J. Oral Rehabil.* **2023**, *50*, 627–634. [CrossRef] [PubMed]
18. Minervini, G.; Franco, R.; Marrapodi, M.M.; Ronsivalle, V.; Shapira, I.; Cicciù, M. Prevalence of Temporomandibular Disorders in Subjects Affected by Parkinson Disease: A Systematic Review and Metanalysis. *J. Oral Rehabil.* **2023**. online. [CrossRef]
19. Ortiz-Gutiérrez, A.L.; Beltrán-Salinas, B.; Cienfuegos, R. Mandibular Condyle Fractures: A Diagnosis with Controversial Treatment. *Cirugía Cir.* **2019**, *87*, 587–594. [CrossRef]
20. Inchingolo, F.; Tatullo, M.; Marrelli, M.; Inchingolo, A.M.; Tarullo, A.; Inchingolo, A.D.; Dipalma, G.; Podo Brunetti, S.; Tarullo, A.; Cagiano, R. Combined Occlusal and Pharmacological Therapy in the Treatment of Temporo-Mandibular Disorders. *Eur. Rev. Med. Pharmacol. Sci.* **2011**, *15*, 1296–1300.
21. Ellis, E.; Throckmorton, G.S. Bite Forces after Open or Closed Treatment of Mandibular Condylar Process Fractures. *J. Oral Maxillofac. Surg.* **2001**, *59*, 389–395. [CrossRef] [PubMed]
22. De Riu, G.; Gamba, U.; Anghinoni, M.; Sesenna, E. A Comparison of Open and Closed Treatment of Condylar Fractures: A Change in Philosophy. *Int. J. Oral Maxillofac. Surg.* **2001**, *30*, 384–389. [CrossRef] [PubMed]
23. Wieczorek, A.; Loster, J.; Loster, B.W. Relationship between Occlusal Force Distribution and the Activity of Masseter and Anterior Temporalis Muscles in Asymptomatic Young Adults. *BioMed Res. Int.* **2012**, *2013*, e354017. [CrossRef] [PubMed]
24. Ferrario, V.F.; Tartaglia, G.M.; Galletta, A.; Grassi, G.P.; Sforza, C. The Influence of Occlusion on Jaw and Neck Muscle Activity: A Surface EMG Study in Healthy Young Adults. *J. Oral Rehabil.* **2006**, *33*, 341–348. [CrossRef]
25. Yang, W.-G.; Chen, C.-T.; Tsay, P.-K.; Chen, Y.-R. Functional Results of Unilateral Mandibular Condylar Process Fractures after Open and Closed Treatment. *J. Trauma* **2002**, *52*, 498–503. [CrossRef]
26. Sybil, D.; Gopalkrishnan, K. Assessment of Masticatory Function Using Bite Force Measurements in Patients Treated for Mandibular Fractures. *Craniomaxillofac. Trauma Reconstr.* **2013**, *6*, 247–250. [CrossRef]
27. Malcangi, G.; Inchingolo, A.D.; Patano, A.; Coloccia, G.; Ceci, S.; Garibaldi, M.; Inchingolo, A.M.; Piras, F.; Cardarelli, F.; Settanni, V.; et al. Impacted Central Incisors in the Upper Jaw in an Adolescent Patient: Orthodontic-Surgical Treatment—A Case Report. *Appl. Sci.* **2022**, *12*, 2657. [CrossRef]
28. Fridrich, K.L.; Pena-Velasco, G.; Olson, R.A.J. Changing Trends with Mandibular Fractures: A Review of 1067 Cases. *J. Oral Maxillofac. Surg.* **1992**, *50*, 586–589. [CrossRef] [PubMed]
29. Vermesan, D.; Prejbeanu, R.; Trocan, I.; Birsasteanu, F.; Florescu, S.; Balanescu, A.; Abbinante, A.; Caprio, M.; Potenza, A.; Dipalma, G.; et al. Reconstructed ACLs Have Different Cross-Sectional Areas Compared to the Native Contralaterals on Postoperative MRIs. A Pilot Study. *Eur. Rev. Med. Pharmacol. Sci.* **2015**, *19*, 1155–1160.

30. Vermesan, D.; Inchingolo, F.; Patrascu, J.M.; Trocan, I.; Prejbeanu, R.; Florescu, S.; Damian, G.; Benagiano, V.; Abbinante, A.; Caprio, M.; et al. Anterior Cruciate Ligament Reconstruction and Determination of Tunnel Size and Graft Obliquity. *Eur. Rev. Med. Pharmacol. Sci.* **2015**, *19*, 357–364.

31. Inchingolo, F.; Tatullo, M.; Abenavoli, F.M.; Marrelli, M.; Inchingolo, A.D.; Corelli, R.; Mingrone, R.; Inchingolo, A.M.; Dipalma, G. Simple Technique for Augmentation of the Facial Soft Tissue. *Sci. World J.* **2012**, *2012*, 262989. [CrossRef]

32. Lindahl, L. Condylar Fractures of the Mandible. IV. Function of the Masticatory System. *Int. J. Oral Surg.* **1977**, *6*, 195–203. [CrossRef]

33. Adina, S.; Dipalma, G.; Bordea, I.R.; Lucaciu, O.; Feurdean, C.; Inchingolo, A.D.; Septimiu, R.; Malcangi, G.; Cantore, S.; Martin, D.; et al. Orthopedic Joint Stability Influences Growth and Maxillary Development: Clinical Aspects. *J. Biol. Regul. Homeost. Agents* **2020**, *34*, 747–756. [CrossRef]

34. Malcangi, G.; Patano, A.; Pezzolla, C.; Riccaldo, L.; Mancini, A.; Di Pede, C.; Inchingolo, A.D.; Inchingolo, F.; Bordea, I.R.; Dipalma, G.; et al. Bruxism and Botulinum Injection: Challenges and Insights. *J. Clin. Med.* **2023**, *12*, 4586. [CrossRef]

35. Jensen, T.; Jensen, J.; Nørholt, S.E.; Dahl, M.; Lenk-Hansen, L.; Svensson, P. Open Reduction and Rigid Internal Fixation of Mandibular Condylar Fractures by an Intraoral Approach: A Long-Term Follow-up Study of 15 Patients. *J. Oral Maxillofac. Surg.* **2006**, *64*, 1771–1779. [CrossRef]

36. Farronato, M.; Farronato, D.; Giannì, A.B.; Inchingolo, F.; Nucci, L.; Tartaglia, G.M.; Maspero, C. Effects on Muscular Activity after Surgically Assisted Rapid Palatal Expansion: A Prospective Observational Study. *Bioengineering* **2022**, *9*, 361. [CrossRef]

37. Suvinen, T.I.; Kemppainen, P. Review of Clinical EMG Studies Related to Muscle and Occlusal Factors in Healthy and TMD Subjects. *J. Oral Rehabil.* **2007**, *34*, 631–644. [CrossRef] [PubMed]

38. Ferrario, V.F.; Tartaglia, G.M.; Luraghi, F.E.; Sforza, C. The Use of Surface Electromyography as a Tool in Differentiating Temporomandibular Disorders from Neck Disorders. *Man. Ther.* **2007**, *12*, 372–379. [CrossRef] [PubMed]

39. Di Paolo, C.; Qorri, E.; Falisi, G.; Gatto, R.; Tari, S.R.; Scarano, A.; Rastelli, S.; Inchingolo, F.; Di Giacomo, P. RA.DI.CA. Splint Therapy in the Management of Temporomandibular Joint Displacement without Reduction. *J. Pers. Med.* **2023**, *13*, 1095. [CrossRef] [PubMed]

40. Handschel, J.; Rüggeberg, T.; Depprich, R.; Schwarz, F.; Meyer, U.; Kübler, N.R.; Naujoks, C. Comparison of Various Approaches for the Treatment of Fractures of the Mandibular Condylar Process. *J. Cranio-Maxillofac. Surg.* **2012**, *40*, e397–e401. [CrossRef]

41. Devlin, M.F.; Hislop, W.S.; Carton, A.T.M. Open Reduction and Internal Fixation of Fractured Mandibular Condyles Bya Retromandibular Approach: Surgical Morbidity and Informed Consent. *Br. J. Oral Maxillofac. Surg.* **2002**, *40*, 23–25. [CrossRef]

42. Eckelt, U.; Hlawitschka, M. Clinical and Radiological Evaluation Following Surgical Treatment of Condylar Neck Fractures with Lag Screws. *J. Craniomaxillofac. Surg.* **1999**, *27*, 235–242. [CrossRef]

43. Singh, P.; Mohanty, S.; Chaudhary, Z.; Sharma, P.; Kumar, J.; Verma, A. Does Mandibular Condylar Morphology after Fracture Healing Predict Functional Outcomes in Patients Treated with Closed Reduction? *J. Oral Maxillofac. Surg.* **2022**, *80*, 691–699. [CrossRef]

44. Berner, T.; Essig, H.; Schumann, P.; Blumer, M.; Lanzer, M.; Rücker, M.; Gander, T. Closed versus Open Treatment of Mandibular Condylar Process Fractures: A Meta-Analysis of Retrospective and Prospective Studies. *J. Craniomaxillofac. Surg.* **2015**, *43*, 1404–1408. [CrossRef] [PubMed]

45. Shakya, S.; Zhang, X.; Liu, L. Key Points in Surgical Management of Mandibular Condylar Fractures. *Chin. J. Traumatol.* **2020**, *23*, 63–70. [CrossRef]

46. Ferrillo, M.; Ammendolia, A.; Paduano, S.; Calafiore, D.; Marotta, N.; Migliario, M.; Fortunato, L.; Giudice, A.; Michelotti, A.; de Sire, A. Efficacy of rehabilitation on reducing pain in muscle-related temporomandibular disorders: A systematic review and meta-analysis of randomized controlled trials. *J. Back Musculoskelet. Rehabil.* **2022**, *35*, 921–936. [CrossRef] [PubMed]

47. Ferrillo, M.; Giudice, A.; Marotta, N.; Fortunato, F.; Di Venere, D.; Ammendolia, A.; Fiore, P.; de Sire, A. Pain Management and Rehabilitation for Central Sensitization in Temporomandibular Disorders: A Comprehensive Review. *Int. J. Mol. Sci.* **2022**, *23*, 12164. [CrossRef]

48. The PRISMA Statement for Reporting Systematic Reviews and Meta-Analyses of Studies That Evaluate Healthcare Interventions: Explanation and Elaboration | The BMJ. Available online: https://www.bmj.com/content/339/bmj.b2700 (accessed on 18 February 2023).

49. Talwar, R.M.; Ellis, E.; Throckmorton, G.S. Adaptations of the Masticatory System after Bilateral Fractures of the Mandibular Condylar Process. *J. Oral Maxillofac. Surg.* **1998**, *56*, 430–439. [CrossRef]

50. Throckmorton, G.S.; Talwar, R.M.; Ellis, E. Changes in Masticatory Patterns after Bilateral Fracture of the Mandibular Condylar Process. *J. Oral Maxillofac. Surg.* **1999**, *57*, 500–508; discussion 508–509. [CrossRef]

51. Choi, B.H. Comparison of Computed Tomography Imaging before and after Functional Treatment of Bilateral Condylar Fractures in Adults. *Int. J. Oral Maxillofac. Surg.* **1996**, *25*, 30–33. [CrossRef] [PubMed]

52. Kahl-Nieke, B.; Fischbach, R. Condylar Restoration after Early TMJ Fractures and Functional Appliance Therapy. Part II: Muscle Evaluation. *J. Orofac. Orthop.* **1999**, *60*, 24–38. [CrossRef] [PubMed]

53. Kuntamukkula, S.; Sinha, R.; Tiwari, P.K.; Paul, D. Dynamic Stability Assessment of the Temporomandibular Joint as a Sequela of Open Reduction and Internal Fixation of Unilateral Condylar Fracture. *J. Oral Maxillofac. Surg.* **2018**, *76*, 2598–2609. [CrossRef] [PubMed]

54. Raustia, A.M.; Oikarinen, K.S.; Pyhtinen, J. Changes in the Main Masticatory Muscles in CT after Mandibular Condyle Fracture. *Rofo* **1990**, *153*, 501–504. [CrossRef] [PubMed]

55. Salunkhe, S.M.; Kadam, H.; Nakhate, M.; Edsor, E.; Kamble, R.; Vadane, A.K. Evaluation of Masticatory Forces in Patients Treated for Mandibular Fractures: A Case-Control Study. *Cureus* **2022**, *14*, e29295. [CrossRef] [PubMed]

56. Sforza, C.; Tartaglia, G.M.; Lovecchio, N.; Ugolini, A.; Monteverdi, R.; Giannì, A.B.; Ferrario, V.F. Mandibular Movements at Maximum Mouth Opening and EMG Activity of Masticatory and Neck Muscles in Patients Rehabilitated after a Mandibular Condyle Fracture. *J. Cranio-Maxillofac. Surg.* **2009**, *37*, 327–333. [CrossRef]

57. Throckmorton, G.S.; Ellis, E.; Hayasaki, H. Masticatory Motion after Surgical or Nonsurgical Treatment for Unilateral Fractures of the Mandibular Condylar Process. *J. Oral Maxillofac. Surg.* **2004**, *62*, 127–138. [CrossRef]

58. Hjorth, T.; Melsen, B.; Møller, E. Masticatory Muscle Function after Unilateral Condylar Fractures: A Prospective and Quantitative Electromyographic Study. *Eur. J. Oral Sci.* **1997**, *105*, 298–304. [CrossRef]

59. Da Silva, A.P.; Sassi, F.C.; de Andrade, C.R.F. Oral-motor and electromyographic characterization of patients submitted to open and closed reductions of mandibular condyle fracture. *Codas* **2016**, *28*, 558–566. [CrossRef]

60. Loukota, R.A.; Eckelt, U.; Bont, L.D.; Rasse, M. Subclassification of Fractures of the Condylar Process of the Mandible. *Br. J. Oral Maxillofac. Surg.* **2005**, *43*, 72–73. [CrossRef]

61. Di Cosola, M.; Cazzolla, A.P.; Charitos, I.A.; Ballini, A.; Inchingolo, F.; Santacroce, L. Candida Albicans and Oral Carcinogenesis. A Brief Review. *J. Fungi* **2021**, *7*, 476. [CrossRef]

62. Nussbaum, M.L.; Laskin, D.M.; Best, A.M. Closed Versus Open Reduction of Mandibular Condylar Fractures in Adults: A Meta-Analysis. *J. Oral Maxillofac. Surg.* **2008**, *66*, 1087–1092. [CrossRef] [PubMed]

63. Santler, G.; Kärcher, H.; Ruda, C.; Köle, E. Fractures of the Condylar Process: Surgical versus Nonsurgical Treatment. *J. Oral Maxillofac. Surg.* **1999**, *57*, 392–397. [CrossRef] [PubMed]

64. Cantore, S.; Mirgaldi, R.; Ballini, A.; Coscia, M.F.; Scacco, S.; Papa, F.; Inchingolo, F.; Dipalma, G.; De Vito, D. Cytokine Gene Polymorphisms Associate with Microbiogical Agents in Periodontal Disease: Our Experience. *Int. J. Med. Sci.* **2014**, *11*, 674–679. [CrossRef]

65. Raustia, A.M.; Pyhtinen, J.; Virtanen, K.K. Examination of the Temporomandibular Joint by Direct Sagittal Computed Tomography. *Clin. Radiol.* **1985**, *36*, 291–296. [CrossRef]

66. Marrelli, M.; Tatullo, M.; Dipalma, G.; Inchingolo, F. Oral Infection by Staphylococcus Aureus in Patients Affected by White Sponge Nevus: A Description of Two Cases Occurred in the Same Family. *Int. J. Med. Sci.* **2012**, *9*, 47–50. [CrossRef]

67. Bulcke, J.A.; Termote, J.L.; Palmers, Y.; Crolla, D. Computed Tomography of the Human Skeletal Muscular System. *Neuroradiology* **1979**, *17*, 127–136. [CrossRef]

68. Häggmark, T.; Jansson, E.; Svane, B. Cross-Sectional Area of the Thigh Muscle in Man Measured by Computed Tomography. *Scand. J. Clin. Lab. Investig.* **1978**, *38*, 355–360. [CrossRef]

69. Inchingolo, F.; Tatullo, M.; Abenavoli, F.M.; Marrelli, M.; Inchingolo, A.D.; Gentile, M.; Inchingolo, A.M.; Dipalma, G. Non-Syndromic Multiple Supernumerary Teeth in a Family Unit with a Normal Karyotype: Case Report. *Int. J. Med. Sci.* **2010**, *7*, 378–384. [CrossRef]

70. Kuwahara, T.; Bessette, R.W.; Maruyama, T. Chewing Pattern Analysis in TMD Patients with and without Internal Derangement: Part I. *Cranio* **1995**, *13*, 8–14. [CrossRef]

71. Kuwahara, T.; Bessette, R.W.; Maruyama, T. Characteristic Chewing Parameters for Specific Types of Temporomandibular Joint Internal Derangements. *Cranio* **1996**, *14*, 12–22. [CrossRef]

72. Youssef, R.E.; Throckmorton, G.S.; Ellis, E.; Sinn, D.P. Comparison of Habitual Masticatory Patterns in Men and Women Using a Custom Computer Program. *J. Prosthet. Dent.* **1997**, *78*, 179–186. [CrossRef] [PubMed]

73. Throckmorton, G.S.; Ellis, E.; Hayasaki, H. Jaw Kinematics during Mastication after Unilateral Fractures of the Mandibular Condylar Process. *Am. J. Orthod. Dentofac. Orthop.* **2003**, *124*, 695–707. [CrossRef] [PubMed]

Journal of
Clinical Medicine

Article

Prevalence of Temporomandibular Disorders Based on a Shortened Symptom Questionnaire of the Diagnostic Criteria for Temporomandibular Disorders and Its Screening Reliability for Children and Adolescents Aged 7–14 Years

Mathias Rentsch [1,2], Aleksandra Zumbrunn Wojczyńska [1], Luigi M. Gallo [1] and Vera Colombo [1,*]

1   Clinic of Masticatory Disorders, Center of Dental Medicine, University of Zurich, 8032 Zurich, Switzerland; mathias.rentsch@uzh.ch (M.R.); aleksandra.zumbrunn@zzm.uzh.ch (A.Z.W.); luigi.gallo@zzm.uzh.ch (L.M.G.)

2   Public-School Dental Services of the City of Zurich, 8002 Zurich, Switzerland

*   Correspondence: vera.colombo@zzm.uzh.ch; Tel.: +41-44-634-33-53

Abstract: The prevalence and adequacy of diagnostic approaches for temporomandibular disorders (TMD) in children and adolescents are still matters of debate. This study aimed to determine the prevalence of TMD and oral habits in children and adolescents aged 7–14 years and evaluate the consistency between self-reported TMD symptoms and clinical findings using a shortened Axis I of Diagnostic Criteria for TMD (DC/TMD). Children (aged 7–10) and adolescents (aged 11–14) of both sexes were invited to participate in this study (n = 1468). Descriptive statistics for all observed variables and Mann–Whitney U-Tests for the clinical examination were performed. A total of 239 subjects participated in the study (response rate 16.3%). The self-reported prevalence of TMD was found to be 18.8%. The most frequently reported oral habit was nail biting (37.7%), followed by clenching (32.2%) and grinding (25.5%). Self-reported headache increased with age, while clenching and grinding decreased. Based on the answers to the DC/TMD Symptom Questionnaire, subgroups of asymptomatic and symptomatic participants (n = 59; 24.7%) were established and randomly selected for the clinical examination (f = 30). The shortened Symptom Questionnaire showed a sensitivity of 0.556 and a specificity of 0.719 for pain during the clinical examination. Although the Symptom Questionnaire exhibited high specificity (0.933), its sensitivity (0.286) for temporomandibular joint sounds was low. Disc displacement with reduction (10.2%) and myalgia (6.8%) were the most common diagnoses. In conclusion, the self-reported prevalence of TMD in children and adolescents in this study was comparable to that reported in the literature for adults. However, the accuracy of the shortened Symptom Questionnaire as a screening tool for TMD-related pain and jaw sounds in children and adolescents was found to be low.

Keywords: adolescents; children; diagnostic criteria for temporomandibular disorders; temporomandibular joint disorders

Citation: Rentsch, M.; Zumbrunn Wojczyńska, A.; Gallo, L.M.; Colombo, V. Prevalence of Temporomandibular Disorders Based on a Shortened Symptom Questionnaire of the Diagnostic Criteria for Temporomandibular Disorders and Its Screening Reliability for Children and Adolescents Aged 7–14 Years. *J. Clin. Med.* **2023**, *12*, 4109. https://doi.org/10.3390/jcm12124109

Academic Editor: Eiji Tanaka

Received: 24 May 2023
Revised: 8 June 2023
Accepted: 16 June 2023
Published: 17 June 2023

1. Introduction

Temporomandibular disorder (TMD) is a collective term describing dysfunction and pain in the masticatory muscles, as well as in the temporomandibular joints (TMJs) and related structures [1]. Subjects with TMD often exhibit a limited range of motion, joint noises, pain, or a combination of these symptoms [2,3]. TMDs are believed to have a complex, multifactorial etiology. According to recent literature, macrotraumas caused by impact injuries to the chin resulting from accidents [4–6], as well as microtraumas due to oral parafunctional habits such as clenching or bruxism [7], are considered etiological factors. Additionally, multiple or frequent oral parafunctions are found to be associated with the incidence of TMD [8]. Psychosocial factors such as stress, anxiety, insomnia, and depression [9–12] can significantly contribute to the development of TMD. For example, anxiety and stress can lead to increased muscle tension, central sensitization, and bruxism, while also reducing coping strategies [12–14]. Furthermore, incorporating psychosocial factors into the treatment strategy has been shown to improve the outcome of TMD treatment [15,16]. Systemic diseases such as rheumatoid arthritis, lupus erythematosus, juvenile idiopathic arthritis, and psoriatic arthritis [17] are also involved in the development of TMD. Comorbid pain, including pre-existing lower back pain or genital pain conditions, sleep disturbance or smoking, has been identified as an important predictor for TMD [18].

TMD prevalence in adults is estimated to be in the range of 5–50% [1,19,20]. TMD may occur at any age; however, the peak occurrence is between 20 and 40 years of age [19], and women are approximately twice as likely to be affected than men [21,22]. Furthermore, the congruence between self-reported TMD symptoms and diagnosed TMD shows a sensitivity of 0.43–0.49 and a specificity of 0.93–0.95 [23,24]. Reported TMD prevalence strongly depends on diagnostic criteria, clinical examination protocols, study population, and training of the investigators [21,25,26]. The Research Diagnostic Criteria for Temporomandibular Disorders (RDC/TMD) clinical examination tool, published in 1992 [27], became a gold standard in the diagnosis of TMD and was used for adults but also for adolescents and children with good reliability [28] until the Diagnostic Criteria for Temporomandibular Disorders (DC/TMD) clinical examination protocol replaced the RDC/TMD guidelines in 2014. DC/TMD is a validated screening tool for detecting TMD as well as for differentiating common pain-related TMDs in adults [29]. However, the DC/TMD examination has not yet been validated for use in children and adolescents.

The prevalence of TMD in children has already been investigated in previous studies using different diagnostic systems and has been shown to be similar to that in adults [30–32]. In contrast to adults, mixed results on sex differences [33–37] were obtained, and no differences in the mandibular range of motion for children with and without TMD were found [38,39]. Oral habits (i.e., biting nails, clenching or grinding teeth) in children and adolescents were as prevalent as in adults [34,40,41]. However, only a few studies were based on the DC/TMD protocol [33,36,42,43]. Only one study evaluated the accuracy of the DC/TMD protocol for TMD diagnosis in children aged 8–12 years, which showed a lower accuracy than in adults [28]. Lately, an international group of TMD experts tried to find a consensus on the DC/TMD Axis I using a Delphi study. It was agreed appropriate questionnaires about general health and demographics should be created for children (<10 years old) and adolescents (between 10 and 19 years old). Three screening questions (3Q/TMD) [44] instead of the TMD pain screener [45] should be used for both age groups. The Symptom Questionnaire should be rephrased and adapted for each group. Recommended revisions of the clinical examination contained the abandonment of mandatory commands, a different number of palpation sites, and a reduced threshold for limited mouth opening [46].

The main objective of this observational study was to conduct a clinical quantitative assessment of representative parameters for TMD in a sample of children and adolescents between the ages of 7 and 14 to evaluate the reliability of a shortened DC/TMD Symptom Questionnaire as a screening tool for clinical examination. The secondary objective was to estimate the self-reported prevalence of TMD in this specific age group using the collected data.

2. Materials and Methods

2.1. Recruitment and Study Participants

The city of Zurich, Switzerland, offers a yearly dental check-up free of charge for all children and adolescents between 2 and 18 years of age. Whole school classes attend one of the six public school dental clinics depending on their location. Recruitment and examinations in the present study were conducted in Zurich between August 2019 and February 2020. This study involved children (aged 7–10) and adolescents (aged 11–14) and was performed in one of the public school dental clinics of the city of Zurich.

For recruitment, two to three weeks prior to the visit to the dental clinic, second-, fourth-, sixth-, and eighth-grade teachers were given envelopes to distribute to the pupils. Each envelope contained the study information, an informed consent form, and a questionnaire about TMD symptoms. Pupils took the envelopes home to decide on participation in accordance with their legal guardians. Teachers were instructed to collect the envelopes and return them to the dental clinic on the day of the visit.

On examination day, two groups were created based on the answers given to the questionnaires. The symptomatic group included subjects with any pain in the jaw, temple, anteriorly to or inside the ear, and/or headache that included the temporal areas, and activities that influenced any headache they had experienced in the last 30 days. The remaining subjects (including subjects with pain-free joint noises) were assigned to the asymptomatic group (Figure 1). Afterward, only one child or adolescent of a class was clinically examined after the normal dental check-up. First, a symptomatic participant was randomly chosen for the test group. If the symptomatic participant was missing in the class, an asymptomatic participant was randomly chosen for the control group (Figure 1).

Recruitment flow chart for the symptomatic / asymptomatic group

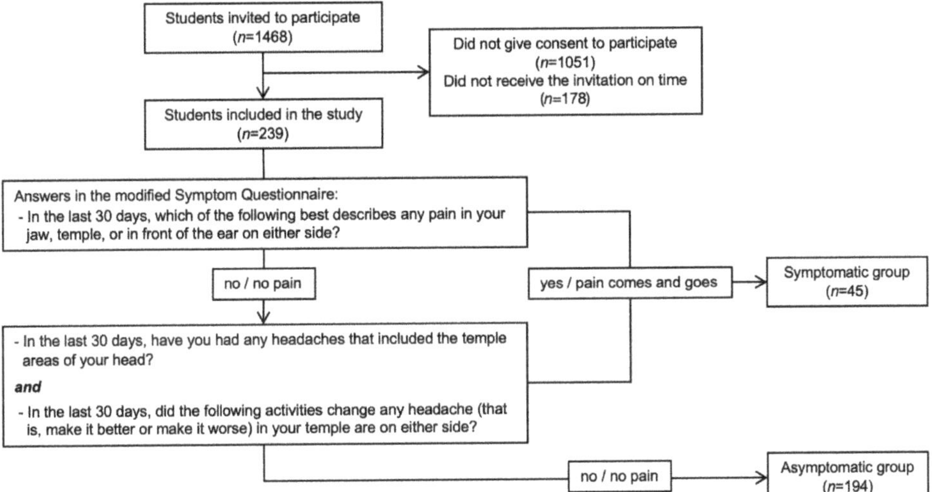

Recruitment flow chart for the clinical examination in each grade

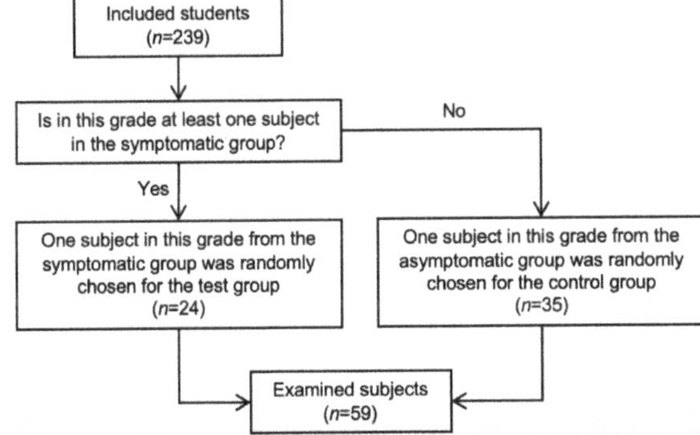

Figure 1. Recruitment flow chart for the test/control sample and for the clinical examination in each class.

2.2. The DC/TMD Symptom Questionnaire

A shortened German version of the DC/TMD questionnaire (Table 1) was used to evaluate the presence of TMD signs or symptoms [47]. All main questions about pain, headache, jaw joint noises, and closed as well as open locking of the jaw were included unaltered. Six questions were skipped due to ease and understanding. In order to evaluate the prevalence of oral habits, five additional questions about oral habits were added. The shortened DC/TMD Symptom Questionnaire and the answers were discussed with the participants prior to the clinical examination to detect divergent answers.

Table 1. Comparison between the original DC/TMD protocol [47] and the shortened version.

DC/TMD Symptom Questionnaire		DC/TMD Examination Protocol	
Original Version	Shortened Version	Original Version	Shortened Version
SQ1	SQ1	E1a	E1a
SQ2	—	E1b	E1b
SQ3	SQ3	E2	E2
SQ4	SQ4	E3	E3
SQ5	SQ5	E4a	E4a
SQ6	—	E4b	E4b
SQ7	SQ7	E4c	—
SQ8	SQ8	E4d	—
SQ9	SQ9	E5	E5
SQ10	—	E6	E6
SQ11	—	E7	E7
SQ12	—	E8	E8
SQ13	SQ13	E9	E9
SQ14	—	E10	—
—	OH *		

Numbers indicate the questionnaire number or the examination step. SQ = Symptom Questionnaire; E = examination; OH = oral habits. * Additional questions about oral habits: A. Do you grind your teeth?; B. Do you clench your teeth?; C. Do you bite your nails?; D. Do you suck a pacifier or your thumb?; E. Do you play a wind instrument?

2.3. Clinical Examiner (Calibration)

The examiner (MR, pediatric dentist employed by the school dental clinic) underwent training by an orofacial pain specialist with expertise in DC/TMD examination at the University of Zurich (AZW). The training included theory and palpation techniques first on adults and then on an eight-year-old boy at the University of Zurich.

2.4. The Clinical DC/TMD Examination

The clinical examination was performed according to the DC/TMD Examination Protocol. Maximum assisted opening, termination of movement, and the examination of supplemental muscle pain with palpation were skipped to ease and avoid discomfort (Table 1). Incisal relationships, maximum opening, and jaw movements were measured with a ruler. TMJ noises were detected bilaterally via palpation. Calibration of the palpation pressure was conducted using commercially available electronic scales [48]. The use of mandatory DC/TMD commands was omitted due to unimpaired results [49] and better understanding for children and adolescents.

2.5. Diagnosis of TMD

TMD diagnosis was made based on the DC/TMD Diagnostic Decision Tree and, accordingly, the Diagnostic Criteria Table. The shortened DC/TMD Symptom Questionnaire (Table 2) and the condensed DC/TMD clinical examination were incorporated into the TMD diagnosis. The diagnoses of intra-articular joint disorders were made based on the clinical findings.

Table 2. Prevalence of self-reported pain, jaw function disturbances, and orals habits as reported in the shortened DC/TMD Symptom Questionnaire.

Answers (*n* = 239)	No/No Pain *n* (%)					Yes/Pain Comes and Goes *n* (%)				
	Total	Girls	Boys	Aged 7–10	Aged 11–14	Total	Girls	Boys	Aged 7–10	Aged 11–14
Ever experienced pain in the temporomandibular region	173 (72.4)	92 (73.6%)	81 (71.1)	100 (73.0)	73 (71.6)	66 (27.6)	33 (26.4)	33 (28.9)	37 (27.0)	29 (28.4)
Description of any pain in the temporomandibular region in last 30 days	207 (86.6)	109 (87.2)	98 (86.0)	122 (89.1)	85 (83.3)	32 (13.4)	16 (12.8)	16 (14.0)	15 (10.9)	17 (16.7)
Pain during jaw activities										
A. Chewing	227 (95.0)	119 (95.2)	108 (94.7)	132 (96.4)	95 (93.1)	12 (5.0)	6 (4.8)	6 (5.3)	5 (3.6)	7 (6.9)
B. Opening/movements to the front or to the side	232 (97.1)	122 (97.6)	110 (96.5)	133 (97.1)	99 (97.1)	7 (2.9)	3 (2.4)	4 (3.5)	4 (2.9)	3 (2.9)
C. Jaw habits	229 (95.8)	120 (96.0)	108 (95.6)	136 (99.3)	93 (91.2)	10 (4.2)	5 (4.0)	5 (4.4)	1 (0.7)	9 (8.8)
D. Other jaw activities (talking, kissing, yawning, …)	235 (98.3)	123 (98.4)	112 (98.2)	137 (100)	98 (96.1)	4 (1.7)	2 (1.6)	2 (1.8)	0 (0.0)	4 (3.9)
Temporal headache in the last 30 days	188 (78.7)	97 (77.6)	91 (79.8)	115 (83.9)	73 (71.6)	51 (21.3)	28 (22.4)	23 (20.2)	22 (16.1)	29 (28.4)
Temporal headache during jaw activities										
A. Chewing	223 (93.3)	115 (92.0)	108 (94.7)	132 (96.4)	91 (89.2)	16 (6.7)	10 (8.0)	6 (5.3)	5 (3.6)	11 (10.8)
B. Opening/movements to the front or to the side	232 (97.1)	121 (96.8)	111 (97.4)	133 (97.1)	99 (97.1)	7 (2.9)	4 (3.2)	3 (2.6)	4 (2.9)	3 (2.9)
C. jaw habits	224 (93.7)	115 (91.2)	109 (96.5)	132 (96.4)	92 (90.2)	15 (6.3)	11 (8.8)	4 (3.5)	5 (3.6)	10 (9.8)
D. Other jaw activities (talking, kissing, yawning, …)	228 (95.4)	118 (94.4)	110 (96.5)	134 (97.8)	94 (92.2)	11 (4.6)	7 (5.6)	4 (3.5)	3 (2.2)	8 (7.8)
Jaw joint noises	213 (89.1)	109 (87.2)	104 (91.2)	121 (88.3)	92 (90.2)	26 (10.9)	16 (12.8)	10 (8.8)	16 (11.7)	10 (9.8)
Closed locking of the jaw	233 (97.5)	121 (96.8)	112 (98.2)	136 (99.3)	97 (95.1)	6 (2.5)	4 (3.2)	2 (1.8)	1 (0.7)	5 (4.9)
Open locking of the jaw	236 (98.7)	123 (98.4)	113 (99.1)	135 (98.5)	101 (99.0)	3 (1.3)	2 (1.6)	1 (0.9)	2 (1.5)	1 (1.0)
Oral habits										
A. Grinding	162 (67.8)	78 (62.4)	84 (73.7)	85 (62.0)	77 (75.5)	77 (32.2)	47 (37.6)	30 (26.3)	52 (38.0)	25 (24.5)
B. Clenching	178 (74.5)	96 (76.8)	82 (71.9)	101 (73.7)	77 (75.5)	61 (25.5)	29 (23.2)	32 (28.1)	36 (26.3)	25 (24.5)

Table 2. *Cont.*

Answers (n = 239)	No/No Pain n (%)					Yes/Pain Comes and Goes n (%)				
C. Nail biting	149 (62.3)	82 (65.6)	67 (58.8)	99 (72.3)	50 (49.0)	90 (37.7)	43 (34.4)	47 (41.2)	38 (27.7)	52 (51.0)
D. Pacifier/thumb sucking	231 (96.7)	121 (96.8)	110 (96.5)	131 (95.6)	100 (98)	8 (3.3)	4 (3.2)	4 (3.5)	6 (4.4)	2 (2.0)
E. Wind instrument	212 (88.7)	109 (87.2)	103 (90.4)	122 (89.1)	90 (88.2)	27 (11.3)	16 (12.8)	11 (9.6)	15 (10.9)	12 (11.8)

2.6. Data Analysis and Statistics

The questionnaire was analyzed with descriptive statistics using crosstabulations (e.g., age group, sex). The data were summarized and visualized with tables and diagrams. In a further step, Pearson's correlation coefficients were used to correlate the data of the questionnaire with sex, age group, and test and control sample. Results of the clinical examination were analyzed with descriptive statistics. Mean values and standard deviations were obtained for all parameters in all conditions observed. Nonparametric Mann–Whitney tests for independent samples were used to assess the differences in the quantitative results of clinical examinations (overjet, overbite, pain-free mouth opening, maximum unassisted mouth opening, laterotrusion, protrusion) between test and control subjects globally in both age groups. If a significant difference was found, pairwise tests for the variable were performed on sex and age group. The level of significance was set at α = 0.05. Microsoft Excel (Version 2019, Microsoft Corporation, Redmond, WA, USA) and SPSS (Version 26.0., IBM Corporation, Armonk, NY, USA) for Windows were used to perform statistical analyses.

3. Results

3.1. Study Participants

A total of 1468 letters were sent out to 71 classes of four different levels. Of these, 239 children and adolescents (16.3%) agreed to participate in the study. In 8 of the 71 invited classes (178 students), the study envelopes were not delivered in time for the evaluation of potential participants by their teachers. In 5 of 71 classes, none of the participants gave consent to participate. Subjects were divided into two age groups for subsequent analysis: 7–10 years (second and fourth grade, n = 137); 11–14 years (sixth and eighth grade, n = 102).

3.2. Symptom Questionnaire

A total of 239 subjects completed the shortened DC/TMD Symptom Questionnaire. The entire study sample consisted of 114 boys (47.7%) and 125 girls (52.3%). The mean age of all participants was 10.0 ± 1.9 (range 7–14 years). Almost 19% of the participants reported that they were currently suffering from TMD symptoms. Sixty-six subjects (27.6%) stated that they had experienced pain in the jaw, temple, or anteriorly to or inside the ear on either side earlier in their life. In addition, 32 subjects (13.4%) described pain that "comes and goes" in the past 30 days, and jaw activities modified the pain in almost a third of the subjects. Temporal headache in the last 30 days was reported by 51 of 239 children and adolescents (21.3%). In almost 30% of these cases, chewing and/or habits modified the headache. Pearson's correlation coefficient of 0.944 showed an increase in the prevalence of headage with age (Figure 2).

TMJ noises occurring in the last 30 days were reported by 26 individuals (10.9%). At least one oral habit was found in 160 subjects (66.9%). The most frequently reported oral habit was nail biting, which was found in 90 pupils (37.7%), followed by clenching in 77 pupils (32.2%), and grinding teeth in 61 pupils (25.5%). Self-reported teeth grinding and clenching decreased with age. Detailed information about the answers to the shortened Symptom Questionnaire and oral habits can be found in Table 2.

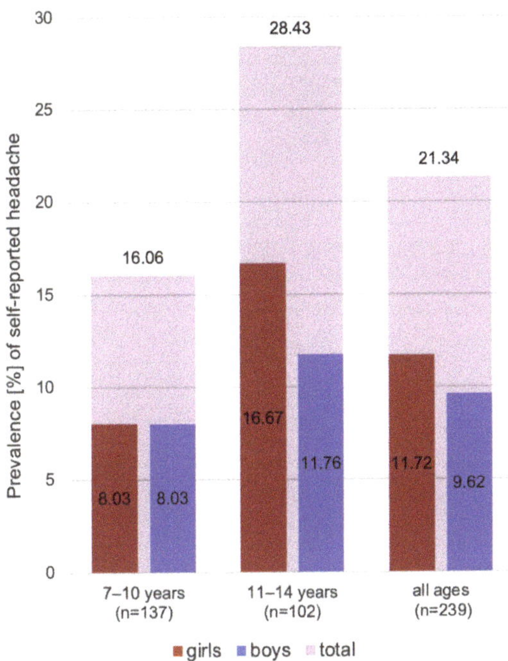

Figure 2. Increase in prevalence of self-reported headache by age and sex.

3.3. Clinical Examination

Of the 59 subjects examined clinically, 29 were boys (49.2%) and 30 were girls (50.8%), with a mean age of 10.0 ± 2.3 years (range 7–14). The age distribution of the subjects was as follows: 35 subjects (59.3%) in the younger age group and 24 subjects (40.7%) in the older age group. Of the whole sample, 35 (59.3%) asymptomatic subjects belonged to the controls and 24 (40.7%) symptomatic subjects to the test group. No subject reported changes in the given answers in the Symptom Questionnaire.

3.3.1. Self-Reported Localization of Pain

A total of 14 (23.7%) participants (control: $n = 0$; test: $n = 14$) reported pain in the temporomandibular region. Among them, the masseter muscle was indicated as a site of pain eight times, the temporal muscle four times, and the TMJs eight times. Headache was indicated by 20 (33.9%) pupils (control: $n = 4$; test: $n = 16$). Nine subjects indicated the temporal region (control: $n = 2$; test = 7), ten (control: $n = 2$; test: $n = 8$) indicated other regions, and one test subject reported headache in both the temporal region and other regions.

3.3.2. Jaw Motion and TMJ Sounds

Both pain-free mouth opening and maximum unassisted mouth opening ranged between 35 and 61 mm. The mean measured values and standard deviation for mouth opening and lateral and protrusive movements divided by age, sex, and control and test sample are shown in Table 3. The Mann–Whitney test revealed significantly larger pain-free mouth opening ($p = 0.025$) and maximum unassisted opening ($p = 0.015$) in the test group than in the control group. Maximum unassisted opening increased significantly with age ($p = 0.014$). Sex did not seem to have an influence on maximum pain-free and unassisted mouth opening. Familiar pain was reported by five children (control: $n = 2$; test: $n = 3$) during opening movements and by three test subjects during horizontal movements.

Table 3. Average values measured in millimeters (mm) and standard deviation (*SD*) for mouth opening, lateral movements, and protrusion divided according to age group, sex, and control/test.

Group	Pain-Free Opening	Maximum Un-Assisted Opening	Right Lateral Movement	Left Lateral Movement	Protrusion
		Age Group (in Years)			
7–10	45.8 (5.76)	46.71 (5.72)	9 (2.11)	9.31 (1.97)	9.4 (2.14)
11–14	48.29 (5.61)	50.17 (5.47) *	9.54 (1.59)	9.58 (2.54)	10.17 (1.27)
		Sex			
Girls	47.07 (5.36)	48.4 (5.46)	8.83 (1.82)	9.4 (1.71)	9.37 (1.87)
Boys	46.55 (6.29)	47.83 (6.27)	9.62 (1.97)	9.45 (2.64)	10.07 (1.83)
		Control/Test			
Control	45.49 (6.1)	46.63 (5.96)	9.09 (2.27)	9.11 (2.61)	9.51 (3.67)
Test	48.75 (4.78) *	50.29 (4.99) *	9.42 (1.28)	9.88 (1.33)	10 (1.79)
		Total			
Total	46.81 (5.79)	48.12 (5.83)	9.22 (1.92)	9.42 (2.2)	9.71 (1.87)

* $p < 0.05$.

Jaw joint noises occurred in 14 subjects (23.7%); in six of these cases (control: $n = 4$; test: $n = 2$), a sound was heard by the child and detected by the examiner, and in eight cases (control: $n = 7$; test: $n = 1$), the sound was only detected by the examiner.

3.3.3. Palpation of Muscles and TMJs

Pain during palpation of the masseter and temporal muscle was indicated by 20 subjects (control: $n = 10$; test: $n = 10$). Additionally, three test subjects (12.5%) reported familiar pain, while one control pupil (2.9%) showed familiar pain during muscle palpation. During palpation of the TMJs, 23 children reported pain (control: $n = 9$; test: $n = 14$). In addition, six of them (control: $n = 1$; test: $n = 5$) indicated that they had familiar pain. The most frequently reported site of pain during palpation was the lateral pole of the TMJs.

3.4. Comparison between Symptom Questionnaire and Self-Reported Localization of Pain

The 41 subjects examined (control: $n = 35$; test: $n = 6$) who stated that they did not experience episodic pain in the Symptom Questionnaire also reported no pain location prior to the examination. However, four subjects who reported episodic pain in the questionnaire did not indicate any pain location. The remaining 14 test subjects localized pain in both the questionnaire and the examination.

Three pupils (5.1%) provided divergent answers between the Symptom Questionnaire and self-reported headache localization on examination day, whereas the other responses were consistent.

3.5. Comparison of Symptom Questionnaire and Clinical Examination

The sensitivity and specificity for the pain answers in the Symptom Questionnaire and the clinical findings were 0.556 and 0.719. During the clinical examination, 34.3% of the controls and 62.5% of the test subjects reported pain. Pain during palpation of the masseter muscle, temporal muscle, and/or TMJs was recorded in 14 of 24 pupils (58.3%) in the test group. In addition to pain during palpation, six children also showed pain during mandibular movements. Only one child presented pain during the opening phases alone. Furthermore, 12 control subjects (34.3%) reported pain at palpation and 5 (14.3%) during jaw movements. No pain during the clinical examination was found in 23 control subjects and 9 test subjects (Figure 3).

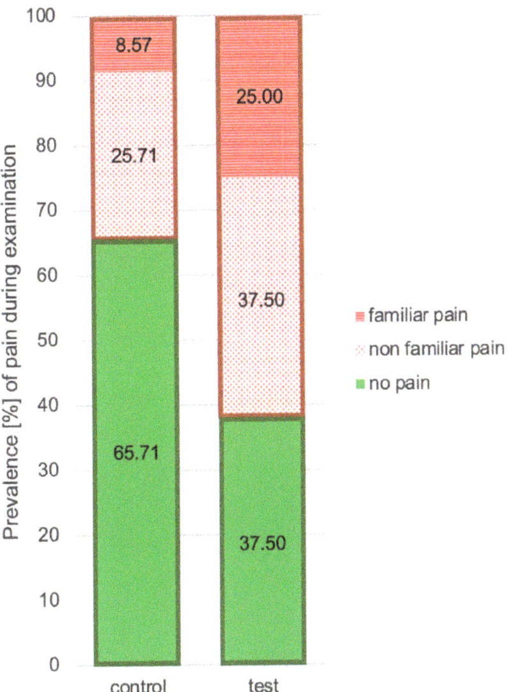

Figure 3. Prevalence of pain during examination (palpation and mouth movements) by control and test.

Sensitivity and specificity for joint sounds in the Symptom Questionnaire were 0.286 and 0.933. Subjects reporting joint sounds indeed showed a clicking sound during movement in four out of seven cases. A sound was also recorded in 10 of 52 pupils (19.2%) that did not report experiencing sounds in the questionnaire. Moreover, 8 of these 10 pupils heard no sound during the examination, even when indicated by the examiner.

3.6. Comparison between Self-Reported Pain Localization and Clinical Examination

The sensitivity and specificity of positive pain localization were 0.407 and 0.906. A total of 64.4% of the pupils who indicated no pain location also felt no pain during the examination. Agreement between self-reported pain localization and clinical examination was achieved in 81% of cases for the masseter and/or TMJ structures and in 25% for the temporal muscle. Additionally, we combined positive pain localization and temporal headache localization to verify the accuracy of the temporal muscle. As a result, we improved sensitivity but weakened specificity. For the combined measure, we observed a sensitivity and specificity of 0.484 and 0.857.

3.7. Diagnoses

Ten individuals had one or more TMD diagnoses. Disc displacement with reduction in five subjects was the most common diagnosis, followed by local myalgia, which was reported in two subjects. Arthralgia was diagnosed in one case, whereas two subjects showed combined diagnoses: in one case local myalgia, arthralgia, and disc displacement with reduction, and in the other myofascial pain with referral combined with arthralgia (Table 4).

Table 4. Individually considered TMD diagnoses according to the DC/TMD diagnostic decision tree among 59 examined children and adolescents.

	Number (%)	Sex		Age Group (Years)	
		Girls	Boys	7–10	11–14
Myalgia	4 (6.8)	2	2	1	3
Local myalgia	3 (5.1)	2	1	1	2
Myofascial pain with referral	1 (1.7)		1		1
Arthralgia	3 (5.1)	1	2	1	2
Disc displacement with reduction [†]	6 (10.2)	2	4	4	2
Total number of examined symptomatic subjects with a TMD diagnosis	10 * (16.9)				

[†] For one student, the diagnosis of disc displacement could not be subdivided due to the shortened Symptom Questionnaire; * Multiple diagnoses are possible.

4. Discussion

4.1. Prevalence of Self-Reported Oral Habits and TMD

The aim of the present study was to estimate the prevalence of oral habits, self-reported TMD symptoms, and TMD diagnosed using a shortened German version of the DC/TMD. The most frequently reported oral habit was nail biting, followed by teeth clenching and grinding, both of which decreased with age. Conversely, self-reported headache was more frequent with increasing age. The prevalence of self-reported TMD was comparable to the adult population or even slightly lower. Furthermore, disc displacement with reduction was the most common diagnosis, followed by myalgia.

4.1.1. Oral Habits

Nail biting was found to be the most frequently mentioned oral habit, followed by teeth grinding and clenching, which decreased with age. Our results agree with those of other studies conducted on similar age samples, which showed a prevalence of nail biting around 44–58% and 10–32% for clenching/grinding [34,35,50]. The importance of oral habits for the onset of TMD and the cause-and-effect relationship need to be re-examined in larger longitudinal studies [51,52].

4.1.2. Headache

Our results indicate an increase in headache with age, which corresponds with a nationwide Austrian study that showed that the incidence of headache is associated with older age in pediatric subjects [53]. Furthermore, seven out of ten subjects diagnosed with TMD according to DC/TMD criteria reported suffering from headache in our study. A prospective study showed that the presence of other pain conditions (e.g., headache) at baseline is a predictive factor for the onset of facial pain and TMD in 11-year-olds [54]. Moreover, comorbidity between TMD and headache is bidirectional for both conditions [55]. Therefore, children and adolescents with headache should be screened early for signs of TMD.

4.1.3. Prevalence of Self-Reported TMD and Diagnosis of TMD

The prevalence of pain-related TMD in our study was 18.8%, in agreement with a Swedish systematic review in which the prevalence ranged from 7.3% to 30.4% [30], and with a Dutch cross-sectional questionnaire survey with a prevalence of self-reported TMD of 21.6% [35]. Compared with adults, the prevalence of TMD in children appears to be similar or slightly lower [21,30,56]. The most frequent diagnosis was disc displacement with reduction, followed by myalgia. Even though our results may be compromised because of the shortened Symptom Questionnaire and the clinical examination, they are consistent with other European studies [33,34,50,57]. In contrast, other studies from Brazil, China, and Saudi Arabia found that myofascial pain was the most common diagnosis [2,37,58]. Therefore, ethnic discrepancies cannot be ruled out, as has also been observed in the adult population in the United States [21]. Nevertheless, because of the different forms of clinical examination and different diagnostic criteria, the results are not completely comparable.

4.2. Reliability of the DC/TMD Symptom Questionnaire

In addition to TMD prevalence, we analyzed the reliability of a self-completed Symptom Questionnaire as a screening tool in children and adolescents aged 7–14 years. Our main results show low sensitivity and specificity of the Symptom Questionnaire for pain. Compared to the given answers in the Symptom Questionnaire, specificity for pain during the clinical examination increased when the subject was directly asked about the location of pain, while sensitivity decreased. Pain during palpation was mainly found in the lateral TMJ pole or in the masseter muscle, while several controls also reported pain. Most of the subjects who indicated pain in the masseter muscle and/or TMJs also experienced pain during the examination, while for the temporal muscle, this was the case in only a quarter of the subjects. Sensitivity for joint noises was low, while specificity was high. TMJ noises occurred in almost a quarter of the sample, whereas the examiner alone heard sounds in half of the subjects.

4.2.1. DC/TMD Symptom Questionnaire

Our findings show that the reliability of the shortened DC/TMD Symptom Questionnaire in children is lower than for the full version in adults. Nevertheless, our findings are comparable with a German study that used the full version for children aged 8–12 [42]. Based on history and clinical examination, the DC/TMD shows good sensitivity and specificity for TMD in adults [29]. The discrepancy between the two groups could be explained by differences in pain perception [59] and in pain memory [60]. Furthermore, the unequal comprehension and language ability of the subjects due to their development [61] could have led to a misunderstanding of the questions or the clinical examination.

Therefore, our findings indicate that the shortened DC/TMD Symptom Questionnaire is not an adequate screening tool for TMD in children and adolescents. In conclusion, the TMD pain screener [45] presumably shows similar results because the Symptom Questionnaire includes its items. The three validated screening questions (3Q/TMD) are used in adults and adolescents with good reliability [44,62,63]. Although the screening tool has not yet been tested for children, its use has recently been recommended for both age groups [46]. Further studies are needed to validate the 3Q/TMD in children.

Despite low reliability, we cautiously recommend the use of the DC/TMD assessment protocol as a reference template in general practice. The current DC/TMD protocol can serve as a scheme for dentists to minimize the number of undiagnosed cases until new guidelines and assessment tools are established. The lack of a standardized protocol demonstrates the need for a validated screening and examination tool for children and adolescents. An international consortium of TMD experts is therefore currently working on developing an adapted and validated DC/TMD Axis I and II protocol specifically designed for this particular age group [46,64].

4.2.2. Mouth Opening and Horizontal Movements

The older age group showed significantly larger unassisted mouth opening than the younger age group. Additionally, test subjects also showed significantly larger unassisted mouth opening than controls. A Swiss study measured maximum unassisted mouth opening in 20,719 pupils (F: 10,060 with a median age range of 9.9 years (3.3–18.3); M: 10,659 with a median age range of 10.0 years (2.8–18.7)). Up to the age of 13, no significant sex differences were found. Later, between the ages of 14 and 17, boys showed greater mouth opening than girls. In summary, their study showed an increase in mouth opening with age, but with a wide range within children of the same age group, most likely due to different craniofacial morphologies and skeletal ages. Therefore, they recommended observing individual changes in maximum mouth opening in children at high risk for TMJ afflictions [65]. Our results of greater unassisted mouth opening in the test group could be explained by the wide mouth opening range within children of the same age group and the differences in craniofacial morphology and skeletal age. Not examining maximum assisted opening might negatively affect the sensitivity and specificity for myalgia and arthralgia.

4.2.3. TMJ Noises

More than half of the children with TMJ noises were not aware of any sounds, although they were present during movements, as identified by the examiner. Our findings are consistent with those of a German research group [42] which found that only 3.2% of children were aware of TMJ sounds. Therefore, they suggest that only examiner confirmation should be considered in cases where an examination is needed to diagnose degenerative joint diseases, although patient confirmation is required for the diagnosis [42]. The sensitivity for the diagnosis of disc displacement with reduction in adults is 0.34, while the specificity is 0.92 [29]. Magnetic resonance imaging (MRI) improves sensitivity by up to 78% [66]. Furthermore, a retrospective study of 56 patients with multiple sclerosis without TMD symptoms showed deviated disc position in 12% of cases (Schuknecht B, Kuhn F, MRI Zürich, unpublished normative data 2017). Therefore, inconsistency between self-reported symptoms and clinical signs should be considered when diagnosing disc displacement with reduction in adults and children.

4.2.4. Palpation of Muscle and TMJs

Pupils in the control and test sample overall reported pain during palpation mainly in the masseter and TMJs, with good congruence between the location of self-reported pain and clinical examination, whereas there seemed to be a discrepancy for the temporal muscle. Therefore, we pooled pain in the temporal muscle and headache in the temporal region to see if we were able to increase sensitivity. The improved results suggest that children and adolescents are unaware of the difference between pain in the temporal region and headache in the temporal region. When defining new questionnaires, this result should be taken into consideration.

Furthermore, although subjects reported no pain in the questionnaire, almost 30% experienced pain during palpation and a non-negligible number of controls reported familiar pain. Similar results were found in a German study, which showed that 36.2% of the children experienced pain during palpation, although they reported no pain previously [42]. Therefore, the question arises whether the amount of pressure applied is too high or whether new examination questions appropriate for children with higher sensitivity and specificity should be formulated.

4.3. Limitations

The main limitation of this study was the use of a shortened version of the existing DC/TMD Symptom Questionnaire and examination protocol, which has not yet been validated for children and adolescents [29]. An international consortium of TMD specialists recently recommended some adjustments to the adult version of the DC/TMD [46]. Although most of the recently published suggestions have already been included in our study, we did not use a modified Symptom Questionnaire adapted for each age group. Instead, we shortened the existing DC/TMD Symptom Questionnaire and examination protocol. Regardless of the alterations, the results should maintain their comparability with other studies, and the impact on the DC/TMD Diagnostic Decision Tree and Table should be minimized. Three of six skipped questions from the DC/TMD Symptom Questionnaire and two of three skipped examinations did not influence the DC/TMD Diagnostic Decision Tree and Table in our study. Intra-articular joint disorders could not be clearly specified in one case (participant number 223) because of the omission of three questions in the Symptom Questionnaire. The other five diagnoses could be correctly made after the DC/TMD Diagnostic Decision Tree and Table. Furthermore, maximum assisted opening was not examined, which might have led to false-negative (undetected) diagnoses concerning myalgia and arthralgia. Therefore, the prevalence of myalgia and arthralgia might be underestimated in our study. However, Katsikogianni et al. showed that there is no significant difference between maximum assisted or unassisted opening in children with or without pain report, and that maximum opening (assisted or unassisted) was painful for 69% of the children with pain and for 41% of the children without pain [42]. Therefore, we decided to omit this examination step to avoid discomfort and additional pain for the pupils although the results may be compromised. Despite our study limitations, our results remain comparable to the current literature [30,33–35,50,57].

Another limitation of this study was the number of participants due to the logistics of the study. In addition, there were some potential limitations due to the recruitment strategy. Participants were recruited indirectly by their teachers and depended on them for the delivery of the study documents on the day of the dental visit. Although the differences between children-reported and parent-reported pain were low [67], bias due to parents' answers in the questionnaire cannot be excluded.

4.4. Implications

Early screening and diagnosis are important to avoid misinformation, to improve prevention of chronic pain, and to reduce overtreatment [68–70]. Moreover, orofacial pain has an individual but also a social and economic burden [71]. TMD has a strong impact on quality of life [72], and TMD patients consult several healthcare providers before seeing a TMD specialist [73,74]. In addition, treatment costs are estimated at CHF 1778 (approximately USD 1950) per patient in Switzerland, which is almost 30% of the average monthly salary. About 45% of the costs are covered by general insurance, while 55% are borne by the patient [75]. On the other hand, TMD patients show reduced productivity at work [71] and increased school absences [76]. Therefore, early screening and proper diagnosis, together with appropriate treatment, are essential for each individual and society. Dentists could play a key role because they are often the first contact and periodically check their patients through regular recall and follow-up systems. The lack of a standardized screening and diagnostic tool with high accuracy for children complicates the work of the general dentist. We hope that our findings will help improve awareness of the signs and symptoms of TMD in children and adolescents among clinicians and emphasize the need for standardized screening and examination tools. In addition, we hope that our results contain useful information for the creation of a new assessment protocol.

Another important aspect of TMD is its large intra-individual fluctuation during the affected person's lifetime [77,78], which demonstrates the need for longitudinal studies during childhood and adolescence. Further follow-up studies can better elucidate the development of TMD in relation to age.

5. Conclusions

In conclusion, the accuracy of the self-completed shortened DC/TMD Symptom Questionnaire is low. The reason for this is either our adjustments, which may have compromised the diagnostic process and the TMD diagnosis itself, or the fact that neither the Symptom Questionnaire or the clinical examination protocol is suitable for children and adolescents. Standardized screening and examination tools are therefore required for these specific age groups. Dentists should be aware of the relatively high prevalence of TMD in children and adolescents. Furthermore, TMJ sounds are often not perceived by subjects in this age group. The masseter and TMJs show good congruence between self-reported pain location and clinical examination. Children and adolescents tend not to be able to distinguish between temporal pain and temporal headache. Disc displacement with reduction is the most common type of TMD among pediatric subjects, followed by myalgia.

Author Contributions: Conceptualization, A.Z.W., L.M.G., M.R. and V.C.; methodology, M.R.; software, V.C.; validation, A.Z.W., L.M.G., M.R. and V.C.; formal analysis, M.R.; investigation, M.R.; resources, M.R. and V.C.; data curation, M.R. and V.C.; writing—original draft preparation, M.R.; writing—review and editing, A.Z.W., L.M.G., M.R. and V.C.; visualization, M.R.; supervision, L.M.G. and V.C.; project administration, L.M.G. and V.C.; All authors have read and agreed to the published version of the manuscript.

Funding: This research received no external funding.

Institutional Review Board Statement: The study was conducted in accordance with the Declaration of Helsinki, and approved by the Review Board of the State of Zurich (KEK-ZH, no. 2019-01361).

Informed Consent Statement: Informed consent was obtained from all subjects (or their legal guardians if the participant was younger than 14 years old) involved in the study.

Data Availability Statement: The data are available upon request from the corresponding author.

Acknowledgments: The authors would like to thank all participants and the City of Zurich, Switzerland, for allowing our study to be performed.

Conflicts of Interest: The authors declare no conflict of interest.

References

1. LeResche, L. Epidemiology of temporomandibular disorders: Implications for the investigation of etiologic factors. *Crit. Rev. Oral Biol. Med.* **1997**, *8*, 291–305. [CrossRef]
2. Al-Khotani, A.; Naimi-Akbar, A.; Albadawi, E.; Ernberg, M.; Hedenberg-Magnusson, B.; Christidis, N. Prevalence of diagnosed temporomandibular disorders among Saudi Arabian children and adolescents. *J. Headache Pain* **2016**, *17*, 41. [CrossRef]
3. Laskin, D.M.; Greene, C.S.; Hylander, W.L. *Temporomandibular Disorders: An Evidence-Based Approach to Diagnosis and Treatment*; Quintessence Pub.: Chicago, IL, USA, 2006.
4. Fischer, D.J.; Mueller, B.A.; Critchlow, C.W.; LeResche, L. The association of temporomandibular disorder pain with history of head and neck injury in adolescents. *J. Orofac. Pain* **2006**, *20*, 191–198. [PubMed]
5. Greco, C.M.; Rudy, T.E.; Turk, D.C.; Herlich, A.; Zaki, H.H. Traumatic onset of temporomandibular disorders: Positive effects of a standardized conservative treatment program. *Clin. J. Pain* **1997**, *13*, 337–347. [CrossRef] [PubMed]
6. Imahara, S.D.; Hopper, R.A.; Wang, J.; Rivara, F.P.; Klein, M.B. Patterns and outcomes of pediatric facial fractures in the United States: A survey of the National Trauma Data Bank. *J. Am. Coll. Surg.* **2008**, *207*, 710–716. [CrossRef] [PubMed]
7. Dym, H.; Israel, H. Diagnosis and treatment of temporomandibular disorders. *Dent. Clin. N. Am.* **2012**, *56*, 149–161. [CrossRef] [PubMed]
8. Ohrbach, R.; Fillingim, R.B.; Mulkey, F.; Gonzalez, Y.; Gordon, S.; Gremillion, H.; Lim, P.F.; Ribeiro-Dasilva, M.; Greenspan, J.D.; Knott, C.; et al. Clinical findings and pain symptoms as potential risk factors for chronic TMD: Descriptive data and empirically identified domains from the OPPERA case-control study. *J. Pain* **2011**, *12*, T27-45. [CrossRef]
9. Karibe, H.; Goddard, G.; Kawakami, T.; Aoyagi, K.; Rudd, P.; McNeill, C. Comparison of subjective symptoms among three diagnostic subgroups of adolescents with temporomandibular disorders. *Int. J. Paediatr. Dent.* **2010**, *20*, 458–465. [CrossRef]
10. de Leeuw, R.; Klasser, G.D. *Orofacial Pain: Guidelines for Assessment, Diagnosis, and Management*; Quintessence Publishing Company, Incorporated: Hanover Park, IL, USA, 2018.
11. List, T.; Wahlund, K.; Larsson, B. Psychosocial functioning and dental factors in adolescents with temporomandibular disorders: A case-control study. *J. Orofac. Pain.* **2001**, *15*, 218–227. [PubMed]
12. Minervini, G.; Franco, R.; Marrapodi, M.M.; Mehta, V.; Fiorillo, L.; Badnjević, A.; Cervino, G.; Cicciù, M. The Association between COVID-19 Related Anxiety, Stress, Depression, Temporomandibular Disorders, and Headaches from Childhood to Adulthood: A Systematic Review. *Brain Sci.* **2023**, *13*, 481. [CrossRef]
13. Fillingim, R.B.; Ohrbach, R.; Greenspan, J.D.; Knott, C.; Dubner, R.; Bair, E.; Baraian, C.; Slade, G.D.; Maixner, W. Potential psychosocial risk factors for chronic TMD: Descriptive data and empirically identified domains from the OPPERA case-control study. *J. Pain* **2011**, *12*, T46-60. [CrossRef]
14. Kindler, S.; Samietz, S.; Houshmand, M.; Grabe, H.J.; Bernhardt, O.; Biffar, R.; Kocher, T.; Meyer, G.; Völzke, H.; Metelmann, H.R.; et al. Depressive and anxiety symptoms as risk factors for temporomandibular joint pain: A prospective cohort study in the general population. *J. Pain* **2012**, *13*, 1188–1197. [CrossRef] [PubMed]
15. Dworkin, S.F.; Huggins, K.H.; Wilson, L.; Mancl, L.; Turner, J.; Massoth, D.; LeResche, L.; Truelove, E. A randomized clinical trial using research diagnostic criteria for temporomandibular disorders-axis II to target clinic cases for a tailored self-care TMD treatment program. *J. Orofac. Pain* **2002**, *16*, 48–63. [PubMed]
16. Costa, Y.M.; Porporatti, A.L.; Stuginski-Barbosa, J.; Bonjardim, L.R.; Conti, P.C. Additional effect of occlusal splints on the improvement of psychological aspects in temporomandibular disorder subjects: A randomized controlled trial. *Arch. Oral Biol.* **2015**, *60*, 738–744. [CrossRef] [PubMed]
17. Howard, J.A. Temporomandibular joint disorders in children. *Dent. Clin. N. Am.* **2013**, *57*, 99–127. [CrossRef]

18. Sanders, A.E.; Slade, G.D.; Bair, E.; Fillingim, R.B.; Knott, C.; Dubner, R.; Greenspan, J.D.; Maixner, W.; Ohrbach, R. General health status and incidence of first-onset temporomandibular disorder: The OPPERA prospective cohort study. *J. Pain* **2013**, *14*, T51-62. [CrossRef]

19. Manfredini, D.; Guarda-Nardini, L.; Winocur, E.; Piccotti, F.; Ahlberg, J.; Lobbezoo, F. Research diagnostic criteria for temporomandibular disorders: A systematic review of axis I epidemiologic findings. *Oral Surg. Oral Med. Oral Pathol. Oral Radiol. Endod.* **2011**, *112*, 453–462. [CrossRef]

20. Prevalence of TMJD and Its Signs and Symptoms. Available online: https://www.nidcr.nih.gov/research/data-statistics/facial-pain/prevalence (accessed on 9 February 2020).

21. Yost, O.; Liverman, C.T.; English, R.; Mackey, S.; Bond, E.C. *National Academies of Sciences, Engineering and Medicine: Temporomandibular Disorders: Priorities for Research and Care*; The National Academies Press: Washington, DC, USA, 2020; p. 426.

22. Slade, G.D.; Bair, E.; Greenspan, J.D.; Dubner, R.; Fillingim, R.B.; Diatchenko, L.; Maixner, W.; Knott, C.; Ohrbach, R. Signs and symptoms of first-onset TMD and sociodemographic predictors of its development: The OPPERA prospective cohort study. *J. Pain* **2013**, *14*, T20–T32.e21–e23. [CrossRef]

23. Bair, E.; Brownstein, N.C.; Ohrbach, R.; Greenspan, J.D.; Dubner, R.; Fillingim, R.B.; Maixner, W.; Smith, S.B.; Diatchenko, L.; Gonzalez, Y.; et al. Study protocol, sample characteristics, and loss to follow-up: The OPPERA prospective cohort study. *J. Pain* **2013**, *14*, T2–T19. [CrossRef] [PubMed]

24. Janal, M.N.; Raphael, K.G.; Nayak, S.; Klausner, J. Prevalence of myofascial temporomandibular disorder in US community women. *J. Oral Rehabil.* **2008**, *35*, 801–809. [CrossRef] [PubMed]

25. List, T.; John, M.T.; Dworkin, S.F.; Svensson, P. Recalibration improves inter-examiner reliability of TMD examination. *Acta Odontol. Scand.* **2006**, *64*, 146–152. [CrossRef] [PubMed]

26. Vainionpää, R.; Kinnunen, T.; Pesonen, P.; Laitala, M.L.; Anttonen, V.; Sipilä, K. Prevalence of temporomandibular disorders (TMD) among Finnish prisoners: Cross-sectional clinical study. *Acta Odontol. Scand.* **2019**, *77*, 264–268. [CrossRef]

27. Dworkin, S.F.; LeResche, L. Research diagnostic criteria for temporomandibular disorders: Review, criteria, examinations and specifications, critique. *J. Craniomandib. Disord.* **1992**, *6*, 301–355. [PubMed]

28. Wahlund, K.; List, T.; Dworkin, S.F. Temporomandibular disorders in children and adolescents: Reliability of a questionnaire, clinical examination, and diagnosis. *J. Orofac. Pain* **1998**, *12*, 42–51.

29. Schiffman, E.; Ohrbach, R.; Truelove, E.; Look, J.; Anderson, G.; Goulet, J.P.; List, T.; Svensson, P.; Gonzalez, Y.; Lobbezoo, F.; et al. Diagnostic Criteria for Temporomandibular Disorders (DC/TMD) for Clinical and Research Applications: Recommendations of the International RDC/TMD Consortium Network* and Orofacial Pain Special Interest Group†. *J. Oral Facial Pain Headache* **2014**, *28*, 6–27. [CrossRef]

30. Christidis, N.; Lindström Ndanshau, E.; Sandberg, A.; Tsilingaridis, G. Prevalence and treatment strategies regarding temporomandibular disorders in children and adolescents-A systematic review. *J. Oral Rehabil.* **2019**, *46*, 291–301. [CrossRef]

31. Tecco, S.; Crincoli, V.; Di Bisceglie, B.; Saccucci, M.; Macrí, M.; Polimeni, A.; Festa, F. Signs and symptoms of temporomandibular joint disorders in Caucasian children and adolescents. *Cranio* **2011**, *29*, 71–79. [CrossRef] [PubMed]

32. Minervini, G.; Franco, R.; Marrapodi, M.M.; Fiorillo, L.; Cervino, G.; Cicciù, M. Prevalence of temporomandibular disorders in children and adolescents evaluated with Diagnostic Criteria for Temporomandibular Disorders: A systematic review with meta-analysis. *J. Oral Rehabil.* **2023**, *50*, 522–530. [CrossRef]

33. Graue, A.M.; Jokstad, A.; Assmus, J.; Skeie, M.S. Prevalence among adolescents in Bergen, Western Norway, of temporomandibular disorders according to the DC/TMD criteria and examination protocol. *Acta Odontol. Scand.* **2016**, *74*, 449–455. [CrossRef]

34. Köhler, A.A.; Helkimo, A.N.; Magnusson, T.; Hugoson, A. Prevalence of symptoms and signs indicative of temporomandibular disorders in children and adolescents. A cross-sectional epidemiological investigation covering two decades. *Eur. Arch. Paediatr. Dent.* **2009**, *10* (Suppl. 1), 16–25. [CrossRef]

35. Marpaung, C.; Lobbezoo, F.; van Selms, M.K.A. Temporomandibular Disorders among Dutch Adolescents: Prevalence and Biological, Psychological, and Social Risk Indicators. *Pain Res. Manag.* **2018**, *2018*, 5053709. [CrossRef] [PubMed]

36. Rauch, A.; Schierz, O.; Körner, A.; Kiess, W.; Hirsch, C. Prevalence of anamnestic symptoms and clinical signs of temporomandibular disorders in adolescents-Results of the epidemiologic LIFE Child Study. *J. Oral Rehabil.* **2020**, *47*, 425–431. [CrossRef] [PubMed]

37. Wu, N.; Hirsch, C. Temporomandibular disorders in German and Chinese adolescents. *J. Orofac. Orthop.* **2010**, *71*, 187–198. [CrossRef] [PubMed]

38. Bonjardim, L.R.; Gavião, M.B.; Pereira, L.J.; Castelo, P.M. Mandibular movements in children with and without signs and symptoms of temporomandibular disorders. *J. Appl. Oral Sci.* **2004**, *12*, 39–44. [CrossRef]

39. Hirsch, C.; John, M.T.; Lautenschläger, C.; List, T. Mandibular jaw movement capacity in 10-17-yr-old children and adolescents: Normative values and the influence of gender, age, and temporomandibular disorders. *Eur. J. Oral Sci.* **2006**, *114*, 465–470. [CrossRef]

40. Almutairi, A.F.; Albesher, N.; Aljohani, M.; Alsinanni, M.; Turkistani, O.; Salam, M. Association of oral parafunctional habits with anxiety and the Big-Five Personality Traits in the Saudi adult population. *Saudi Dent. J.* **2021**, *33*, 90–98. [CrossRef]

41. Winocur, E.; Littner, D.; Adams, I.; Gavish, A. Oral habits and their association with signs and symptoms of temporomandibular disorders in adolescents: A gender comparison. *Oral Surg. Oral Med. Oral Pathol. Oral Radiol. Endod.* **2006**, *102*, 482–487. [CrossRef]

42. Katsikogianni, E.; Schweigert-Gabler, S.; Krisam, J.; Orhan, G.; Bissar, A.; Lux, C.J.; Schmitter, M.; Giannakopoulos, N.N. Diagnostic accuracy of the Diagnostic Criteria for Temporomandibular Disorders for children aged 8–12 years. *J. Oral Rehabil.* **2021**, *48*, 18–27. [CrossRef]

43. Restrepo, C.C.; Suarez, N.; Moratto, N.; Manrique, R. Content and construct validity of the Diagnostic Criteria for Temporomandibular Disorders Axis I for children. *J. Oral Rehabil.* **2020**, *47*, 809–819. [CrossRef]

44. Lövgren, A.; Visscher, C.M.; Häggman-Henrikson, B.; Lobbezoo, F.; Marklund, S.; Wänman, A. Validity of three screening questions (3Q/TMD) in relation to the DC/TMD. *J. Oral Rehabil.* **2016**, *43*, 729–736. [CrossRef]

45. Gonzalez, Y.M.; Schiffman, E.; Gordon, S.M.; Seago, B.; Truelove, E.L.; Slade, G.; Ohrbach, R. Development of a brief and effective temporomandibular disorder pain screening questionnaire: Reliability and validity. *J. Am. Dent. Assoc.* **2011**, *142*, 1183–1191. [CrossRef] [PubMed]

46. Rongo, R.; Ekberg, E.; Nilsson, I.M.; Al-Khotani, A.; Alstergren, P.; Conti, P.C.R.; Durham, J.; Goulet, J.P.; Hirsch, C.; Kalaykova, S.I.; et al. Diagnostic criteria for temporomandibular disorders (DC/TMD) for children and adolescents: An international Delphi study-Part 1-Development of Axis I. *J. Oral Rehabil.* **2021**, *48*, 836–845. [CrossRef]

47. See Diagnostic Criteria for Temporomandibular Disorders: Assessment Instruments (German). Available online: https://ubwp.buffalo.edu/rdc-tmdinternational/tmd-assessmentdiagnosis/dc-tmd-translations/ (accessed on 23 May 2023).

48. Herpich, C.M.; Gomes, C.; Dibai-Filho, A.V.; Politti, F.; Souza, C.D.S.; Biasotto-Gonzalez, D.A. Correlation Between Severity of Temporomandibular Disorder, Pain Intensity, and Pressure Pain Threshold. *J. Manip. Physiol. Ther.* **2018**, *41*, 47–51. [CrossRef] [PubMed]

49. Österlund, C.; Berglund, H.; Åkerman, M.; Nilsson, E.; Petersson, H.; Lam, J.; Alstergren, P. Diagnostic criteria for temporomandibular disorders: Diagnostic accuracy for general dentistry procedure without mandatory commands regarding myalgia, arthralgia and headache attributed to temporomandibular disorder. *J. Oral Rehabil.* **2018**, *45*, 497–503. [CrossRef] [PubMed]

50. Paduano, S.; Bucci, R.; Rongo, R.; Silva, R.; Michelotti, A. Prevalence of temporomandibular disorders and oral parafunctions in adolescents from public schools in Southern Italy. *Cranio* **2020**, *38*, 370–375. [CrossRef]

51. Maixner, W.; Diatchenko, L.; Dubner, R.; Fillingim, R.B.; Greenspan, J.D.; Knott, C.; Ohrbach, R.; Weir, B.; Slade, G.D. Orofacial pain prospective evaluation and risk assessment study—The OPPERA study. *J. Pain* **2011**, *12*, T4–T11.e11–e12. [CrossRef]

52. Ohrbach, R.; Bair, E.; Fillingim, R.B.; Gonzalez, Y.; Gordon, S.M.; Lim, P.F.; Ribeiro-Dasilva, M.; Diatchenko, L.; Dubner, R.; Greenspan, J.D.; et al. Clinical orofacial characteristics associated with risk of first-onset TMD: The OPPERA prospective cohort study. *J. Pain* **2013**, *14*, T33–T50. [CrossRef]

53. Philipp, J.; Zeiler, M.; Wöber, C.; Wagner, G.; Karwautz, A.F.K.; Steiner, T.J.; Wöber-Bingöl, Ç. Prevalence and burden of headache in children and adolescents in Austria—A nationwide study in a representative sample of pupils aged 10–18 years. *J. Headache Pain* **2019**, *20*, 101. [CrossRef]

54. LeResche, L.; Mancl, L.A.; Drangsholt, M.T.; Huang, G.; Von Korff, M. Predictors of onset of facial pain and temporomandibular disorders in early adolescence. *Pain* **2007**, *129*, 269–278. [CrossRef]

55. Speciali, J.G.; Dach, F. Temporomandibular dysfunction and headache disorder. *Headache* **2015**, *55* (Suppl. 1), 72–83. [CrossRef]

56. De Stefano, A.A.; Guercio-Mónaco, E.; Uzcátegui, A.; Boboc, A.M.; Barbato, E.; Galluccio, G. Temporomandibular disorders in Venezuelan and Italian adolescents. *Cranio* **2020**, *40*, 517–523. [CrossRef] [PubMed]

57. Hirsch, C.; Hoffmann, J.; Türp, J.C. Are temporomandibular disorder symptoms and diagnoses associated with pubertal development in adolescents? An epidemiological study. *J. Orofac. Orthop.* **2012**, *73*, 6–18. [CrossRef] [PubMed]

58. Franco-Micheloni, A.L.; Fernandes, G.; de Godoi Gonçalves, D.A.; Camparis, C.M. Temporomandibular Disorders in a Young Adolescent Brazilian Population: Epidemiologic Characterization and Associated Factors. *J. Oral Facial Pain Headache* **2015**, *29*, 242–249. [CrossRef] [PubMed]

59. El Tumi, H.; Johnson, M.I.; Dantas, P.B.F.; Maynard, M.J.; Tashani, O.A. Age-related changes in pain sensitivity in healthy humans: A systematic review with meta-analysis. *Eur. J. Pain* **2017**, *21*, 955–964. [CrossRef]

60. von Baeyer, C.L.; Marche, T.A.; Rocha, E.M.; Salmon, K. Children's memory for pain: Overview and implications for practice. *J. Pain* **2004**, *5*, 241–249. [CrossRef]

61. Nell, K.D.; Joanne, C. The Development of Comprehension. In *Handbook of Reading Research: Volume IV*; Routledge: London, UK, 2010.

62. Lövgren, A.; Parvaneh, H.; Lobbezoo, F.; Häggman-Henrikson, B.; Wänman, A.; Visscher, C.M. Diagnostic accuracy of three screening questions (3Q/TMD) in relation to the DC/TMD in a specialized orofacial pain clinic. *Acta Odontol. Scand.* **2018**, *76*, 380–386. [CrossRef]

63. Nilsson, I.M.; List, T.; Drangsholt, M. The reliability and validity of self-reported temporomandibular disorder pain in adolescents. *J. Orofac. Pain* **2006**, *20*, 138–144.

64. Rongo, R.; Ekberg, E.; Nilsson, I.M.; Al-Khotani, A.; Alstergren, P.; Rodrigues Conti, P.C.; Durham, J.; Goulet, J.P.; Hirsch, C.; Kalaykova, S.I.; et al. Diagnostic criteria for temporomandibular disorders in children and adolescents: An international Delphi study-Part 2-Development of Axis II. *J. Oral Rehabil.* **2022**, *49*, 541–552. [CrossRef]

65. Müller, L.; van Waes, H.; Langerweger, C.; Molinari, L.; Saurenmann, R.K. Maximal mouth opening capacity: Percentiles for healthy children 4-17 years of age. *Pediatr. Rheumatol. Online J.* **2013**, *11*, 17. [CrossRef]

56. Vogl, T.J.; Lauer, H.C.; Lehnert, T.; Naguib, N.N.; Ottl, P.; Filmann, N.; Soekamto, H.; Nour-Eldin, N.E. The value of MRI in patients with temporomandibular joint dysfunction: Correlation of MRI and clinical findings. *Eur. J. Radiol.* **2016**, *85*, 714–719. [CrossRef]

57. Lifland, B.E.; Mangione-Smith, R.; Palermo, T.M.; Rabbitts, J.A. Agreement Between Parent Proxy Report and Child Self-Report of Pain Intensity and Health-Related Quality of Life After Surgery. *Acad. Pediatr.* **2018**, *18*, 376–383. [CrossRef] [PubMed]

58. Durham, J.; Steele, J.; Moufti, M.A.; Wassell, R.; Robinson, P.; Exley, C. Temporomandibular disorder patients' journey through care. *Community Dent. Oral Epidemiol.* **2011**, *39*, 532–541. [CrossRef] [PubMed]

59. Greene, C.S.; Manfredini, D. Transitioning to chronic temporomandibular disorder pain: A combination of patient vulnerabilities and iatrogenesis. *J. Oral Rehabil.* **2021**, *48*, 1077–1088. [CrossRef] [PubMed]

70. Macfarlane, G.J. The epidemiology of chronic pain. *Pain* **2016**, *157*, 2158–2159. [CrossRef]

71. Slade, G.; Durham, J. Prevalence, impact, and costs of treatment for temporomandibular disorders. Paper commissioned by the Committee on Temporomandibular Disorders (TMDs): From Research Discoveries to Clinical Treatment. In *Temporomandibular Disorders: Priorities for Research and Care*; see Appendix C; The National Academies Press: Washington, DC, USA, 2020.

72. de Magalhães Barros, V.; Seraidarian, P.I.; de Souza Côrtes, M.I.; de Paula, L.V. The impact of orofacial pain on the quality of life of patients with temporomandibular disorder. *J. Orofac. Pain* **2009**, *23*, 28–37.

73. Durham, J.; Shen, J.; Breckons, M.; Steele, J.G.; Araujo-Soares, V.; Exley, C.; Vale, L. Healthcare Cost and Impact of Persistent Orofacial Pain: The DEEP Study Cohort. *J. Dent. Res.* **2016**, *95*, 1147–1154. [CrossRef]

74. Glaros, A.G.; Glass, E.G.; Hayden, W.J. History of treatment received by patients with TMD: A preliminary investigation. *J. Orofac. Pain* **1995**, *9*, 147–151.

75. Katsoulis, K.; Bassetti, R.; Windecker-Gétaz, I.; Mericske-Stern, R.; Katsoulis, J. Temporomandibular disorders/myoarthropathy of the masticatory system. Costs of dental treatment and reimbursement by Swiss federal insurance agencies according to the Health Care Benefits Ordinance (KLV). *Schweiz. Monatsschr Zahnmed.* **2012**, *122*, 510–526.

76. Nilsson, I.M. Reliability, validity, incidence and impact of temporormandibular pain disorders in adolescents. *Swed. Dent. J. Suppl.* **2007**, *183*, 7–86.

77. Egermark, I.; Carlsson, G.E.; Magnusson, T. A 20-year longitudinal study of subjective symptoms of temporomandibular disorders from childhood to adulthood. *Acta Odontol. Scand.* **2001**, *59*, 40–48. [CrossRef]

78. Könönen, M.; Nyström, M. A longitudinal study of craniomandibular disorders in Finnish adolescents. *J. Orofac. Pain* **1993**, *7*, 329–336. [PubMed]

Journal of
Clinical Medicine

Article

Prevalence of Clinical Signs and Symptoms of Temporomandibular Joint Disorders Registered in the EUROTMJ Database: A Prospective Study in a Portuguese Center

David Faustino Ângelo [1,2,*], Beatriz Mota [3], Ricardo São João [4,5], David Sanz [1] and Henrique José Cardoso [1]

1 Instituto Português da Face, 1050-227 Lisboa, Portugal
2 Faculdade de Medicina, Universidade de Lisboa, 1649-028 Lisboa, Portugal
3 Serviço de Estomatologia, Centro Hospitalar Universitário Lisboa Norte (CHULN), 1649-035 Lisboa, Portugal
4 Department of Computer Science and Quantitative Methods, School of Management and Technology, Polytechnic Institute of Santarém, 2001-904 Santarém, Portugal
5 CEAUL—Centro de Estatística e Aplicações, Faculdade de Ciências, Universidade de Lisboa, 1749-016 Lisboa, Portugal
* Correspondence: david.angelo@ipface.pt

Abstract: Temporomandibular joint disorders (TMDs) are characterized by their multifactorial etiology and pathogenesis. A 3-year prospective study was conducted in a Portuguese TMDs department to study the prevalence of different TMDs signs and symptoms and their association with risk factors and comorbidities. Five hundred ninety-five patients were included using an online database: EUROTMJ. Most patients were female (80.50%), with a mean age of 38.20 ± 15.73 years. The main complaints were: (1) temporomandibular joint (TMJ) clicking (13.26%); (2) TMJ pain (12.49%); (3) masticatory muscle tension (12.15%). The main clinical findings were myalgia (74%), TMJ clicking (60–62%), and TMJ arthralgia (31–36%). Risk factors such as clenching (60%) and bruxism (30%) were positively associated with TMJ pain and myalgia. Orthodontic treatment (20%) and wisdom tooth removal (19%) were positively associated with TMJ clicking, while jaw trauma (6%), tracheal intubation (4%) and orthognathic surgery (1%) were positively associated with TMJ crepitus, limited mandibular range of motion, and TMJ pain, respectively. In total, 42.88% of TMDs patients had other associated chronic diseases, most of them were mental behavioral or neurodevelopmental disorders (33.76%), namely, anxiety (20%) and depression (13%). The authors also observed a positive association of mental disorders with the degree of TMJ pain and myalgia. The online database seems to be a relevant scientific instrument for healthcare providers who treat TMDs. The authors expect that the EUROTMJ database can serve as a milestone for other TMDs departments.

Keywords: temporomandibular joint disorders; parafunctional habits; risk factors; chronic diseases

Citation: Ângelo, D.F.; Mota, B.; João, R.S.; Sanz, D.; Cardoso, H.J. Prevalence of Clinical Signs and Symptoms of Temporomandibular Joint Disorders Registered in the EUROTMJ Database: A Prospective Study in a Portuguese Center. *J. Clin. Med.* **2023**, *12*, 3553. https://doi.org/10.3390/jcm12103553

Academic Editor: Mieszko Wieckiewicz

Received: 5 May 2023
Revised: 11 May 2023
Accepted: 16 May 2023
Published: 18 May 2023

1. Introduction

Temporomandibular joint disorders (TMDs) are a group of dysfunctions that appear to be of multifactorial origin, affecting the temporomandibular joint (TMJ), masticatory muscles, and adjacent structures [1,2]. Nowadays, TMDs are considered the most frequent cause of chronic orofacial pain of non-dental origin and the third stomatological disorder leading to pain and disability. In addition, myofascial TMDs are documented as the most frequent subtype, followed by internal derangements such as disc displacement and TMJ arthralgia [3,4].

It is estimated that TMDs affect about 31% of the adult population and 11% of children/adolescents [5]. TMDs prevalence is significantly higher in women (female/male ratio 5:1) and younger subjects [6]. However, TMD signs seem to increase with age [1,2,7].

Psychosocial, environmental, biological, and neurophysiological factors are considered etiological entities significantly associated with TMD symptoms. Factors such as emotional stress (anxiety and depression), bruxism, occlusal disharmony, orthodontic treatment, masticatory dysfunction, and postural deviation are reported to increase the risk for TMDs [1,4,7].

The most prominent symptoms are restricted joint function with alteration of the mandibular physiological dynamics, muscular or articular pain that intensifies with mastication, headache, and TMJ noises. Limited mandibular range of motion, pain, crepitation, or clicking in TMJ palpation are all common signs of TMDs assessed during a clinical examination [1,2].

This study attempts to evaluate and analyze the prevalence of TMD clinical signs and symptoms in a Portuguese TMD department and their association with various sociodemographic and individual factors such as age, gender, oral behaviors, risk factors, and other comorbidities.

2. Materials and Methods

2.1. Database Description

European temporomandibular joint (EUROTMJ) is an online electronic medical database that allows data collection from patients with TMDs (Figure 1). The patient's data is encrypted and only accessible through a password attributed to the clinician. The database works in the English language. EUROTMJ displays a tree format table of contents (Figure 1). The database is constituted with different menus: general data (patient ID, employment status, comorbidities, daily medication, drug allergy); TMJ history (complaints, risk factors, past treatments, VAS Pain, TMJ click, life impact/habits; clinical evaluation; exams, diagnosis, clinical resume, notes); questionnaires (SF-36, HADS, EQ-5D, OHIP-14, WHO, Fonseca, Pain Screener, GAD-2, PHQ-2); treatments (Proposed treatments); and clinical evaluation (evolution charts, evolution table).

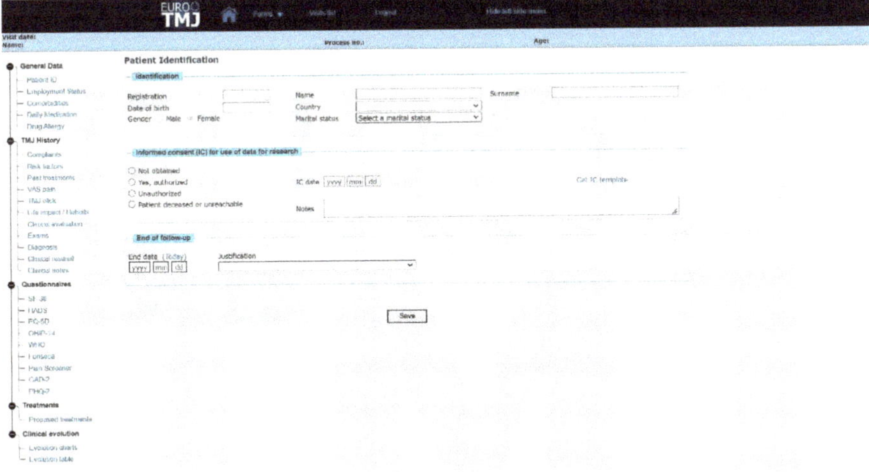

Figure 1. EUROTMJ database screen.

2.2. Study Design and Data Collection

A 3-year prospective study was conducted in a Portuguese TMD department from 1 August 2019 to 1 August 2022. This study was approved by the Instituto Português da Face ethics committee (PT/IPFace//RCT/0822/01). All enrolled patients gave their informed consent in writing, following current legislation. The inclusion criteria were: (1) registration of all the variables under study in a first consultation and (2) clinical diagnosis of TMDs. The exclusion criteria were severe medical problems or impaired cognitive capacity. All patients with inclusion criteria over these three years were included. Descriptive data and clinical outcomes were registered in EUROTMJ. Demographic data (date of birth and gender) was recorded. In the first consultation, patients were instructed to answer questions regarding their complaints (TMJ pain, TMJ clicking, TMJ crepitus, limited mouth opening, masticatory muscles tension, cervical muscles tension, tinnitus, TMJ edema, and vertigo), laterality (when applicable), and duration of the symptoms (<3M; 3M–1Y; 1Y–5Y > 5Y). They were also asked about parafunctional habits in their daily life, such as bruxism and clenching. A clinical record registered other comorbidities and potential risk factors for TMDs. The TMJ pain recording was accessed in the right and left joints through the Visual Analog Scale (VAS, 0–10, with 0 being no pain; 1–3: mild, 4–6: moderate, and 7–10 severe pain). The VASLIFE was accessed by asking: "If you could give a life impact score to your TMJ problem on a 0 to 10 scale, where 0 means no impact, and 10 means the maximum impact possible, what would be your score?" [8]. In clinical evaluation, maximum mouth opening (MMO, mm), the presence of clicks, crepitus, arthralgia (right and left joint), and myalgia were recorded. MMO was accessed using a certified ruler between the incisor teeth (TheraBite Jaw ROM Scale). Limited mouth opening was registered when MMO < 40 mm [9,10]. Myalgia was diagnosed according to a positive clinical history for: (1) in the past 30 days, pain in the jaw, in front, or directly in the ear, with confirmation of pain through palpation of the masticatory muscles by the examiner and (2) accompanied pain with jaw movement, function or parafunction, and a positive clinical evaluation for palpation pressure (5 s/1 kg pressure) in masseter and temporalis muscles as defined in DC/TMD [9,10]. Myalgia was graded accordingly with pain intensity in each muscle: 0 = no pain/pressure only; 1 = mild pain; 2 = moderate pain; 3 = severe pain [11]. Arthralgia was reported if verified: (1) history of pain in the TMJ area and (2) accompanied pain with jaw movement, function, or parafunction [9,10]. The level of TMJ arthralgia was registered through the pain on palpation of the lateral pole or around or pain on maximum unassisted or assisted opening, lateral, or protrusive movements. The same clinician (D.F.A.) performed the clinical evaluation for all patients.

2.3. Statistical Analysis

Data were analyzed using the GraphPad Prism (v9, (Boston, MA, USA) and IBM SPSS (v26, Armonk, NY, USA) software. The variables were expressed as the mean ($\pm$standard deviation (SD)) or frequency (%). The biserial correlation Pearson Test (rpb) assessed the variables' correlation. The non-parametric Chi-square test (χ^2) was used to determine the associations' presence, and its intensity was measured using Cramér's V Coefficient (φc). A p-value < 0.05 was considered statistically significant. For a graphic representation of risk factors and other comorbidities, a percentage >1% was assumed.

3. Results

A total of 595 patients were registered in the EUROTMJ database. The mean age was 38.20 ± 15.73 years at the first visit, 479 (80.50%) of whom were female gender (Table 1). 86.39% of the patients presented symptoms bilaterally, while 6.89% and 6.72% presented only on the right and left, respectively.

Table 1. Demographic data and side of the TMJ symptoms.

Variables		n (%), or Mean $\pm$ SD
Number of patients		595
Sex	Female	479 (80.50%)
	Male	116 (19.50%)
Age		38.20 ± 15.73
Side of the joint with symptoms	Only Right	41 (6.89%)
	Only Left	40 (6.72%)
	Bilateral	514 (86.39%)

The mean global TMJ pain in the right and left joints was 3.42 ± 3.01 and 3.34 ± 3.01, respectively (Figure 2a). The impact on life (VASLIFE) was 6.38 ± 2.51 (Figure 2a).

Global Pain in first visit	
Variables	mean ± SD
TMJ pain (R) (VAS, 0–10)	3.42 ± 3.01
TMJ pain (L) (VAS, 0–10)	3.34 ± 3.01
Pain (VASLIFE)	6.38 ± 2.51

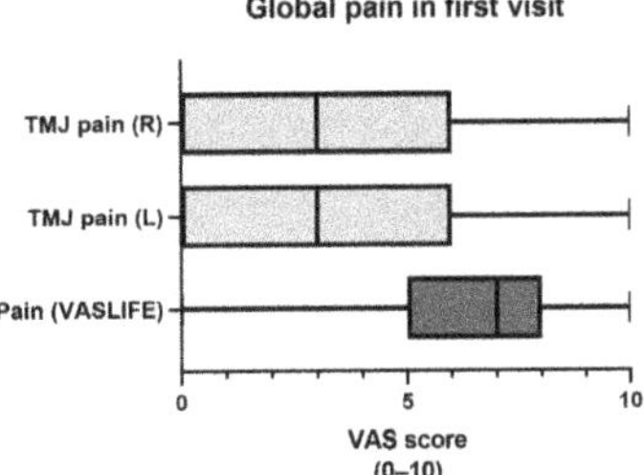

(a)

Patients who report pain	
Variables	mean ± SD
TMJ pain (R) (VAS, 0–10)	4.25 ± 2.83
TMJ pain (L) (VAS, 0–10)	4.34 ± 2.73
Pain (VASLIFE)	6.67 ± 2.34

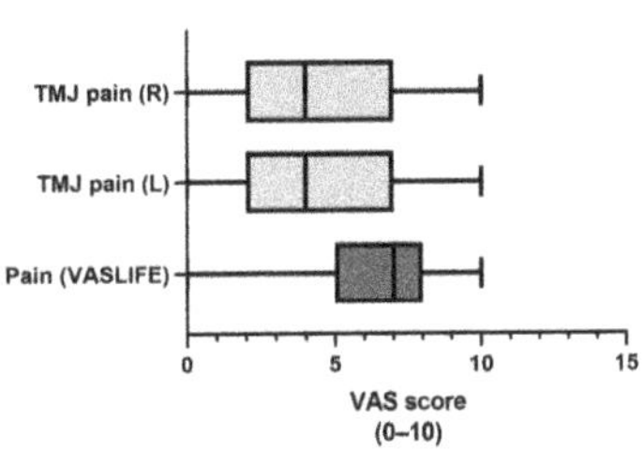

(b)

Figure 2. Global pain in the first visit (**a**) and Visual Analogue Score (VAS) in the patients that reported pain (**b**). The representative median VAS values in the range [Q1 − 1.5IQR; Q3 + 1.5IQR] are shown in the right graphs where IQR = interquartile range and Q1 and Q3 are, respectively, the 1st and 3rd quartiles (*n* = 595).

Considering only those patients who presented pain, pain right and left was 4.25 ± 2.83 and 4.34 ± 2.73, whereas VASLIFE was 6.67 ± 2.34 (Figure 2b). The main complaints of the patients were TMJ clicking (13.26%), TMJ pain (12.49%), and masticatory muscle tension (12.15%), while the least frequent were vertigo (4.43%), crepitus (4.20%), and edema (2.17%) (Figure 3a). However, considering the patient's main complaint, the one which presented the higher predominance was TMJ pain (22.78%) (Figure 3b). There was also a high prevalence of symptom duration between 1–5 years (32–39%) and over 5 years (24–33%), while a low frequency of symptoms under 3 months was verified (5–12%) (Figure 3c).

The correlation of complaints was analyzed in Figure 4. TMJ pain was moderately correlated with limited mouth opening, masticatory muscle tension, and headache (rpb = 0.3). TMJ pain also had a small correlation with other complaints (rpb = 0.1–0.2). Limitation of mouth opening was moderately correlated with TMJ locking and masticatory muscle tension (rpb = 0.3). Masticatory muscle tension was moderately correlated with headache and cervical muscle tension (rpb = 0.3). Additionally, headache was a moderate correlation with vertigo (rpb = 0.3).

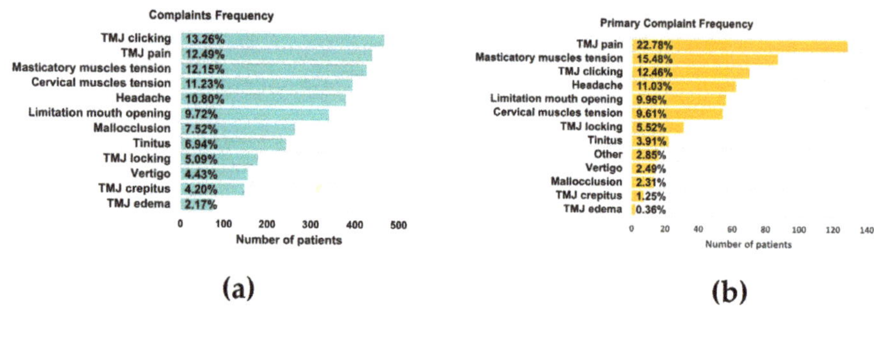

(a)

(b)

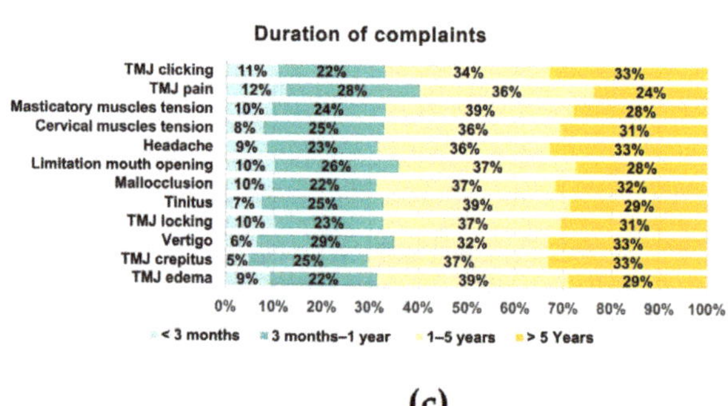

(c)

Figure 3. Global Complaints. (**a**) Complaints frequency; (**b**) Primary complaint frequency; (**c**) Duration of each complaint ($n = 595$).

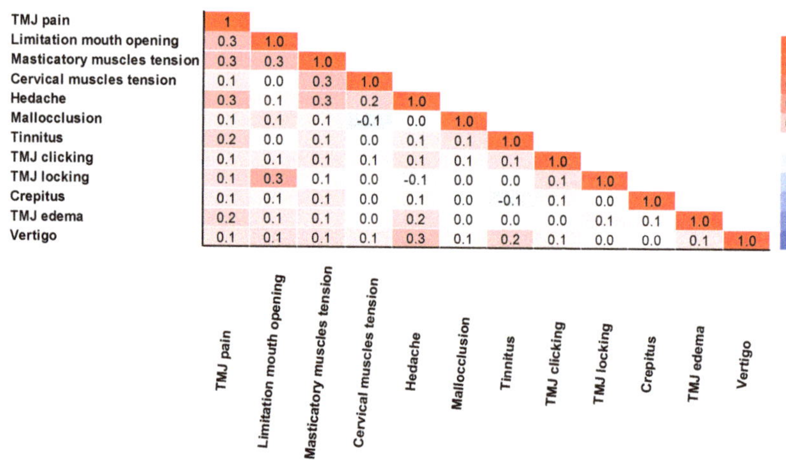

Figure 4. Correlation matrix of complaints obtained by biserial correlation Pearson Test ($n = 595$).

Regarding parafunctional habits, 180 (30%) of the patients reported positively to bruxism, 138 (23%) at night, 35 (6%) at day and night, and 7 (1%) during the day. On the contrary, 212 (36%) said that they do not have bruxism habits, and 203 (34%) do not know (Table 2). Regarding clenching, 359 (60%) of the patients answered that they do it, while 159 (27%) did not, and 77 (13%) did not know.

Table 2. Parafunctional habits. Frequency of Clenching and bruxism habits. *n* = 595.

Parafunctional Habits	
Variables	*n* **(%)**
Bruxism	
Day	7 (1%)
Night	138 (23%)
Day and Night	35 (6%)
No	212 (36%)
Does not know	203 (34%)
Clenching	
Yes	359 (60%)
No	159 (27%)
Does not know	77 (13%)

In clinical evaluation, patients presented 38.02 ± 9.31 mm of MMO (Figure 5). The mean myalgia degree was 2.03 ± 1.07 and 1.87 ± 1.09 on the right and left sides, respectively. A total of 370 (~62%) and 356 (~60%) patients presented clicks in the right and left TMJ, while 84 (~14%) and 72 (~12%) patients were verified crepitus (Figure 5). A suggestive pain indicative of TMJ arthralgia was detected in 213 (~36%) and 184 (~31%) patients in the right and left TMJ (Figure 5).

Clinical Evaluation	
Variables	**mean ± SD or N (%)**
MMO (mm)	38.02 ± 9.31
Myalgia Degree (R) (0–3)	2.03 ± 1.07
Myalgia Degree (L) (0–3)	1.87 ± 1.09
Clicks (R)	370 (62.18%)
Clicks (L)	356 (59.83%)
Crepitus (R)	84 (14.12%)
Crepitus (L)	72 (12.10%)
Arthralgia (R)	213 (35.80%)
Arthralgia (L)	184 (30.92%)

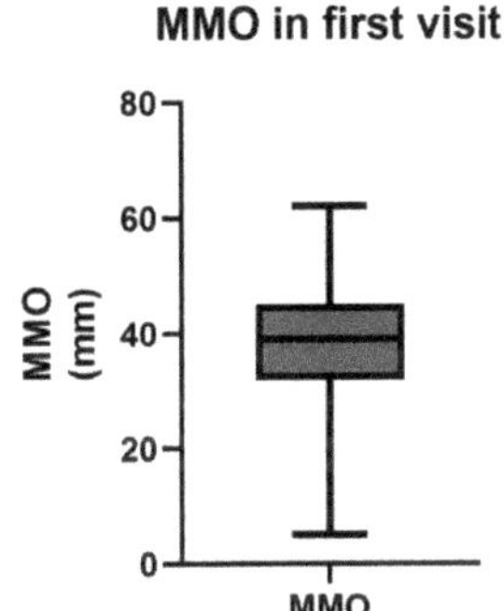

Figure 5. Clinical evaluation in the first visit. Maximum mouth opening (MMO), muscle tenderness, clicks, crepitus, and arthralgia were registered. The right graph shows the representative median MMO value in the range [Q1 − 1.5IQR; Q3 + 1.5IQR] where IQR = interquartile range and Q1 and Q3 are, respectively, the 1st and 3rd quartiles (*n* = 595).

Five hundred twenty-seven patients were also questioned regarding the diagnosis of other diseases, and it was found that 226 (42.88%) of the patients had another illness, with 111 (21.06%) with one condition and 115 (21.82%) with two or more diseases (Figure 6a). In a total of 545 reported medical conditions, there was a significant predominance of mental, behavioral, or neurodevelopmental disorders (184 patients, 33.76%) (Figure 6b). These values were supported by the diagnosis of anxiety and depression, verified in 110 (20%) and 69 (13%) patients, respectively (Figure 6c). The authors observed an essential percentage of patients with respiratory and endocrine/metabolic diseases, namely thyroid pathology (48 patients, 9%) (Figure 6c).

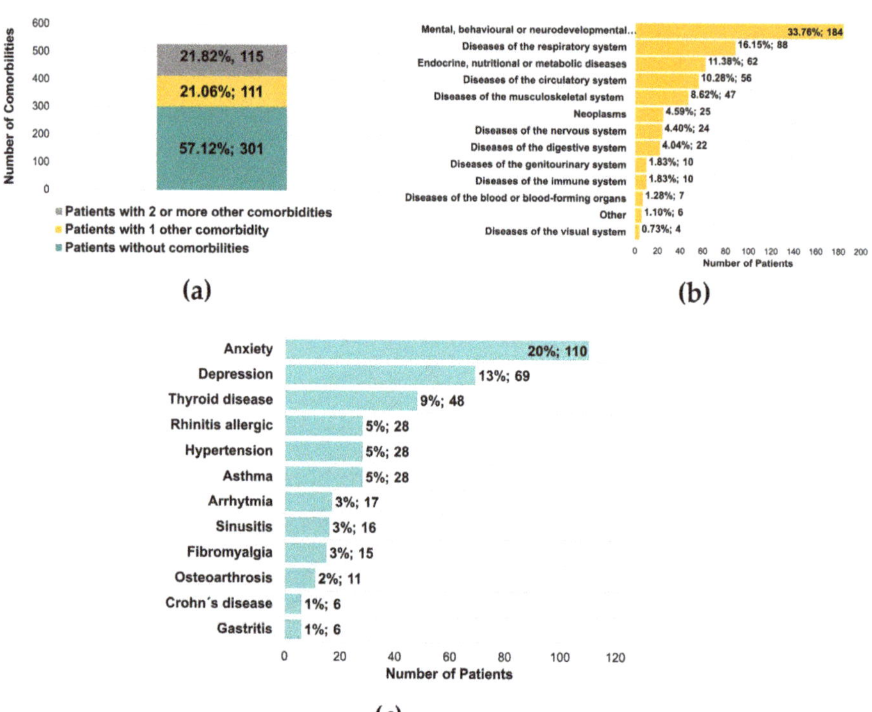

Figure 6. Other comorbidities identified in patients with temporomandibular disorders. (**a**) Number of patients without or with at least 1 or 2 comorbidities (*n* = 527) (**b**) Number of patients for a group of comorbidities (*n* = 545) (**c**) Number of patients with specific comorbidities (*n* = 545).

Potential risk factors described in the literature for triggering the onset of TMDs were also identified (Figure 7). In total, 53% of the patients presented at least one risk factor (Figure 7). Orthodontic treatment (119, 20%), wisdom tooth removal (116, 19%), dental treatment (86, 14%), jaw trauma (33, 6%), intubation (23, 4%), and orthognathic surgery (8, 1%) represent a large proportion of the risk factors identified (Figure 7).

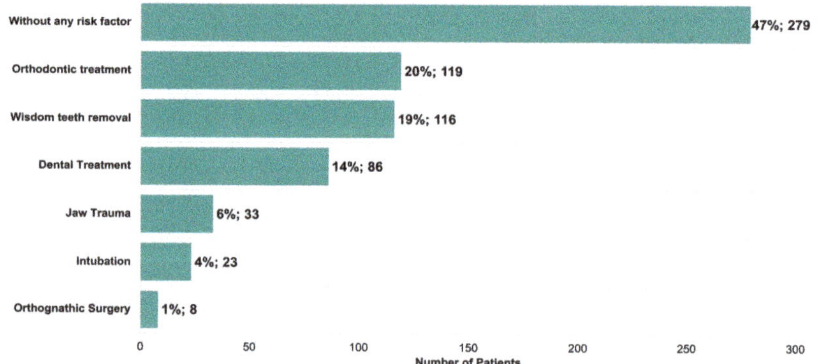

Figure 7. Risk factors identified in patients with symptoms of temporomandibular disease (*n* = 595).

The association of clinical variables and TMJ pain with demographic data, parafunctional habits, risk factors, and other comorbidities were then studied (Table S1).

In the demographic data, sex has a strong association with pain ($p < 0.001$, φc = 0.189, Table 3) and limitation of mouth opening ($p < 0.001$, φc = 0.154, Table 3), and a very strong association with the degree of myalgia ($p < 0.001$, φc = 0.277, Table 3).

Table 3. Association table with demographic data and TMJ pain and clinical variables. Significant values are in bold.

	TMJ Pain and Clinical Variables					
Demographic Data	**TMJ Pain**	**Limitation of Mouth Opening**	**Myalgia Degree**	**TMJ Clicks**	**TMJ Crepitus**	**Arthralgia**
	p-Value, Cramer's V					
Sex	**<0.001, 0.189**	**<0.001, 0.154**	**<0.001, 0.277**	0.698, 0.016	0.179, 0.060	0.143, 0.064
Age	0.889, 0.060	0.543, 0.072	0.456, 0.088	**0.002, 0.177**	**<0.001, 0.229**	**0.032, 0.142**

This association was corroborated by a higher degree of TMJ pain and myalgia and a lower MMO average in females ($p < 0.001$, Figure 8). At younger ages, the presence of clicks was more common ($p = 0.002$, φc = 0.177, Table 3), while at older ages, the presence of crepitus was more common ($p < 0.001$, φc = 0.229, Table 3).

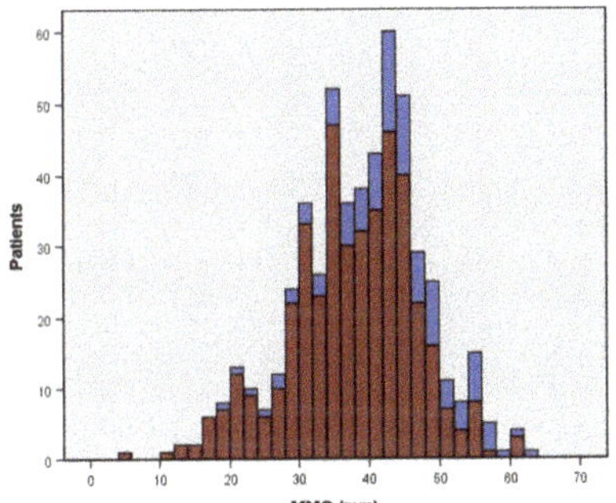

Clinical Outcome	Female mean ± SD	Male mean ± SD	p-value
TMJ pain (VAS, 0–10)	3.70 ± 2.19	2.51 ± 3.70	<0.001***
Maximum mouth opening (MMO)	36.86 ± 9.04	42.85 ± 8.89	<0.001***
Myalgia Degree (0–3)	2.11 ± 0.99	1.38 ± 1.1	<0.001***

Figure 8. Distribution of maximum mouth opening (MMO) and statistical difference between TMJ pain, MMO, and myalgia degree by sex. F-female; M-male. *** $p < 0.001$.

Parafunctional habits have also been shown to have an association with the clinical variables: clenching was strongly associated with myalgia degree ($p = 0.008$, φc = 0.159, Table 4); bruxism was strongly and moderately associated with myalgia degree and TMJ pain ($p = 0.006$, φc = 0.186 and $p = 0.035$, φc = 0.148, Table 4).

Table 4. Association table with parafunctional habits and TMJ pain and clinical variables. Significant values are in bold.

	TMJ Pain and Clinical Variables					
Parafunctional Habits	**TMJ Pain**	**Limitation of Mouth Opening**	**Myalgia Degree**	**TMJ Clicks**	**TMJ Crepitus**	**Arthralgia**
	p-Value, Cramer's V					
Clenching	0.203, 0.094	0.340, 0.042	**0.008, 0.159**	0.760, 0.013	0.571, 0.026	0.677, 0.020
Bruxism	**0.035, 0.148**	0.732, 0.017	**0.006, 0.186**	0.890, 0.012	0.496, 0.036	0.364, 0.048

Evaluating the association of risk factors with clinical variables revealed: past intubation was weakly related to limited mouth opening ($p = 0.047$, $\varphi c = 0.082$, Table 5), while past orthognathic surgery showed a moderate association with TMJ pain ($p = 0.011$, $\varphi c = 0.137$, Table 5); orthodontic treatment and wisdom teeth removal were weakly associated with TMJ clicks ($p = 0.026$, $\varphi c = 0.094$; $p = 0.038$, $\varphi c = 0.088$, Table 5) and jaw trauma with TMJ crepitus ($p = 0.037$, $\varphi c = 0.093$, Table 5).

Table 5. Association table with risk factors and TMJ pain and clinical variables. Significant values are in bold.

	Global Pain and Clinical Variables					
Risk Factors	**TMJ Pain**	**Open Mouth Limitation**	**Myalgia Degree**	**TMJ Clicks**	**TMJ Crepitus**	**Arthralgia**
	p-Value, Cramer's V					
Dental treatment	0.343, 0.075	0.168, 0.057	0.891, 0.035	0.698, 0.016	0.270, 0.049	0.065, 0.080
Orthodontic treatment	0.419, 0.069	0.116, 0.064	0.386, 0.077	**0.026, 0.094**	0.238, 0.053	0.523, 0.028
Intubation	0.489, 0.064	**0.047, 0.082**	0.323, 0.082	0.169, 0.058	0.135, 0.067	0.731, 0.015
Orthognathic surgery	**0.011, 0.137**	0.234, 0.049	0.286, 0.086	0.858, 0.008	0.313, 0.045	0.524, 0.028
Wisdom teeth removal	0.537, 0.060	0.115, 0.065	0.879, 0.036	**0.038, 0.088**	0.824, 0.010	0.729, 0.015
Jaw trauma	0.216, 0.087	0.326, 0.040	0.179, 0.098	0.732, 0.014	**0.037, 0.093**	0.201, 0.056

The presence of other comorbidities has been shown to have a strong association with the degree of TMJ pain and myalgia ($p = 0.002$, $\varphi c = 0.131$ and $p < 0.001$, $\varphi c = 0.185$, Table 6). Within the identified disease classes, mental, behavioral, or neurodevelopmental disorders were strongly associated with the degree of myalgia and a moderate association with TMJ pain intensity ($p < 0.001$, $\varphi c = 0.202$ and $p = 0.008$, $\varphi c = 0.142$, Table 6). Circulatory system diseases were moderately associated with the absence of TMJ clicks ($p = 0.003$, $\varphi c = 0.125$, Table 6).

Table 6. Association table of other comorbidities and TMJ pain and clinical variables. Significant values are in bold.

	TMJ Pain and Clinical Variables					
Other Comorbidities	**TMJ Pain**	**Limitation of Mouth Opening**	**Myalgia Degree**	**TMJ Clicks**	**TMJ Crepitus**	**Arthralgia**
	p-Value, Cramer's V					
n (no, 1, $\geq$2)	**0.002, 0.131**	0.354, 0.059	**<0.001, 0.185**	0.125, 0.086	0.490, 0.054	0.786, 0.030
Group of other Comorbidities						
Mental, behavioral, or neurodevelopmental disorders	**0.008, 0.142**	0.678, 0.017	**<0.001, 0.202**	0.708, 0.016	0.677, 0.019	0.645, 0.020
Diseases of the respiratory system	0.492, 0.064	0.475, 0.029	0.138, 0.103	0.381, 0.037	0.236, 0.053	0.835, 0.009
Endocrine, nutritional, or metabolic diseases	0.154, 0.094	0.798, 0.011	0.195, 0.096	0.264, 0.047	0.370, 0.040	0.487, 0.030
Diseases of the circulatory system	0.990, 0.014	0.679, 0.017	0.956, 0.025	**0.003, 0.125**	0.066, 0.083	0.416, 0.035
Diseases of the musculoskeletal system	0.228, 0.085	0.584, 0.022	0.719, 0.051	0.979, 0.001	0.125, 0.069	0.253, 0.050

4. Discussion

TMDs are one of the principal causes of chronic facial pain, affecting a considerable part of the population [12]. This 3-year prospective study provides a comprehensive and detailed characterization of the TMDs population in this research and demonstrates possible correlations with other comorbidities or risk factors. In this study, 80% of our patients were females with a mean age of 38 years. The age pattern and proportion of females were consistent with the results of other studies [3,13,14]. Iodice, et al. [3] revealed that this phenomenon might result from the biological, behavioral, psychological, and/or social factors associated with the female gender [3].

In our current study, the most common TMDs symptom was TMJ clicking (13.26%), as reported in other studies [3,15,16], followed by TMJ pain (12.49%) and masticatory muscle tension (12.15%). However, considering only the patient's main complaint, we found that 22.78% referred to TMJ pain as their leading symptom. Additionally, in our study, the duration of complaints was mainly 1–5 and over 5 years, with a low occurrence of symptoms under 3 months. Therefore, we hypothesize that most patients take some time to associate the symptoms with TMDs and find adequate medical opinion. This correlates with the fact that overall, the frequency of seeking treatment increases as the symptoms interfere with day-to-day activities. In addition, we also observed that TMDs pain occurs as part of a group of symptoms rather than as a single entity [3].

In our study, TMDs signs are more frequent than symptoms [2,17]. This can partly be explained by the difference between the prevalence of TMJ clicking as a complaint (13.26%) or clinical signs during the medical evaluation. Despite being a TMDs symptom, TMJ clicking can also be a sign of TMDs. Nevertheless, the most common TMDs sign was TMJ sounds (~62% and ~60% clicking in the right and left TMJ; ~14% and ~12 crepitus in the right and left TMJ). Suggestive pain is indicative of possible TMJ arthralgia (~36% and ~31% in the right and left TMJ), masticatory muscles myalgia (2.03 ± 1.07 and 1.87 ± 1.09 on the right and left sides, respectively), and MMO (38 ± 9.31 mm) was also evaluated. This data suggests that an expert clinical evaluation can sometimes valorize signs that patients are unaware of. Previous studies correlate the association between TMDs signs and symptoms with age, gender, and TMDs diagnosis. TMJ clicking has been shown to be common in younger subjects, while TMJ crepitation showed a stronger association with higher age [3,18]. This result is in accordance with our study, as our population was mainly younger subjects and the predominant clinical sign was TMJ clicking. At the same time, crepitus was more prevalent in older patients. This result is in agreement that the presence of crepitus is related to the diagnosis of osteoarthritis, which is more common in older patients [19]. Additionally, TMJ pain and myalgia were significantly higher in females, while MMO was significantly lower. This result shows that besides the prevalence of the disease in women, the symptoms are also more exacerbated. The correlation between signs and symptoms can contribute to an appropriate TMDs diagnosis. Previous studies demonstrate that muscle pain and tenderness may indicate myofascial disorders, while TMJ pain on palpation and limited mouth opening (LMO) can suggest intra-articular conditions. In our research, this was observed in pain indicative of arthralgia (~36% and ~31% in the right and left TMJ) and a mean MMO of 38 ± 9.31 mm, encountering a limitation of mouth opening as a cut-off of MMO < 40 mm was defined in our study.

Furthermore, TMJ pain was moderately correlated with LMO, while masticatory muscle tension had a moderate correlation with headache and cervical muscle tension. Almoznino, et al. [20] also showed higher cervical tenderness scores in myogenous TMDs patients and a positive association with TMJ pain and headaches. This connection can be related to myofascial trigger points in the trapezius that potentiate electromyographic changes in the masticatory muscles [21]. LMO was moderately associated with TMJ locking. On the other hand, TMJ clicking is considered a potential sign of disc displacement with reduction. Otherwise, LMO can be a typical sign of disc displacement without reduction [2]. In the present study, ~62% and ~60% of the patients presented TMJ clicking in the right and left TMJ, respectively, and the mean of MMO measured was 38 ± 9.31 mm. By analyzing these results, we observed that over one-half of the patients evaluated had signs of disc displacement with reduction. However, on average, the majority of the patients presented LMO. Previous studies support these findings as they linked LMO with other entities besides disc displacement without reduction, such as myalgia and degenerative joint disease associated with the normal aging process, possibly explaining the high percentage of individuals with LMO in this study [16]. In addition, some authors described chronic disc displacement without reduction with the absence of LMO, while acute disease with the presence of LMO [22,23].

The most frequent possible causes for TMDs described by the patients included: (1) parafunctional habits (clenching (60%), bruxism (30%)); (2) previous orthodontic treatment (20%); (3) wisdom tooth removal (19%), (4) general dental treatment (14%), (5) jaw trauma (6%), (6) tracheal intubation (4%), and (7) orthognathic surgery (1%). Our study shows a strong association between clenching and bruxism with the degree of myalgia and a moderate association between clenching and TMJ pain. In addition, previous tracheal intubation, orthognathic surgery, and wisdom tooth removal were positively associated with a limited mandibular range of motion, TMJ pain, and TMJ clicking, respectively. These findings are in accordance with previous studies, such as Marklund et al. [24], which performed a 2-year prospective observational study and concluded that biomechanical factors, such as bruxism, mandibular instability, and malocclusion, were linked to the incidence and persistence of TMJ signs and symptoms. Moreover, trauma and long-standing load may also contribute to the development and course of TMDs [24,25].

Regarding other comorbidities, it is unclear to which extent TMDs may reflect symptoms or manifestations of underlying diseases. In the current study, although most of the patients evaluated had no comorbidities (57.12%), 42.88% of TMDs had other conditions and 21.06% presented only one disease. In comparison, 21.82% of the patients referred to two or more diseases.

Almost one-third of these patients reported mental behavior and neurodevelopmental diseases, a pertinent factor for many patients with TMDs. A total of 20% had positive results for anxiety and 13% for depression. This study shows a correlation between mental behavioral and neurodevelopmental diseases with TMJ pain and myalgia. This result is consistent with other studies that demonstrated significant associations between stress, anxiety, depression, and TMDs [26,27]. Higher levels of stress and depression are related to changes in electrical potentials and asymmetry of the masticatory muscles during clenching [28,29]. In fact, higher levels of masticatory muscle pain perception were demonstrated in response to psychological stress in Sprague-Dawley rats [30]. Previous and present results strongly support an association and should be aware that specialists to keep psychosocial modulators as possible treatment-associated modalities. In addition, other chronic diseases summarized in Figure 6, such as respiratory and endocrine alterations, might be related to TMDs. This is corroborated in other population studies as there is evidence for an association between impaired general health and TMDs, suggesting that TMDs symptoms may share characteristics with other chronic conditions [31,32].

One of the strengths of this study was the consistency of a questionnaire and clinical examination to obtain reliable results for TMJ signs and symptoms in the first appointment. In addition, the EUROTMJ database allows the collection of data and evaluation of TMDs-related symptoms, correlating the findings with TMDs signs. This can be a valuable and applicable platform for TMDs screening and treatment decisions. Additionally, only one specialist examined all the patients in a TMDs-specialized appointment, making it an advantage of this study. However, as with most studies, the present analysis has limitations. Firstly, the database and questionnaire are not validated; furthermore, information bias might be present since the parameters evaluated using the VAS scale could be over or underestimated by the patients. Additionally, the evaluation of the correlation between TMDs signs and symptoms and TMJ diagnosis and proposed treatment was not included in this study, making it a limitation. Moreover, this study is single-center, which should be reproduced in a multi-center study.

5. Conclusions

TMDs are a group of dysfunctions that affect a considerable part of the population nowadays, mainly younger female patients. In the subjects included in this study, TMDs were shown to be a group of disorders with a broad spectrum of clinical manifestations, pathophysiology, and associated comorbid conditions. Significant associations between TMDs signs and symptoms with intrinsic characteristics, such as age, gender, and para-functional habits, such as clenching and bruxism, have been made. Positive correlations were also made with mental behavior and neurodevelopmental diseases, such as anxiety and depression, and other comorbidities, such as respiratory and endocrine alterations. Interestingly, risk factors such as wisdom tooth removal, orthodontic treatment, jaw trauma, tracheal intubation, and orthognathic surgery may increase the susceptibility to developing TMDs clinical symptoms. However, more studies are needed to understand such associations. We believe this data will serve as a milestone in providing helpful information for researchers and healthcare providers treating patients with TMDs.

Supplementary Materials: The following supporting information can be downloaded at: https://www.mdpi.com/article/10.3390/jcm12103553/s1, Table S1: Relative frequency (%) of the TMJ pain and clinical variables.

Author Contributions: Conceptualization, D.F.Â., D.S., B.M. and H.J.C.; methodology, D.F.Â., R.S.J. and H.J.C.; validation, D.F.Â., D.S. and B.M. formal analysis, R.S.J. and H.J.C.; investigation, D.F.Â., D.S., B.M., R.S.J. and H.J.C.; resources, D.F.Â., D.S.; data curation, R.S.J. and H.J.C.; writing—original draft preparation, D.F.Â., B.M. and H.J.C.; writing—review and editing, D.F.Â., D.S., B.M., R.S.J. and H.J.C.; supervision, D.F.Â.; project administration, D.F.Â. and H.J.C. All authors have read and agreed to the published version of the manuscript.

Funding: This work was supported by Research Grant SORG 2019 assigned to David Ângelo—project: "EUROTMJ—recording TMJ outcomes in a central database." We also thank the sub-group created by the European Society of TMJ Surgeons (ESTMJS) for their contribution to the creation of the database, namely to: Florencio Monje; Raúl González-García; Vladimir Machoň; Mattias Ulmner; Niall McLeod; Jacinto Fernández Sanromán. The funding for publication is supported by CEAUL—Centro de Estatística e Aplicações, Faculdade de Ciências, Universidade de Lisboa, Portugal financed by national funds through FCT—Fundação para a Ciência e a Tecnologia under the project UIDB/00006/2020.

Institutional Review Board Statement: The study was conducted in accordance with the Declaration of Helsinki, and approved by the Instituto Português da Face ethics committee (PT/IPFace// RCT/0822/01).

Informed Consent Statement: Informed consent was obtained from all subjects involved in the study. Written informed consent has been obtained from the patient(s) to publish this paper.

Data Availability Statement: Not applicable.

Conflicts of Interest: The authors declare no conflict of interest. The funders had no role in the design of the study; in the collection, analyses, or interpretation of data; in the writing of the manuscript; or in the decision to publish the results.

References

1. Chaurasia, A.; Ishrat, S.; Katheriya, G.; Chaudhary, P.K.; Dhingra, K.; Nagar, A. Temporomandibular disorders in North Indian population visiting a tertiary care dental hospital. *Natl. J. Maxillofac. Surg.* **2020**, *11*, 106–109. [CrossRef] [PubMed]
2. Qvintus, V.; Sipilä, K.; Le Bell, Y.; Suominen, A.L. Prevalence of clinical signs and pain symptoms of temporomandibular disorders and associated factors in adult Finns. *Acta Odontol. Scand.* **2020**, *78*, 515–521. [CrossRef] [PubMed]
3. Iodice, G.; Cimino, R.; Vollaro, S.; Lobbezoo, F.; Michelotti, A. Prevalence of temporomandibular disorder pain, jaw noises and oral behaviours in an adult Italian population sample. *J. Oral. Rehabil.* **2019**, *46*, 691–698. [CrossRef]
4. Kmeid, E.; Nacouzi, M.; Hallit, S.; Rohayem, Z. Prevalence of temporomandibular joint disorder in the Lebanese population, and its association with depression, anxiety, and stress. *Head Face Med.* **2020**, *16*, 19. [CrossRef] [PubMed]
5. Valesan, L.F.; Da-Cas, C.D.; Réus, J.C.; Denardin, A.C.S.; Garanhani, R.R.; Bonotto, D.; Januzzi, E.; de Souza, B.D.M. Prevalence of temporomandibular joint disorders: A systematic review and meta-analysis. *Clin. Oral Investig.* **2021**, *25*, 441–453. [CrossRef] [PubMed]

6. Manfredini, D.; Piccotti, F.; Ferronato, G.; Guarda-Nardini, L. Age peaks of different RDC/TMD diagnoses in a patient population. *J. Dent.* **2010**, *38*, 392–399. [CrossRef]

7. Johansson, A.; Unell, L.; Carlsson, G.E.; Söderfeldt, B.; Halling, A. Risk factors associated with symptoms of temporomandibular disorders in a population of 50- and 60-year-old subjects. *J. Oral. Rehabil.* **2006**, *33*, 473–481. [CrossRef]

8. de Boer, A.G.; van Lanschot, J.J.; Stalmeier, P.F.; van Sandick, J.W.; Hulscher, J.B.; de Haes, J.C.; Sprangers, M.A. Is a single-item visual analogue scale as valid, reliable and responsive as multi-item scales in measuring quality of life? *Qual. Life Res. Int. J. Qual. Life Asp. Treat. Care Rehabil.* **2004**, *13*, 311–320. [CrossRef]

9. Schiffman, E.L.; Ohrbach, R.; Truelove, E.L.; Tai, F.; Anderson, G.C.; Pan, W.; Gonzalez, Y.M.; John, M.T.; Sommers, E.; List, T.; et al. The Research Diagnostic Criteria for Temporomandibular Disorders. V: Methods used to establish and validate revised Axis I diagnostic algorithms. *J. Orofac. Pain* **2010**, *24*, 63–78.

10. Schiffman, E.; Ohrbach, R.; Truelove, E.; Look, J.; Anderson, G.; Goulet, J.-P.; List, T.; Svensson, P.; Gonzalez, Y.; Lobbezoo, F.; et al. Diagnostic Criteria for Temporomandibular Disorders (DC/TMD) for Clinical and Research Applications: Recommendations of the International RDC/TMD Consortium Network* and Orofacial Pain Special Interest Group. *J. Oral Facial Pain Headache* **2014**, *28*, 6–27. [CrossRef]

11. Goiato, M.C.; Zuim, P.R.J.; Moreno, A.; Dos Santos, D.M.; da Silva, E.V.F.; de Caxias, F.P.; Turcio, K.H.L. Does pain in the masseter and anterior temporal muscles influence maximal bite force? *Arch. Oral Biol.* **2017**, *83*, 1–6. [CrossRef]

12. Kothari, K.; Jayakumar, N.; Razzaque, A. Multidisciplinary management of temporomandibular joint ankylosis in an adult: Journey from arthroplasty to oral rehabilitation. *BMJ Case Rep.* **2021**, *14*, e245120. [CrossRef]

13. Gesch, D.; Bernhardt, O.; Alte, D.; Schwahn, C.; Kocher, T.; John, U.; Hensel, E. Prevalence of signs and symptoms of temporomandibular disorders in an urban and rural German population: Results of a population-based Study of Health in Pomerania. *Quintessence Int.* **2004**, *35*, 143–150.

14. Feteih, R.M. Signs and symptoms of temporomandibular disorders and oral parafunctions in urban Saudi Arabian adolescents: A research report. *Head Face Med.* **2006**, *2*, 25. [CrossRef] [PubMed]

15. Gonçalves, D.A.; Dal Fabbro, A.L.; Campos, J.A.; Bigal, M.E.; Speciali, J.G. Symptoms of temporomandibular disorders in the population: An epidemiological study. *J. Orofac. Pain* **2010**, *24*, 270–278. [PubMed]

16. Mobilio, N.; Casetta, I.; Cesnik, E.; Catapano, S. Prevalence of self-reported symptoms related to temporomandibular disorders in an Italian population. *J. Oral. Rehabil.* **2011**, *38*, 884–890. [CrossRef] [PubMed]

17. Könönen, M.; Waltimo, A.; Nyström, M. Does clicking in adolescence lead to painful temporomandibular joint locking? *Lancet* **1996**, *347*, 1080–1081. [CrossRef] [PubMed]

18. Lamot, U.; Strojan, P.; Šurlan Popovič, K. Magnetic resonance imaging of temporomandibular joint dysfunction-correlation with clinical symptoms, age, and gender. *Oral Surg. Oral Med. Oral Pathol. Oral Radiol.* **2013**, *116*, 258–263. [CrossRef]

19. Abrahamsson, A.K.; Kristensen, M.; Arvidsson, L.Z.; Kvien, T.K.; Larheim, T.A.; Haugen, I.K. Frequency of temporomandibular joint osteoarthritis and related symptoms in a hand osteoarthritis cohort. *Osteoarthr. Cartil.* **2017**, *25*, 654–657. [CrossRef]

20. Almoznino, G.; Zini, A.; Zakuto, A.; Zlutzky, H.; Bekker, S.; Shay, B.; Haviv, Y.; Sharav, Y.; Benoliel, R. Cervical Muscle Tenderness in Temporomandibular Disorders and Its Associations with Diagnosis, Disease-Related Outcomes, and Comorbid Pain Conditions. *J. Oral Facial Pain Headache* **2020**, *34*, 67–76. [CrossRef]

21. Zieliński, G.; Byś, A.; Szkutnik, J.; Majcher, P.; Ginszt, M. Electromyographic Patterns of Masticatory Muscles in Relation to Active Myofascial Trigger Points of the Upper Trapezius and Temporomandibular Disorders. *Diagnostics* **2021**, *11*, 580. [CrossRef] [PubMed]

22. Al-Baghdadi, M.; Durham, J.; Araujo-Soares, V.; Robalino, S.; Errington, L.; Steele, J. TMJ Disc Displacement without Reduction Management: A Systematic Review. *J. Dent. Res.* **2014**, *93*, 37s–51s. [CrossRef]

23. Ângelo, D.F.; Sousa, R.; Pinto, I.; Sanz, D.; Gil, F.M.; Salvado, F. Early magnetic resonance imaging control after temporomandibular joint arthrocentesis. *Ann. Maxillofac. Surg.* **2015**, *5*, 255–257. [CrossRef] [PubMed]

24. Marklund, S.; Wänman, A. Risk factors associated with incidence and persistence of signs and symptoms of temporomandibular disorders. *Acta Odontol. Scand.* **2010**, *68*, 289–299. [CrossRef]

25. Johansson, A.; Unell, L.; Carlsson, G.; Söderfeldt, B.; Halling, A.; Widar, F. Associations between social and general health factors and symptoms related to temporomandibular disorders and bruxism in a population of 50-year-old subjects. *Acta Odontol. Scand.* **2004**, *62*, 231–237. [CrossRef] [PubMed]

26. De La Torre Canales, G.; Câmara-Souza, M.B.; Muñoz Lora, V.R.M.; Guarda-Nardini, L.; Conti, P.C.R.; Rodrigues Garcia, R.M.; Del Bel Cury, A.A.; Manfredini, D. Prevalence of psychosocial impairment in temporomandibular disorder patients: A systematic review. *J. Oral. Rehabil.* **2018**, *45*, 881–889. [CrossRef]

27. Rauch, A.; Hahnel, S.; Kloss-Brandstätter, A.; Schierz, O. Patients referred to a German TMD-specialized consultation hour-a retrospective on patients without a diagnosis according to RDC/TMD decision trees. *Clin. Oral Investig.* **2021**, *25*, 5641–5647. [CrossRef]

28. Zieliński, G.; Ginszt, M.; Zawadka, M.; Rutkowska, K.; Podstawka, Z.; Szkutnik, J.; Majcher, P.; Gawda, P. The Relationship between Stress and Masticatory Muscle Activity in Female Students. *J. Clin. Med.* **2021**, *10*, 3459. [CrossRef]

29. Stocka, A.; Sierpinska, T.; Kuc, J.; Golebiewska, M. Relationship between depression and masticatory muscles function in a group of adolescents. *Cranio J. Craniomandib. Pract.* **2018**, *36*, 390–395. [CrossRef]

30. Huang, F.; Zhang, M.; Chen, Y.-J.; Li, Q.; Wu, A.-Z. Psychological Stress Induces Temporary Masticatory Muscle Mechanical Sensitivity in Rats. *J. Biomed. Biotechnol.* **2011**, *2011*, 720603. [CrossRef]
31. Dworkin, S.F.; Huggins, K.H.; LeResche, L.; Von Korff, M.; Howard, J.; Truelove, E.; Sommers, E. Epidemiology of signs and symptoms in temporomandibular disorders: Clinical signs in cases and controls. *J. Am. Dent. Assoc.* **1990**, *120*, 273–281. [CrossRef] [PubMed]
32. Hoffmann, R.G.; Kotchen, J.M.; Kotchen, T.A.; Cowley, T.; Dasgupta, M.; Cowley, A.W., Jr. Temporomandibular disorders and associated clinical comorbidities. *Clin. J. Pain* **2011**, *27*, 268–274. [CrossRef] [PubMed]

Journal of
Clinical Medicine

Systematic Review

Correlation between Temporomandibular Disorders (TMD) and Posture Evaluated trough the Diagnostic Criteria for Temporomandibular Disorders (DC/TMD): A Systematic Review with Meta-Analysis

Giuseppe Minervini [1,*], Rocco Franco [2], Maria Maddalena Marrapodi [3,*], Salvatore Crimi [4,*], Almir Badnjević [5], Gabriele Cervino [6], Alberto Bianchi [4] and Marco Cicciù [4]

1 Multidisciplinary Department of Medical-Surgical and Odontostomatological Specialties, University of Campania "Luigi Vanvitelli", 80121 Naples, Italy
2 Department of Biomedicine and Prevention, University of Rome "Tor Vergata", 00100 Rome, Italy
3 Department of Woman, Child and General and Specialist Surgery, University of Campania "Luigi Vanvitelli", 80121 Naples, Italy
4 Department of General Surgery and Medical-Surgical Specialties, School of Dentistry, University of Catania, 95124 Catania, Italy
5 Verlab Research Institute for Biomedical Engineering, Medical Devices and Artificial Intelligence, 71000 Sarajevo, Bosnia and Herzegovina
6 Department of Biomedical and Dental Sciences and Morphofunctional Imaging, School of Dentistry, University of Messina, Via Consolare Valeria, 1, 98125 Messina, Italy
* Correspondence: giuseppe.minervini@unicampania.it (G.M.); mariamaddalena.marrapodi@studenti.unicampania.it (M.M.M.); torecrimi@gmail.com (S.C.)

Citation: Minervini, G.; Franco, R.; Marrapodi, M.M.; Crimi, S.; Badnjević, A.; Cervino, G.; Bianchi, A.; Cicciù, M. Correlation between Temporomandibular Disorders (TMD) and Posture Evaluated trough the Diagnostic Criteria for Temporomandibular Disorders (DC/TMD): A Systematic Review with Meta-Analysis. *J. Clin. Med.* **2023**, *12*, 2652. https://doi.org/10.3390/jcm12072652

Academic Editor: Eiji Tanaka

Received: 24 February 2023
Revised: 27 March 2023
Accepted: 30 March 2023
Published: 2 April 2023

Abstract: Background: Temporomandibular disorders (TMDs) are a series of disorders that affect the muscles and joint. Symptoms include joint pain, muscle pain, and limitation of mouth opening. One of several multifactorial diseases, temporomandibular dysfunction has mostly been linked to five etiological factors: occlusion, trauma, severe pain stimuli, parafunctional activities, and psychological elements, including stress, anxiety, and depression. The position of the human body as it is displayed in space is referred to as posture. Several nerve pathways regulate posture, and through ligaments, TMD and posture affect each other. The purpose of this study is to evaluate the possible correlation between posture and TMD through a meta-analysis of the literature; Methods: A literature search was performed on PubMed, Lilacs, and Web of science, and articles published from 2000 to 31 December 2022 were considered, according to the keywords entered. The term "temporomandibular disorders" has been combined with "posture", using the Boolean connector AND; Results: At the end of the research, 896 studies were identified from the search conducted on the 3 engines. Only three were chosen to draw up the present systematic study summarizing the article's main findings. The meta-analysis showed through forest plot analysis a correlation between posture and TMD Conclusions: This literature meta-analysis showed a correlation between posture and TMD. Nerve pathways probably regulate both body posture and mandibular posture. Further clinical studies will be needed to confirm this hypothesis and to indicate the main conclusions or interpretations.

Keywords: temporomandibular disorders; posture; TMD; diagnostic criteria for temporomandibular disorders; DC/TMD

1. Introduction

Temporomandibular joint (TMJ), which provides essential biological activities, including chewing and speaking, is one of the most intricate and frequently utilized joints in the human body [1]. Avascular, non-innervated fibrocartilage with a strong capacity for regeneration covers the articular surfaces. The motion of the joint is controlled by the masseter, temporalis muscles, internal pterygoid, the external pterygoid, and the digastric muscle [2]. Temporomandibular disorders (TMD) affect the articulation and masticatory muscles, or both. The following are the most typical symptoms and signs: joint sounds (clicks or crepitus), pre-auricular and/or masticatory muscle soreness, and restrictions or deviation during the mandibular opening. One of several multifactorial diseases, temporomandibular dysfunction has mostly been linked to five etiological factors: trauma, severe pain stimuli [3–5], parafunctional activities, and psychological elements, including stress, anxiety, and depression [6–14]. The position of the human body as it is displayed in space is referred to as posture. The central nervous system-controlled posture by muscle activation enables modifications due to a sophisticated mechanism integrated by the view and hearing [15]. TMD has ligaments and muscle connections with the cervical area, so these connections have led to speculation that posture problems may influence the development of TMD [16–20]. The masticatory cycles should be balanced since unilateral mastication might throw off the body's postural equilibrium while standing by creating an imbalance in the neck muscles and anterior muscle chains. The most popular method for studying postural control involves measuring the oscillation of the body while it is in an upright, resting position using a force platform [21]. In patients with TMD, the position of the head is depressed as the masticatory muscles change the position of the jaw. Proprioceptive afferents may exhibit changes in the mandibular position, which may have an impact on gait stability and the center of pressure of the foot. In the case of TMD, there are frequently significant differences between therapists regarding the best kind of occlusal splint to utilize [22]. Occlusal splints, which are frequently employed for TMD-related pain alleviation, appear to perform a significant role in this scenario. In more depth, there are many occlusal splint kinds, such as bite plates, with various indications and purposes. The stabilization splints are hard acrylic appliances that offer TMD sufferers a perfect occlusion temporarily and removable, relieving orofacial pain by loosening masticatory muscles [23]. The stabilization splints may result in a neuromuscular balance, removing posterior interferences and supplying a stable occlusal relationship and centric relation. The relationship between craniometrical posture and TMDs has been studied; however, despite the huge number of studies, clinicians and academics remain unconvinced [24]. There is evidence from certain studies that people with TMDs have altered head and cervical spine posture, while no such link is found in other investigations. The skull, mandible, and cervical spine exhibit neurological and biomechanical connections, generating a functional complex that may be referred to as the "craniocervical mandibular system," which is related to the cervical area via muscles and ligament [25]. The aim of this literature review with meta-analysis was to evaluate the correlation between posture and TMD. In addition, the meta-analysis performed evaluated the effects of occlusal therapy on posture.

2. Materials and Methods

2.1. Eligibility Criteria

All documents were assessed for eligibility based on the following Population (including animal species), Exposure, Comparator, and Outcomes (PECO):

(P) Participants consisted of patients.

(E) Exposure consists of patients with temporomandibular dysfunction treated with occlusal splint therapy.

(C) Comparison consists of patients with TMD not treated with occlusal splint therapy.

(O) The outcome is to evaluate the effectiveness of bite therapy on posture, as the second outcome is to evaluate the interference and correlation between temporomandibular disorders and posture.

The following inclusion criteria were employed for this meta-analysis: (1) randomized clinical trial (RCT); (2) TMD patients treated and evaluated by DC/TMD; (3) diagnosis of myofascial pain, myofascial pain with a limited opening; (4) disc displacement with reduction; (5) disc displacement without reduction with limited opening; (6) disc displacement without reduction without limited opening; and (7) papers published in English.

Exclusion criteria were: (1) full-text unavailability (i.e., posters and conference abstracts); (2) studies involving animals; (3) review article; (4) case reports; (5) lack of effective statistical analysis; (6) degenerative joint disease (osteoarthrosis and osteoarthritis); (7) loss of more than five teeth, with the exception of the third molars; (8) medical history of motor or neurological disorders; (9) facial or head trauma; and (10) orthopedic and orthodontic treatment.

2.2. Search Strategy

A literature search was performed on PubMed, Lilacs, and Web of science, and articles published from 2000 to 31 December 2022 were considered, according to the keywords entered. The term "Temporomandibular disorders" and "TMD" united with OR has been combined with "Posture", using the Boolean connector AND. The web search was assisted using MESH (Medical Subjects Headings) (Table 1). The criteria for this review are described in the PRISMA and by the following flowchart (Figure 1). Additionally, a manual scan of earlier systematic reviews on the same subject was completed in their references. The Cochrane Handbook for Systematic Reviews of Interventions and Preferred Reporting Items for Systematic Reviews (PRISMA) standards were followed in conducting this systematic review. The International Prospective Register of Systematic Reviews (PROSPERO) has recorded the systematic review procedure under the code CRD42022315350 as of 12 April 2022.

Table 1. Search strategy.

PubMed ((temporomandibular disorders) OR (TMD)) AND (POSTURE)
Lilacs temporomandibular disorders [Palavras] or tmd [Palavras] and posture [Palavras]
Web of Science ((ALL=(temporomandibular disorders)) OR ALL=(tmd)) AND ALL=(posture)

2.3. Data Extraction

Using a specialized data extraction on a Microsoft Excel sheet, two reviewers (GM and RF) separately extracted data from the included studies. When there was disagreement, a third reviewer helped to achieve a consensus (MC).

The following data was obtained: (1) First Author; (2) Year; (3) Sample; (4) Diagnostic criteria; (5) Type of bite; (6) Treatment duration; and (7) Exams to evaluate the effect on posture; (8) Results of therapy.

2.4. Quality Assessment

Using the Cochrane risk-of-bias tool for randomized trials, Version 2, two reviewers evaluated the articles' bias risk (RoB 2). Any discrepancy was discussed with a third reviewer until an agreement was achieved.

2.5. Statistical Analysis

The pooled analyses were carried out utilizing Review Manager 5.2.8 software (Cochrane Collaboration, Copenhagen, Denmark; 2014). The study compared TMD patients receiving biting therapy to a control group to assess the impact on posture. The difference in risk between the two groups was calculated. Low heterogeneity (30%), medium heterogeneity (30–60%), and high heterogeneity (> 60%) were used to measure and categorize study heterogeneity using the Higgins Index (I2) and the chi-square test.

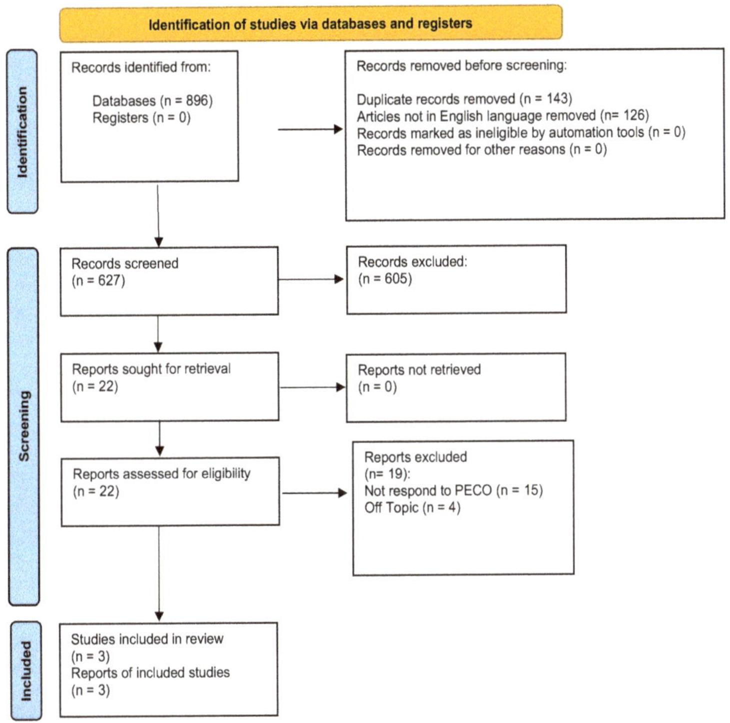

Figure 1. Prisma statement. From: Page, M.J.; McKenzie, J.E.; Bossuyt, P.M.; Boutron l Hoffmann, T.C.; Mulrow, C.D., et al. The PRISMA 2020 statement: an updated guideline for reporting systematic reviews. *BMJ* **2021**, *372*, n71. doi:10.1136/bmi.n71. For more information, visit: http://www.prisma-statement.org/ on 5 March 2023.

3. Results

3.1. Study Characteristics

At the end of the research, 896 studies were identified from the search conducted on the 3 engines. During the initial phase, 143 items were excluded because they were duplicates and 126 because are not in English. During the initial screening phase, 605 articles were excluded from both search engines because they were systematic reviews of the literature, and therefore did not meet the inclusion criteria. In addition, the filter was included in which only randomized clinical trials were considered. During the final screening phase, the abstracts and the full text of 22 articles were evaluated. Only 3 were chosen to draw up the present systematic study, as illustrated by the PRISMA 2020 flowchart in Figure 1; 19 articles were excluded; 15 did not meet PECO; and 4 are off topic. The remaining articles were selected for the title and abstract screening according to the PECO model. The studies considered have a time frame from 2014 to 2018. The studies analyzed were conducted in various parts of the world: Brazil, Italy, and Poland. A total of 154 subjects were analyzed. De Giorgi's study analyzed 45 women with TMD, evaluated by DC/TMD; however, the patients were all female, and all had joint dislocation with or without reduction. The pain was assessed by VAS scale and the effects of bite therapy on posture were by cephalometric examination and by rasterstereography, after which postural parameters were assessed at time 0 and after 1 month and 3 months. The patients were randomly divided between 24 in the study group and 21 in the control group. Patients in the study group were fitted with a 2 mm splint with posterior contact. Oliveira's study evaluated 49 patients with TMD assessed through DCs/TMDs, in which case he selected and evaluated all TMD subgroups.

The patients were randomly divided as follows: 36 patients in the study group to whom a splint was applied to be worn more than 8 h a day for a duration of 12 weeks; a control group of 13 patients to whom nothing but physiotherapy was applied. After that, postural parameters were evaluated through a stabilometric platform, and the parameters were assessed with eyes closed and open. It was mainly evaluated through speed to re-establish the Center of Pressure (COP). A 1.5 mm splint with simultaneous bilateral contacts was used in this study. The WalczyNska-Dragon study took 60 patients with TMD, diagnosed with DC/TMD. The patients considered in this study had TMD and neck pain with movement limitation. The patients were randomly divided as follows: 30 patients in the study group to whom a splint was applied; and 30 patients as the control group. Postural parameters were assessed by a Jaw motion tracking (JMT) which evaluated mandibular movements and a cervical spine motion (CMS), which evaluated improvements at the cervical and postural levels. In addition, during therapy, pain level was reported and assessed by the VAS scale. Patients were evaluated at time 0, and after 3 weeks and 3 months. The type of appliance was a plate with anterior contacts only.

3.2. Main Findings

The purpose of De Giorgi's study was to assess how an occlusal splint affected individuals with intra-articular temporomandibular joint (TMJ) problems' body posture. The study included 45 women with TMJ disorders who were divided into two groups: those who used occlusal splints and those who did not. Rasterstereographic recordings were made at baseline, 1 month, 3 months, and 6 months later in order to examine the following postural parameters: pelvic tilt and torsion, kyphotic and lordotic angles, lumbar and cervical arrows, trunk imbalance, and trunk inclination. In the intragroup analysis, no significant differences were found for the postural parameters. Significant differences between the two groups were found when the cervical arrow, kyphotic angle, and lordotic angle were analyzed. There were no discernible differences between T0 and T1, or T0 and T2 in the postural characteristics of the occlusal splint and control group in the intragroup study. The occlusal splint group between T0 and T3 showed no significant differences either. Significant variations between the two groups were found in the analyses for various postural measures. At T1, the cervical arrow evaluation revealed a statistically significant difference between the control group and the occlusal splint group (61.96 mm 17.73 vs. 58.39 mm 13.73; $p = 0.001$). In terms of the kyphotic angle in the resting position, there was a statistically significant difference at T1 between the control group and the occlusal splint group (54.17o 8.97 vs. 55.25o 10.16; $p = 0.012$), and at T2 between the control group and the occlusal splint group (54.00 vs. 54.84). In terms of the lordotic angle a statistically significant difference was discovered between the control group and the occlusal splint group at rest position at T2 (49.30) ($p = 0.017$) [26]. The study of Oliveira was to investigate how using an occlusal splint influenced postural balance. A prospective, controlled, randomized clinical trial was carried out. A questionnaire developed by the R DC/TD and magnetic resonance imaging of TMJ was used to diagnose 49 patients—36 in the test group and 13 in the control group—who ranged in age from 18 to 75 and were of both sexes. Orientations for physiotherapeutic exercises and an occlusal splint were given to the test group, while just physiotherapeutic exercises were given to the control group. A force plate was used to assess postural balance. The groups were reevaluated after 12 weeks. With their eyes closed, patients from both groups showed a statistically significant increase in antero-posterior speed: test group (p 0.001) and control group ($p = 0.046$). With their eyes open, only patients in the test group showed a statistically significant increase in antero-posterior speed ($p = 0.023$) [27]. The purpose of this study was to assess how TMD therapy affected the cervical spine's range of motion (ROM) and spinal pain relief. Sixty individuals with TMD, cervical spine discomfort, and restricted cervical spine range of motion made up the study group. The subjects completed a questionnaire that asked about their neck pain and TMD symptoms. They also had their masticatory motor system physically evaluated (in accordance with RDC-TMD) and analyzed by a JMA ultrasound instrument. An MCS device was used to analyze the mobility of the cervical spine. The whole group displayed cervical spine discomfort. The treated group's cervical spine pain on the VAS scale considerably decreased over the course of three months of therapy. Cervical spine pain decreased during therapy; it returned after three weeks for 39% of patients, and it was only present in 8% of subjects from the treated group after 3 months (2 subjects). A statistically significant difference existed between the treatment and control groups (*p*= 0.0001). The flexion movement, which only 22% of patients had on the first assessment, was within normative values, and was where the progress could be detected. A total of 70% of the patients in the treated group's flexion movement adhered to the standard at the third examination. There were more subjects in the experimental group who had anteflexion movement outcomes that were consistent with the norm, which was very significant ($p = 0.0006$). The results for the retroflexion movement improved by a highly significant factor ($p = 0.0082$); more subjects in the experimental group had results that were in line with the norm. No significant ($p > 0.05$) improvements were identified in the control group, indicating that the cervical spine's range of motion did not improve in relation to normative values [28] (Table 2).

Table 2. Principal elements of the studies which formed part of the present systematic analysis.

Author	Year	Sample	Diagnostic Criteria	Type of Byte	Treatment Duration	Exams to Evaluate Effect on Posture	Results
De Giorgi et al. [26]	2020	45 women: 24 test 21 control	DC/TMD	2 mm with posterior contact	1, 2, 3 months	Evaluation of VAS Rasterstereography Cephalometric analysis	Significant differences concerning the cervical arrow, the kyphotic and lordotic angles
Oliveira et al. [27]	2019	49 patients: 36 test 13 control	DC/TMD	1.5 mm with simultaneous bilateral contact	12 weeks	Stabilometry test with the eye open and closed	Study group had increased anteroposterior velocity with eyes closed and eyes open
WalczyNska-Dragon et al. [28]	2014	60 patients: 30 test 30 control	DC/TMD	Byte only with anterior contact	Evaluation after 3 weeks and 3 months	Evaluation of VAS Evaluation of mandibular movement with JMT Evaluation of cervical spine movement with MCS	Test group: improvement in TMJ movement and cervical spinal movement Diminutions of VAS

3.3. Metanalysis

The meta-analysis was conducted by fixed model effect because of the low heterogeneity ($I^2 = 0\%$) between the three included studies. The overall effect, reported in the forest plot (Figure 2), shows that showed that bite therapy and TMD are related to posture and that a change in chewing causes effects on posture (RR 1.65; 95%; CI 1.18–2.53).

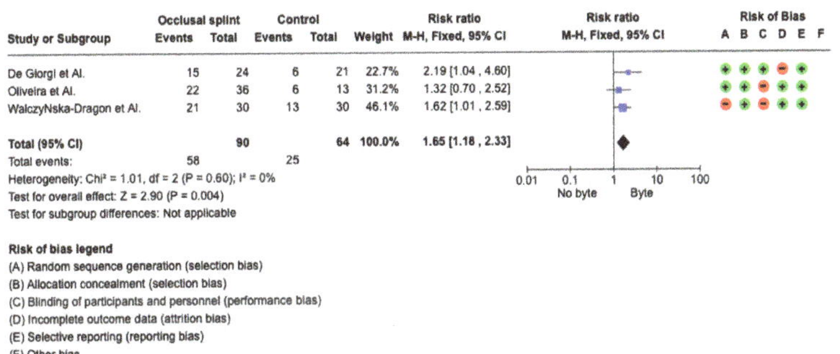

Figure 2. Forest plot of the meta-analysis [26–28].

3.4. Quality Assessment and Risk of Bias

Using RoB 2, the risk of bias was estimated and reported in Figure 3. Regarding the randomization process, 75% of the studies ensured a low risk of bias. However, 25% of the studies excluded performance bias, but 75% reported all outcome data, and 100% of the included studies adequately excluded bias in the selection of reported outcomes, while 25% excluded bias in self-reported outcomes. Overall, all studies were shown to have a low risk of experiencing bias.

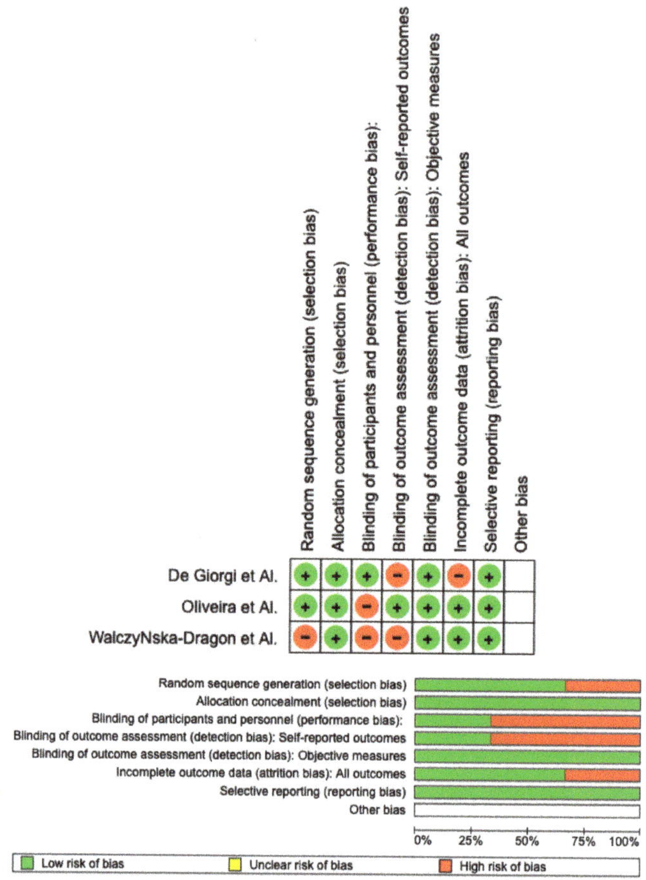

Figure 3. Risk of bias domains of the included studies [26–28].

4. Discussion

The findings and their interpretation in light of prior research and the working hypotheses should be discussed by the authors. It is important to discuss the results and their implications in the widest context possible. It may also highlight potential directions for future study [29].

There is still disagreement in the literature on the relationship between TMJ, muscle and posture, and it is suggested that better controlled trials with thorough TMD diagnoses, larger sample sizes, and objective posture measurements are required [30]. Important gaps in our understanding of this relationship remain as a result of the intricacy of the contributing components. The connection between posture and the TMJ is supported by certain research in the literature, but not all of them [31]. The link between TMD myogenic and posture is heavy but weak between TMD arthrogenus and another type of TMD [32].

De giorgi's study evaluated through different postural analyses their change. In particular, the use of posturography as a diagnostic tool was not supported by the evidence because these analytic techniques did not make any noteworthy advancements. A dependable, non-invasive technique for examining 3D spine anatomy is rasterstereography. This method enables a radiation-free assessment of the body's back surface and has proven to be accurate for determining pelvic and spinal alignment. The intragroup analyses revealed no statistically significant differences, proving that treatment with an occlusal splint had no effect on patients' postural measures. At T1 and T2, however, some variations between the control and occlusal splint groups were discovered, suggesting that some alterations had taken place. However, the results of De Giorgi were not noteworthy from a therapeutic perspective due to the low range of statistical significance [26]. Oliveira et al. [27] found that both with eyes open and closed, the use of an occlusion splint, and a presentation of therapeutic exercises significantly increased the anteroposterior velocity from the COP [33]. In synthesis, the occlusion splint added to the effects of postural control. Studying TMJD and postural balance requires keeping track of patients' anthropometric and clinical parameters. Age, sex, weight, and height are variables that may have an impact on the effectiveness of the treatment and the postural balance. Weight and height did not significantly differ across the groups. Additionally, there was no difference in terms of age, despite the fact that a higher percentage of TMJD patients were between the ages of 20 and 40. There is a significant amount of research supporting the signs and symptoms of TMD, thus they might not differ from those discovered in our study. An attempt at postural readjustment can be concluded from the findings of the use of an occlusal splint and therapeutic exercise guidelines in subjects with TMD over the course of 12 weeks, which included an increase in the velocity of center gravity recovery. It is possible to interpret the acceleration of the postural sway as an increase in the frequency of postural balance corrections made by the individuals [34–41]. The adjustment of the head and neck posture carried on using the occlusal splint may have caused this increase in the frequency of corrections. The new question is whether these findings point to a transient shift in how the body is perceived, indicating that postural balance will change in the future, or perhaps revert to its initial state.

The results obtained by WalczyNska-Dragon have shown a correlation between the diseases and the beneficial effects of treatment for cervical pain, even in people experiencing such discomfort for a long time. In order to treat patients more effectively and efficiently at the beginning, when painful symptoms first appear, and when treating them is possible and much quicker and more efficiently, it is crucial to understand the intricate relationships between posture and TMD. The cause of cervical spine pain is frequently yet unknown, but from these studies, they seem to be related. Numerous scientific studies support the efforts of many researchers to investigate the effects of problems in the "upper quarter" on posture and pain in different body regions. However, the main goal of studies undertaken thus far has been to demonstrate the existence or lack of a relationship between cervical spine discomfort and stomatognathic system dysfunction. The most often used method involved questionnaires with questions on complaints of the motor aspect of the stomatognathic system and pain in the cervical spine. Based on this, researchers would hunt for a connection between the cervical spine pain and the dysfunction of the stomatognathic system's motor component. However, in our investigation, occlusal appliance therapy was used; no intrusive treatment strategies were employed. Most of the participants in the experimental group who received therapy with an occlusal splint reported improvements and the complete removal or significant reduction in cervical spine discomfort and TMD symptoms, while the cervical spine's mobility also increased significantly. When treating TMD, there are frequently significant differences between therapists regarding the best kind of occlusal splint to utilize [28].

Additionally, studies have shown a connection between posture and TMDs. TMD patients have more pronounced alterations in the body's center of gravity. Numerous investigations have demonstrated that patients with TMD frequently exhibit an overly forward head position that is accompanied by shortening of the sternocleidomastoid and the posterior cervical extensor muscles [42,43]. When the head is tilted anteriorly, the field of vision is reduced, and cervical lordosis rises to extend the field of view. The center of gravity is similarly affected by the head's anterior position, supporting the link between TMD and body posture. The cervical region's postural changes can also lead to TMD by altering the position of the mandible and the head's orientation. Previous posturography studies have shown that in participants with unilaterally anaesthetized trigeminal afferents, there is a link between body posture and gaze balance [44]. According to the findings mentioned above, the trigeminal system has a significant impact on how posture and sight are coordinated. The hypothesized neurological basis is still being developed; however, there are several hypotheses [45]. There have been many documented anatomical links between trigeminal systems and the neurological systems that regulate posture. The mesencephalic nucleus of the trigeminus (MNT), a sensory nucleus with distinctive properties, stretches from the caudal region of the superior colliculus to the dorsal portion of the spinal trigeminal nucleus. These nucleus cells are protoneurons with ganglionic cell functions rather than central neurons [46].

The presence of muscle-fascial chains is another fundamental component of the relationship between the stomatognatic system and human posture. Organs are held in place, protected, and fed by fasciae. Fasciae have three layers: superficial, deep, and visceral. Myofibroblasts and several different receptors are heavily concentrated in the deep fasciae, which surrounds muscles, bones, nerves, and blood vessels (nociceptors, proprioceptors, mechanoreceptors, chemoreceptors, and thermoreceptors). Fascial cells called myofibroblasts are produced in response to mechanical stress and actively contract smoothly and resemble muscles [47]. The fascial system contains mechanoreceptors that regulate contractile capacity and can regulate it when stressed, and therefore, distribute muscle tension to adjacent muscles. These tensions move along the muscle fascial chain and affect how the body is positioned in general [48,49]. All these analyzed studies, although they treated patients with different types of splints, showed an intercorrelation between temporomandibular disorders and posture, both in terms of pain and posture [50,51]. All these studies used different methods to be able to evaluate the effects of bite therapy on posture and cervical movements, and this turns out to be a confounding factor in the performed meta-analysis. However, based on the results obtained, there are effects of bite treatment on posture compared to untreated patients, and this shows an intercorrelation between posture and temporomandibular disorders. In addition, a limitation of the study is given by De Giorgi, as he recruited only patients with TMD who have joint disc dislocation. In fact, the limitations of these articles are because the change or readjustment of posture after completing bite therapy is not evaluated, and therefore, the effects over time of this therapy on posture are not evaluated.

5. Conclusions

This meta-analysis and review of the literature showed the correlation between posture and temporomandibular disorders. In fact, in a statistically significant manner, the application of bite therapy had positive effects on what is posture. Although with the limitations of these analyzed studies, we can say that this correlation exists and is present.

Funding: This research received no external funding.

Institutional Review Board Statement: Not applicable.

Informed Consent Statement: Not applicable.

Data Availability Statement: All data described in the study are presented in the manuscript. The datasets analyzed are available from the corresponding author on reasonable request.

Conflicts of Interest: The authors declare no conflict of interest.

Abbreviations

TMD	temporomandibular disorders
TMJ	temporomandibular joint
PECO	patients, exposure, comparison, outcome
DC/TMD	diagnostic criteria for temporomandibular disorders
RCT	randomized clinical trial
MeSh	Medical Subject Headings
ROM	range of motion

References

1. Di Francesco, F.; Lanza, A.; Di Blasio, M.; Vaienti, B.; Cafferata, E.A.; Cervino, G.; Cicciù, M.; Minervini, G. Application of Botulinum Toxin in Temporomandibular Disorders: A Systematic Review of Randomized Controlled Trials (RCTs). *Appl. Sci.* **2022**, *12*, 12409. [CrossRef]
2. la Touche, R.; París-Alemany, A.; von Piekartz, H.; Mannheimer, J.S.; Fernández-Carnero, J.; Rocabado, M. The Influence of Cranio-cervical Posture on Maximal Mouth Opening and Pressure Pain Threshold in Patients with Myofascial Temporomandibular Pain Disorders. *Clin. J. Pain* **2011**, *27*, 48–55. [CrossRef]
3. Tallarico, M.; Meloni, S.M.; Park, C.-J.; Zadrożny; Scrascia, R.; Cicciù, M. Implant Fracture: A Narrative Literature Review. *Prosthesis* **2021**, *3*, 267–279. [CrossRef]
4. Fiorillo, L.; De Stefano, R.; Cervino, G.; Crimi, S.; Bianchi, A.; Campagna, P.; Herford, A.S.; Laino, L.; Cicciù, M. Oral and Psychological Alterations in Haemophiliac Patients. *Biomedicines* **2019**, *7*, 33. [CrossRef]
5. Saccomanno, S.; Quinzi, V.; D'Andrea, N.; Albani, A.; Paskay, L.C.; Marzo, G. Traumatic Events and Eagle Syndrome: Is There Any Correlation? A Systematic Review. *Healthcare* **2021**, *9*, 825. [CrossRef] [PubMed]
6. Minervini, G.; D'amico, C.; Cicciù, M.; Fiorillo, L. Temporomandibular Joint Disk Displacement: Etiology, Diagnosis, Imaging, and Therapeutic Approaches. *J. Craniofacial Surg.* **2022**, 1097. [CrossRef] [PubMed]
7. Campus, G.; Diaz-Betancourt, M.; Cagetti, M.; Carvalho, J.; Carvalho, T.; Cortés-Martinicorena, J.; Deschner, J.; Douglas, G.; Giacaman, R.; Machiulskiene, V.; et al. Study Protocol for an Online Questionnaire Survey on Symptoms/Signs, Protective Measures, Level of Awareness and Perception Regarding COVID-19 Outbreak among Dentists. A Global Survey. *Int. J. Environ. Res. Public Health* **2020**, *17*, 5598. [CrossRef]
8. Chakraborty, T.; Jamal, R.; Battineni, G.; Teja, K.; Marto, C.; Spagnuolo, G. A Review of Prolonged Post-COVID-19 Symptoms and Their Implications on Dental Management. *Int. J. Environ. Res. Public Health* **2021**, *18*, 5131. [CrossRef]
9. Soltani, P.; Baghaei, K.; Tafti, K.T.; Spagnuolo, G. Science Mapping Analysis of COVID-19 Articles Published in Dental Journals. *Int. J. Environ. Res. Public Health* **2021**, *18*, 2110. [CrossRef] [PubMed]
10. Sycinska-Dziarnowska, M.; Maglitto, M.; Woźniak, K.; Spagnuolo, G. Oral Health and Teledentistry Interest during the COVID-19 Pandemic. *J. Clin. Med.* **2021**, *10*, 3532. [CrossRef]
11. Minervini, G.; Franco, R.; Marrapodi, M.M.; Fiorillo, L.; Cervino, G.; Cicciù, M. Prevalence of Temporomandibular Disorders in Children and Adolescents Evaluated with Diagnostic Criteria for Temporomandibular Disorders (DC/TMD): A Systematic Review with Meta-analysis. *J. Oral Rehabil.* **2023**. [CrossRef] [PubMed]
12. Cicciù, M.; Laino, L.; Fiorillo, L. Oral signs and symptoms of COVID-19 affected patients: Dental practice as prevention method. *Minerva Dent. Oral Sci.* **2020**, *71*, 3–6. [CrossRef] [PubMed]
13. Qazi, N.; Pawar, M.; Padhly, P.P.; Pawar, V.; Mehta, V.; D'Amico, C.; Nicita, F.; Fiorillo, L.; Alushi, A.; Minervini, G.; et al. Teledentistry: Evaluation of Instagram Posts Related to Bruxism. *Technol. Health Care* **2023**, 1–12. [CrossRef] [PubMed]
14. Bellini, P.; Iani, C.; Zucchelli, G.; Franchi, M.; Mattioli, A.V.; Consolo, U. Impact of the COVID-19 pandemic on dental hygiene students in the Italian region of Emilia-Romagna. *Minerva Dent. Oral Sci.* **2022**, *71*, 180–191. [CrossRef]
15. Wadhokar, O.C.; Patil, D.S. Current Trends in the Management of Temporomandibular Joint Dysfunction: A Review. *Cureus* **2022**, *14*, e29314. [CrossRef]
16. An, J.-S.; Jeon, D.-M.; Jung, W.-S.; Yang, I.-H.; Lim, W.H.; Ahn, S.-J. Influence of temporomandibular joint disc displacement on craniocervical posture and hyoid bone position. *Am. J. Orthod. Dentofac. Orthop.* **2015**, *147*, 72–79. [CrossRef]
17. Lee, W.Y.; Okeson, J.P.; Lindroth, J. The relationship between forward head posture and temporomandibular disorders. *J. Orofac. Pain* **1995**, *9*.
18. Minervini, G.; Mariani, P.; Fiorillo, L.; Cervino, G.; Cicciù, M.; Laino, L. Prevalence of temporomandibular disorders in people with multiple sclerosis: A systematic review and meta-analysis. *CRANIO®* **2022**, 1–9. [CrossRef] [PubMed]
19. Minervini, G.D.; Del Mondo, D.D.; Russo, D.D.; Cervino, G.D.; D'Amico, C.D.; Fiorillo, L.D. Stem Cells in Temporomandibular Joint Engineering: State of Art and Future Persectives. *J. Craniofacial Surg.* **2022**, *33*, 2181–2187. [CrossRef]
20. Crescente, G.; Minervini, G.; Spagnuolo, C.; Moccia, S. Cannabis Bioactive Compound-Based Formulations: New Per-spectives for the Management of Orofacial Pain. *Molecules* **2022**, *28*, 106. [CrossRef] [PubMed]
21. Kang, J.-H. Effects on migraine, neck pain, and head and neck posture, of temporomandibular disorder treatment: Study of a retrospective cohort. *Arch. Oral Biol.* **2020**, *114*, 104718. [CrossRef]

22. Perinetti, G. Correlations Between the Stomatognathic System and Body Posture: Biological or Clinical Impli-cations? *Clinics* **2009**, *64*, 77–78. [CrossRef]
23. Quinzi, V.; Paskay, L.C.; Manenti, R.J.; Giancaspro, S.; Marzo, G.; Saccomanno, S. Telemedicine for a multidisciplinary as-sessment of orofacial pain in a patient affected by eagle's syndrome: A clinical case report. *Open Dent. J.* **2021**, *15*, 102–110. [CrossRef]
24. Abe, S.; Kawano, F.; Matsuka, Y.; Masuda, T.; Okawa, T.; Tanaka, E. Relationship between Oral Parafunctional and Postural Habits and the Symptoms of Temporomandibular Disorders: A Survey-Based Cross-Sectional Cohort Study Using Propensity Score Matching Analysis. *J. Clin. Med.* **2022**, *11*, 6396. [CrossRef]
25. Miranda, L.S.; Graciosa, M.D.; Puel, A.N.; de Oliveira, L.R.; Sonza, A. Masticatory muscles electrical activity, stress and posture in preadolescents and adolescents with and without temporomandibular dysfunction. *Int. J. Pediatr. Otorhinolaryngol.* **2020**, *141*, 110562. [CrossRef]
26. de Giorgi, I.; Castroflorio, T.; Cugliari, G.; Deregibus, A. Does occlusal splint affect posture? A randomized controlled trial. *CRANIO®* **2020**, *38*, 264–272. [CrossRef]
27. Oliveira, S.S.I.; Pannuti, C.M.; Paranhos, K.S.; Tanganeli, J.P.C.; Laganá, D.C.; Sesma, N.; Duarte, M.; Frigerio, M.L.M.A.; Cho, S. Effect of occlusal splint and therapeutic exercises on postural balance of patients with signs and symptoms of temporo-mandibular disorder. *Clin. Exp. Den.t Res.* **2019**, *5*, 109–115. [CrossRef] [PubMed]
28. Walczyńska-Dragon, K.; Baron, S.; Nitecka-Buchta, A.; Tkacz, E. Correlation between TMD and Cervical Spine Pain and Mobility: Is the Whole Body Balance TMJ Related? *Biomed Res. Int.* **2014**, *2014*, 1–7. [CrossRef]
29. Faulin, E.F.; Guedes, C.G.; Feltrin, P.P.; Joffiley, C.M.M.S.C. Association between temporomandibular disorders and abnormal head postures. *Braz. Oral Res.* **2015**, *29*, 1–6. [CrossRef]
30. Mehta, N.R.; Correa, L.P. Oral Appliance Therapy and Temporomandibular Disorders. *Sleep Med. Clin.* **2018**, *13*, 513–519. [CrossRef] [PubMed]
31. Franco, R.; Basili, M.; Venditti, A.; Chiaramonte, C.; Ottria, L.; Barlattani, A.; Bollero, P. Statistical analysis of the frequency distribution of signs and symptoms of patients with temporomandibular disorders. *Oral Implant.* **2016**, *9*, 190–201. [CrossRef]
32. Saddu, S.C. The Evaluation of Head and Craniocervical Posture among Patients with and without Temporomandibular Joint Disorders- A Comparative Study. *J. Clin. Diagn. Res.* **2015**, *9*, ZC55–ZC58. [CrossRef]
33. Fagundes, N.C.F.; Minervini, G.; Alonso, B.F.; Nucci, L.; Grassia, V.; d'Apuzzo, F.; Puigdollers, A.; Perillo, L.; Flores-Mir, C. Patient-reported outcomes while managing obstructive sleep apnea with oral appliances: A scoping review. *J. Evid. -Based Dent. Pract.* **2022**, *23*, 101786. [CrossRef] [PubMed]
34. Ferrillo, M.; Lippi, L.; Giudice, A.; Calafiore, D.; Paolucci, T.; Renò, F.; Migliario, M.; Fortunato, L.; Invernizzi, M.; de Sire, A. Temporomandibular Disorders and Vitamin D Deficiency: What Is the Linkage between These Conditions? A Sys-tematic Review. *J. Clin. Med.* **2022**, *11*, 6231. [CrossRef] [PubMed]
35. Minervini, G.; Franco, R.; Marrapodi, M.M.; Mehta, V.; Fiorillo, L.; Badnjević, A.; Cervino, G.; Cicciù, M. The Association between COVID-19 Related Anxiety, Stress, Depression, Temporomandibular Disorders, and Headaches from Childhood to Adulthood: A Systematic Review. *Brain Sci.* **2023**, *13*, 481. [CrossRef] [PubMed]
36. Câmara-Souza, M.B.; Bracci, A.; Colonna, A.; Ferrari, M.; Garcia, R.C.M.R.; Manfredini, D. Ecological Momentary Assessment of Awake Bruxism Frequency in Patients with Different Temporomandibular Disorders. *J. Clin. Med.* **2023**, *12*, 501. [CrossRef] [PubMed]
37. Arribas-Pascual, M.; Hernández-Hernández, S.; Jiménez-Arranz, C.; Grande-Alonso, M.; Angulo-Díaz-Parreño, S.; La Touche, R.; Paris-Alemany, A. Effects of Physiotherapy on Pain and Mouth Opening in Temporomandibular Disorders: An Umbrella and Mapping Systematic Review with Meta-Meta-Analysis. *J. Clin. Med.* **2023**, *12*, 788. [CrossRef]
38. Almeida, L.E.; Doetzer, A.; Beck, M.L. Immunohistochemical Markers of Temporomandibular Disorders: A Review of the Literature. *J. Clin. Med.* **2023**, *12*, 789. [CrossRef]
39. Winocur, E.; Wieckiewicz, M. Temporomandibular Disorders Related Pain among Sleep & Awake Bruxers: A Comparison among Sexes and Age. *J. Clin. Med.* **2023**, *12*, 1364. [CrossRef]
40. Chęciński, M.; Chęcińska, K.; Turosz, N.; Sikora, M.; Chlubek, D. Intra-Articular Injections into the Inferior versus Superior Compartment of the Temporomandibular Joint: A Systematic Review and Meta-Analysis. *J. Clin. Med.* **2023**, *12*, 1664. [CrossRef]
41. Colonna, A.; Bracci, A.; Ahlberg, J.; Câmara-Souza, M.B.; Bucci, R.; Conti, P.C.R.; Dias, R.; Emodi-Perlmam, A.; Favero, R.; Häggmän-Henrikson, B.; et al. Ecological Momentary Assessment of Awake Bruxism Behaviors: A Scoping Review of Findings from Smartphone-Based Studies in Healthy Young Adults. *J. Clin. Med.* **2023**, *12*, 1904. [CrossRef] [PubMed]
42. Nota, A.; Tecco, S.; Ehsani, S.; Padulo, J.; Baldini, A. Postural stability in subjects with temporomandibular disorders and healthy controls: A comparative assessment. *J. Electromyogr. Kinesiol.* **2017**, *37*, 21–24. [CrossRef] [PubMed]
43. di Giacomo, P.; Ferrara, V.; Accivile, E.; Ferrato, G.; Polimeni, A.; di Paolo, C. Relationship between Cervical Spine and Skeletal Class II in Subjects with and without Temporomandibular Disorders. *Pain Res. Manag.* **2018**, *2018*, 1–7. [CrossRef] [PubMed]
44. de Chaves, P.J.; de Oliveira, F.E.M.; Damázio, L.C.M. Incidence of postural changes and temporoman-dibular disorders in students. *Acta Ortop. Bras.* **2017**, *25*, 162–164. [CrossRef]
45. Ferreira, M.C.; Grossi, D.; Dach, F.; Speciali, J.G.; Gonçalves, M.C.; Chaves, T.C. Body posture changes in women with migraine with or without temporomandibular disorders. *Braz. J. Phys. Ther.* **2014**, *18*, 19–29. [CrossRef]
46. Paço, M.; Duarte, J.; Pinho, T. Orthodontic Treatment and Craniocervical Posture in Patients with Temporomandibular Disorders: An Observational Study. *Int. J. Environ. Res. Public Health* **2021**, *18*, 3295. [CrossRef]

47. Ekici Camcı, H. Relationship of temporomandibular joint disorders with cervical posture and hyoid bone position. *Cranio®* **2021**, 1–10. [CrossRef]
48. Cortese, S.; Mondello, A.; Galarza, R.; Biondi, A. Postural alterations as a risk factor for temporomandibular disorders. *Acta Odontol. Latinoam.* **2017**, *30*, 57–61.
49. Cuccia, A.; Caradonna, C. The Relationship Between the Stomatognathic System and Body Posture. *Clinics* **2009**, *64*, 61–66. [CrossRef]
50. Manfredini, D.; Castroflorio, T.; Perinetti, G.; Guardanardini, L. Dental occlusion, body posture and temporomandibular disorders: Where we are now and where we are heading for. *J. Oral Rehabil.* **2012**, *39*, 463–471. [CrossRef]
51. Carini, F.; Mazzola, M.; Fici, C.; Palmeri, S.; Messina, M.; Damiani, P.; Tomasello, G. Posture and posturology, anatomical and physiological profiles: Overview and current state of art. *Acta Biomed.* **2017**, *88*, 11–16. [CrossRef] [PubMed]

MDPI
St. Alban-Anlage 66
4052 Basel
Switzerland
www.mdpi.com

Journal of Clinical Medicine Editorial Office
E-mail: jcm@mdpi.com
www.mdpi.com/journal/jcm